Communications in Computer and Information Science 2740

Rationale

The CCIS series is devoted to the publication of proceedings of computer science conferences. Its aim is to efficiently disseminate original research results in informatics in printed and electronic form. While the focus is on publication of peer-reviewed full papers presenting mature work, inclusion of reviewed short papers reporting on work in progress is welcome, too. Besides globally relevant meetings with internationally representative program committees guaranteeing a strict peer-reviewing and paper selection process, conferences run by societies or of high regional or national relevance are also considered for publication.

Topics

The topical scope of CCIS spans the entire spectrum of informatics ranging from foundational topics in the theory of computing to information and communications science and technology and a broad variety of interdisciplinary application fields.

Information for Volume Editors and Authors

Publication in CCIS is free of charge. No royalties are paid, however, we offer registered conference participants temporary free access to the online version of the conference proceedings on SpringerLink (http://link.springer.com) by means of an http referrer from the conference website and/or a number of complimentary printed copies, as specified in the official acceptance email of the event.

CCIS proceedings can be published in time for distribution at conferences or as post-proceedings, and delivered in the form of printed books and/or electronically as USBs and/or e-content licenses for accessing proceedings at SpringerLink. Furthermore, CCIS proceedings are included in the CCIS electronic book series hosted in the SpringerLink digital library at http://link.springer.com/bookseries/7899. Conferences publishing in CCIS are allowed to use our online conference service (Meteor) for managing the whole proceedings lifecycle (from submission and reviewing to preparing for publication) free of charge.

Publication process

The language of publication is exclusively English. Authors publishing in CCIS have to sign the Springer CCIS copyright transfer form, however, they are free to use their material published in CCIS for substantially changed, more elaborate subsequent publications elsewhere. For the preparation of the camera-ready papers/files, authors have to strictly adhere to the Springer CCIS Authors' Instructions and are strongly encouraged to use the CCIS LaTeX style files or templates.

Abstracting/Indexing

CCIS is abstracted/indexed in DBLP, Google Scholar, EI-Compendex, Mathematical Reviews, SCImago, Scopus. CCIS volumes are also submitted for the inclusion in ISI Proceedings.

How to start

To start the evaluation of your proposal for inclusion in the CCIS series, please send an e-mail to ccis@springer.com

Indranil Roy · Krishna K. Kambhampaty ·
Takaki Goto · Gongzhu Hu
Editors

Computer Applications in Industry and Engineering

38th International Conference, CAINE 2025 New Orleans, LA, USA, October 20–21, 2025 Proceedings

Editors
Indranil Roy
Southeast Missouri State University
Cape Girardeau, MT, USA

Krishna K. Kambhampaty
Pennsylvania State University
University Park, PA, USA

Takaki Goto
Toyo University
Tokyo, Japan

Gongzhu Hu
Central Michigan University
Mount Pleasant, MI, USA

ISSN 1865-0929 ISSN 1865-0937 (electronic)
Communications in Computer and Information Science
ISBN 978-3-032-14892-6 ISBN 978-3-032-14893-3 (eBook)
https://doi.org/10.1007/978-3-032-14893-3

This Springer imprint is published by the registered company Springer Nature Switzerland AG
The registered company address is: Gewerbestrasse 11, 6330 Cham, Switzerland

Preface

The 38th International Conference on Computer Applications in Industry and Engineering (CAINE 2025), held October 20–21, 2025, in New Orleans, Louisiana, continued its long-standing tradition as a premier international forum for presenting and exchanging advances in computer technologies across industries and engineering domains. Organized by the International Society for Computers and Their Applications (ISCA), the conference serves as a bridge between academic research and industrial innovation, offering participants a platform to share knowledge, foster collaboration, and explore emerging trends.

For CAINE 2025, 24 full papers were accepted from 48 submissions following a rigorous double-blind peer review process. Each submission was reviewed by an average of three distinguished Program Committee members, ensuring fairness, quality, and objectivity. The selected contributions reflect both the diversity and depth of current research directions, ranging from foundational theories to novel applications in industry and engineering.

This year's technical program was structured around four tracks that capture some of the most dynamic areas in computing today:

- Intelligent Signals & Smart Learning Systems – covering advances in signal processing, learning algorithms, and adaptive systems for real-time decision-making.
- Cloud Futures, Blockchain Trust & P2P Innovation – focusing on scalable cloud computing, decentralized trust mechanisms, and innovative peer-to-peer frameworks.
- Cybersecurity Frontiers & Resilient IoT – addressing the challenges of securing interconnected devices, infrastructures, and networks against emerging cyber threats.
- AI Horizons — Models, Methods & Applications – highlighting advances in artificial intelligence theories, methods, and real-world applications across diverse domains.

Beyond the paper presentations, CAINE 2025 featured keynote lectures by internationally recognized experts, who provided valuable perspectives on current challenges and future directions. The conference also recognized excellence through the Best Paper Award, underscoring CAINE's commitment to celebrating high-quality research and impactful contributions.

We extend our heartfelt appreciation to the Program Committee and reviewers for their tireless efforts in ensuring the high standards of the conference, and to the authors for contributing their research and insights. Most importantly, we thank the participants for their active engagement, which continues to sustain the vitality and success of CAINE.

September 2025

Indranil Roy
Takaki Goto
Krishna K. Kambhampaty

Organization

General Chair

Gongzhu Hu	Central Michigan University, USA

Program Committee Chairs

Krishna K Kambhampaty	Pennsylvania State University, USA
Indranil Roy	Southeast Missouri State University, USA
Takaki Goto	Toyo University, Japan

Program Committee / Reviewers

Abdullah Al-Shoshan	Qassim University, Saudi Arabia
Feng Yu	Youngstown State University, USA
Antoine Bossard	Kanagawa University, Japan
Charitha Hettiarachchi	University of Houston-Clear Lake, USA
Bo-Chao Cheng	National Chung Cheng University, Taiwan
Christoph Wunck	Emden/Leer University of Applied Sciences, Germany
Kendall Nygard	North Dakota State University, USA
Masashi Toda	Kumamoto University, Japan
Naohiro Ishii	Aichi Institute of Technology, Japan
Norhaslinda Kamaruddin	MARA University of Technology, Malaysia
Mario Guimaraes	Southeast Missouri State University, USA
P.K. Wong	University of Macau, China
Said Ghoniemy	Ain Shams University, Egypt
Reshmi Mitra	Southeast Missouri State University, USA
Sultan Aljahdali	Taif University, Saudi Arabia
Bidyut Gupta	Southern Illinois University-Carbondale, USA
Yan Shi	University of Wisconsin-Platteville, USA
Anirban Mitra	Amity University, Kolkata, India
Ansuman Bhattacharya	Southern Illinois University-Carbondale, USA

Somnath Chakroborty	West Virginia University Institute of Technology, USA
Deepak Dawar	Miami University, USA
Juefei Yuan	Southeast Missouri State University, USA

Contents

Intelligent Signals and Smart Learning Systems

An Investigation of Initial Dropout and Effects of Engagement with Course Content in Massive Open Online Courses (MOOCs) 3
Pavankumar Gurazada and Moutusy Maity

Ground Reaction Force Patterns Analysis with Deep Learning 24
Zhanxin Sha, Zhouzhou Li, Yicai Yao, and Juefei Yuan

Real-Time Music and Speech Segregation Using Evolutionary Voting Algorithms 32
Abdullah I. Alshoshan

Analysis of the Transfer Function of an LTI System for a Non-stationary Process 42
Abdullah I. Alshoshan

A Comparative Analysis of RAG and Non-RAG Models to Improve Access to Service Provider Information for Older Adults in Mississippi 54
Saviz Saei and Sujan Anreddy

Cloud Futures, Blockchain Trust and P2P Innovation

Explainable AI-Augmented DevSecOps for Secure and Reproducible Cloud-Native Research Software 69
Akshay Mittal

Cost-Performance Analysis of Cloud-Based Retail Point-of-Sale Systems: A Comparative Study of Google Cloud Platform and Microsoft Azure 87
Ravi Teja Pagidoju

ZK-Pyramid: A Pyramid Tree-Based P2P Framework for Trustworthy and Private Data Sharing Using Zero-Knowledge Succinct Proofs 100
Mehmuna Haque, Samana Dahal, Indranil Roy, Reshmi Mitra, and Bidyut Gupta

Governing Blockchain-Based Decentralized Systems 112
Nipesh Pant, Saurav Ghosh, Hemant Upadhyay, and Reshmi Mitra

Optimization of Outdoor Air Utilization in Industrial Air Conditioning Using Support Vector Machines (SVM) 126
Narayan Debnath, Reinaldo Gomez, Corina Abdelahad, Daniel Riesco, and German Montejano

Cybersecurity Frontiers and Resilient IoT

Cyber Defense by Access Control Hardening: Dynamic Web-Security Based on User Access Behavior 145
Cameron J. Wohlbach and Md Minhaz Chowdhury

Traffic Congestion Sybil Attack Detection in VANET 160
Sarthak Khanduja and Arunita Jaekel

Authentication for IoT Devices to Access Satellite Networks 171
Dharani Parvathaneni, Krishnendu Kottakkal Sugathan, Prasanta Kumar Roy, Ansuman Bhattacharya, and Indranil Roy

STRIKE: A Structured Taxonomy of Cybercrime for Risk, Impact, Knowledge, and Evolution 189
Melissa Pappy, Linh Nguyen, Suman Kumar, Byungkwan Jung, and Bernard Chen

A New Pseudo-Random Binary Generator Based on the Combination of Self-Shrinking Generators and Modified Self-Shrinking Generators Using Boolean Functions 212
Narayan Debnath, Andrés Francisco Farías, Andrés Alejandro Farías, Germán Antonio Montejano, Ana Gabriela Garis, and Daniel Riesco

LiveTest: An Open-Source OMR Testing Software 223
Garrett Mathers, Terry Griffin, and Catherine Stringfellow

Securing E-Commerce: A Framework for Risk Management and Compliance 244
Derek Taylor, Mehdi Barati, and Nick Rahimi

AI Horizons — Models, Methods and Applications

Prediction of Adverse Side Effects of Antidepressants Using an Expert System 265
Paulina Becerra-Vivanco, Lucila I. Castro-Pastrana, and Roberto Rosas-Romero

Elitist Country-Segmented Memetic Algorithm: Integrating Multithreading and SIMD for Efficient Solution Search 277
Piero Pastor, Manuel Tupia, and Rony Cueva

Bert_DPRCNN: A Novel Approach to Text Classification Using Hybrid Deep Learning Models 299
Yogeshvar Reddy Kallam, Lovely Yeswanth Panchumarthi, Lavanya Parchuri, Sriya Padmanabhuni, Satya Krishna Nunna, and Krishna Kambhampaty

Bug Seeding in Open-Source: An OpenStack Study 309
Praveen Kumar Sayani

Analyzing Darknet Address Information Using Machine Learning 321
Trenton Ward, Cheryl Hinds, and Jonathan Graham

Detecting Deviation in User Activity Using Supervised Learning 336
Ciera Miller, Kemon Bynum, and Cheryl Hinds

An Obstacle-Adaptive A^* Heuristic for Path Planning: A Comparative Analysis with Repulsion-Augmented BFGS in Complex Environments 349
Vomsheendhur Raju, Deepak Dawar, and Akhlaqur Rahman

Author Index 365

Intelligent Signals and Smart Learning Systems

An Investigation of Initial Dropout and Effects of Engagement with Course Content in Massive Open Online Courses (MOOCs)

Pavankumar Gurazada[1] and Moutusy Maity[2](✉)

[1] Great Learning, Bengaluru 560102, Karnataka, India
pavankumar.gurazada@greatlearning.in
[2] Marketing Department, HCBC, Southeast Missouri State University, Cape Girardeau, MO, USA
m.maity@semo.edu

Abstract. In this research our objectives include the following: a) in the context of early dropouts of MOOC registrants, what are some factors that predict the drop-out; and b) in the context of those who attend MOOC courses, what is the relationship between engagement with the course material and the grade earned at the end of the course. We focus on the behaviour and outcomes of two distinct groups of participants – early dropouts (Study 1) and participants who appear for exams to obtain certification (Study 2) in MOOCs. Uses and gratification theory, user impatience and engagement theory provide the theoretical underpinnings for the two studies. Study 1 based on the analysis of a data set of $n = 641,138$ participants who enrolled for 16 Harvard and MIT MOOCs for the period 2012–2013. In study 2, we focus on the participants from 15 courses who have completed the course with a grade and have been awarded a certificate ($n = 12,615$). Our findings on initial dropout from Study 1 indicate that early registration is a strong signal that may be used for monitoring participants that are susceptible to dropping out. The findings of Study 2 reveal that the positive effect of increased engagement on grade for the MOOCs indicates that a better approach to enhance tangible outcomes for participants (e.g., final grades) is to focus on increasing interaction/engagement with the course. We suggest that the roles of gratification and belongingness may be further investigated to understand learner motivation to be part of MOOCs.

Keywords: MOOCs · Digital divide · Early dropouts · Certification · Uses and gratification theory · Engagement theory · User impatience

1 Introduction

Massive Open Online Courses (MOOCs) have received widespread attention since their launch in 2012. Since then, MOOCs evolved from being largely free to access to a pay-for-certification model. For example, 78 million participants took part in MOOCs in 2017, with the proportion of participants paying for courses increasing over previous

I. Roy et al. (Eds.): CAINE 2025, CCIS 2740, pp. 3–23, 2026.
https://doi.org/10.1007/978-3-032-14893-3_1

years [1]. Several local universities and governments have also stepped into the fray by offering several of their courses online, where participants might get a certificate based on their performance in the course for a fee. Over the past couple of years MOOCs have even evolved into public policy initiatives with the aim of upskilling the labour force. Some examples of such an effort are Coursera and EdX and Swayam.

MOOCs are designed to have low barriers to both entry and exit unlike traditional education leading to a wide range of participant behaviour. A defining feature of MOOCs is the noticeably low certification rate (usually less than 4%) across courses and providers [2]. There are two arguments proposed by scholars to explain this observation. First, there is a wide gamut of participants who enrol in MOOCs and looking only at the certification rate of a course would not do justice to the utility gained by a participant from a MOOC. Second, even though the number of certifications is less in terms of percentages, absolute numbers are still many multiples of the number of students who complete a typical university course. This fact justifies the investments made into creating and promoting MOOCs. Much of extant research has focused on predicting the grades earned by participants who earn certificates. This focus on the paying segment is understandable since MOOC platforms (e.g., Coursera, EdX) have limited resources that are exhausted once this segment is catered to. However, there is paucity of research regarding the drivers of course completion as well as drivers of grade obtained by participants in the context of MOOCs. In this paper, we focus on the behaviour and outcomes of two distinct groups of participants – early dropouts (Study 1) and participants who appear for exams to obtain certification (Study 2). For early dropouts, that is, participants who register for a MOOC but do not watch a single video, we are interested in crystallizing features that predict the drop-out probability. We note that drop-out rates are deal-breakers for the promise that MOOCs hold. Similar is the case of small private online courses offered to corporates by universities, where high initial drop-out rate can be devastating. In this context, a guiding theme for our investigation in Study 1 is to highlight how MOOC organisers can predict and flag participants who might drop out early.

The second group of participants we study in this paper are participants who complete the course and earn a grade (Study 2). For this group, we are interested in understanding the relationship between their engagement with the course material and the grade earned at the end of the course. We ask the question – if a participant engages more with the course, does it contribute to a higher grade? In particular, we probe whether active engagement with the course through multiple 'events' (for e.g., browsing the course material, posting on forums, attempting the assignments listed in the chapters) leads to a higher grade for these participants. We submit that this analysis is helpful since stimulating engagement by providing participants more 'events' to interact (e.g., short quizzes within videos, links to more material, linking between chapters) is one of the key tactics that can be easily implemented by MOOC organisers.

To understand the effect of higher engagement on grade, we group the participants into two groups – high engagement learners and low engagement participants – using the number of events logged by the participants in the course. Then, we estimate the effect of shifting the low engagement cohort to a high engagement cohort using two meta learning algorithms (T-Learner and X-Learner). Our meta learning algorithms rely on two base machine learning models (each of which is a Gradient Boosted Tree) trained to predict grade for the two groups of participants (high and low engagement).

We note that evaluation of participants in MOOCs is often conducted from a monitoring perspective, that is, course organisers want to ensure that as many participants stay in the course for as long as possible (they should 'survive' in the course). Our findings on initial dropout from Study 1 indicate that early registration is a strong signal that may be used for monitoring participants that are susceptible for dropout. Moreover, the positive effect of increased engagement on grade for the 15 MOOCs we study in Study 2 indicates that a better approach to enhance tangible outcomes for participants (e.g., final grades) is to focus on increasing interaction with the course.

The rest of the paper is organised as follows. Section 2 presents a review of existing literature, and our central hypotheses on predictors of initial drop-out. Section 3 presents our hypotheses concerning the effect of increased engagement on grade. Section 4 presents the details of our predictive models and results for initial dropout (Study 1). Section 5 presents the data, models and results for the effect of engagement on grade for participants who appear for exams to obtain certification (Study 2). The paper concludes with a discussion on the implications of the results of Study 1 and Study 2 and directions for further research.

2 Study 1 - Initial Dropout

In this study, we conceptualise the MOOC participant as a consumer of course information. The purchase (i.e., course registration) is done in advance but the product (i.e., the course) is delivered at a later date. Next, we review the relevant literature and present our hypotheses on factors that drive initial dropout.

2.1 User Impatience and Uses and Gratification Theory

Prior research suggests that as omni-channel retailing and e-commerce become dominant, they are impacting customer expectations on service quality [3]. In this context, an emerging paradigm in customer service literature is the notion of customer impatience where once a purchase is made, customers expect quick product delivery. Consequently, informed, fully visible and on-time delivery process is the norm [4]. Customers are willing to accept delayed outcomes only if a longer waiting time results in a higher future value [5]. Further, for lower-valued outcomes, time sensitivity (i.e., the value placed on time) is higher [6].

Translating these findings to our conceptualization of the MOOC environment, we argue that there are two aspects of MOOC design that might result in customer impatience. First, while participants are aware of the timelines of course delivery during registration, activity commences only close to the launch date of the course. This lack of visible activity might lead to an unresolved impatience. The issue is exacerbated by the fact that growth spurts and attrition characterise the run-up to start of a course. For example, in an analysis of factors that predict completion of a course, [7] argue that MOOCs have a unique development history. Starting from a small participant base upon announcement, new groups of participants join in week after week. Consequently, the authors assert that if supportive communities do not evolve as the course progresses, participants might feel overwhelmed and dropout. These findings have implications for course design and pedagogy, and have been reported in literature [8].

Second, prior research shows that the utility of participating in a MOOC goes beyond the attainment of certification and encompasses a quest to understand a subject, fun, convenience or even exploration of a new learning medium [9]. Uses and gratification theory [10, 11] has also often been used to explain user interaction with specific media. In the context of Internet use, research has indicated that there are three main categories of gratification: content gratification, process gratification, and social gratification [12]. For such participants, the innate desire that drives them to register, which is akin to process gratification, might already be quenched by the act of registration. When the course starts at a later stage, this event has much lesser utility, leading to impatience. Mere access to a media/channel is often identified as a gratification and use of a medium [13].

Following these arguments, we operationalise customer impatience as the waiting time of participants till the commencement of the course and compute this as the lead time between the date of registration of the participant and the start date of the course. We expect that the date of registration, that is, whether the participant registered early or late (relative to the course start date) might be strongly predictive of initial dropout.

We summarise this discussion in the first hypothesis:

H1: Early registration is strongly predictive of initial dropout.

2.2 MOOC Participant Characteristics

As the initial euphoria on MOOCs settles down, significant research interest is being focused on peculiar aspects of participant behaviour in these courses [14]. A typical method used in predictive models of non-completion is to utilise observable aspects of a participants behaviour drawn from click stream data as features [15] along with the demographics of the participants [16]. Following previous studies on grade prediction and course non-completion, we expect participant demographics to be reflective of the infrastructure available in order to successfully engage with a MOOC. For example, a participant from the United States is expected to have access to better internet facilities compared to a participant from Rwanda. On the basis of this discussion, we formulate the second research question:

H2: Demographics of a participant are strongly predictive of initial dropout.

2.3 Diversity in Course Content

Another unique aspect of MOOCs is the variety of course content that draws a wide range of participants [17]. However, prior research on non-completion indicates that dropout rates on MOOCs are high across different course types [2]. Following these arguments, we propose that the course content has no effect on the willingness of the participants to attend the course.

H3: The subject of the course is not an important predictor of initial dropout.

2.4 Data and Descriptive Statistics

We test our hypotheses in Study 1 based on the analysis of a data set of $n = 641,138$ participants who enrolled for 16 Harvard and MIT MOOCs for the period 2012–2013 [17]. Descriptive statistics of the data (depicted in Fig. 1) indicate that majority of the participants are male and North America and Asia contribute most to the number of participants (as mentioned in Table 1, we group participants in the data set by continent).

Table 1. Descriptive statistics of numeric predictor variables employed for the gradient boosted model in Study 1

Variable	Description	Mean	SD	Min	Max
Male	Coded 1 if reported gender is male and 0 otherwise	0.724	0.447	0	1
Age	Age of participant when starting the course. Computed by subtracting the year of birth reported by the participant from course start year (taken as 2013)	28.560	9.602	0	82
Joined early or late	Coded 1 if course start date is greater than the participant's first event date, 0 otherwise	0.435	0.496	0	1
Continent	One of: Africa, Asia, Australia, Europe, North America, South America or Other. Computed by mapping the country reported by the participant to the respective continent	-	-	-	-

To characterise initial dropout on MOOCs, we define a participant as 'engaged' if they register and view at least one video from the course (it follows that those who browsed more than one video and those who earned certificates are also classified as 'engaged'). Similarly, a participant is classified as 'not engaged' if they register but never turn up, that is, they dropout even before starting the course. We note from that for every course, the 'not engaged' category presents a significant challenge. On an average, the initial drop-out rate on the Harvard and MIT MOOCs is 36.9% (Fig. 2 presents the activity status for all the participants across courses).

Since our objective is to predict initial dropout, our data is limited to the behaviour of participants before the course begins. We note that this is in contrast to the predictive model of grades or dropout in prior research, where significantly more data is utilised (e.g., number of chapters accessed, number of videos, discussion forum activity). While other studies do consider dropout overall, that is, investigating why participants do not complete a course, our focus in this study is on an investigation of why they do not even begin the course. While the lack of rich data poses challenges in model building, our endeavour is to select a model that is most accurate in predicting dropout given the data.

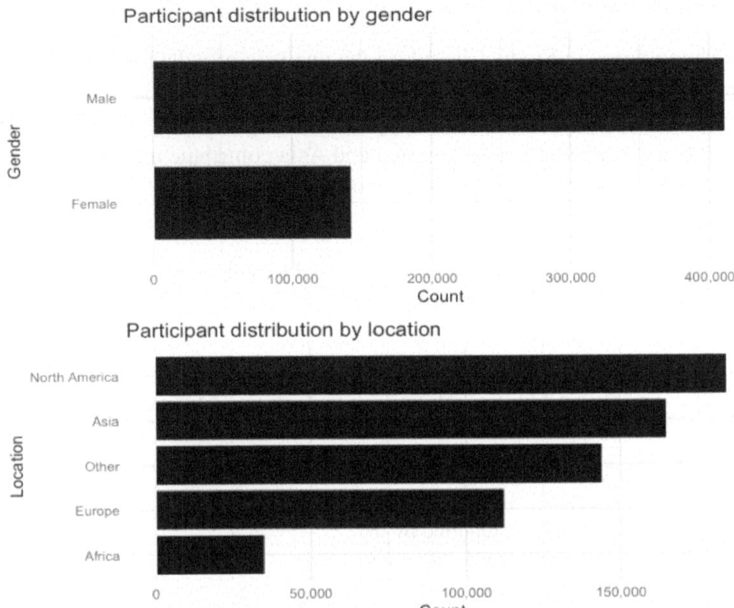

Fig. 1. Distribution of the gender and location of participants on the full data set, $n = 641,138$. After excluding users who registered for more than one course, we build a gradient boosted model to predict initial dropout on a dataset $n = 316,917$ users

2.5 Method

Engagement status was used as the label (coded 1 for 'engaged' and 0 for 'not engaged) to be predicted in all our models. The features used to predict engagement status were extracted from age, gender, country and date of registration. Following our discussion leading to H1, we added a new variable joined_early_or_late by subtracting the start date of a course and the date of registration by the participant. We expect this variable to be the key predictor in our model (H1). The other key features included were one-hot encoded representations of course number (H3), age, gender and continent (H2). Finally, we excluded the participants who registered for more than one course (this allows us to justify the independent samples assumption underlying the models). The final sample size on which the models were evaluated is $n = 316,917$. Table 1 presents descriptive statistics of the numeric variables considered in Study 1.

To develop a predictive model for initial dropout, we used 3 methods - Logistic Regression (base model), Gradient Boosted Trees and Neural Networks. Since we want our models to be strongly predictive of dropout, we attach prime importance to misclassified participants. In particular, we want the number of participants who were misclassified as non-dropouts to be minimum. To enforce this, we scored our models using AUC of the ROC curve as the metric and selected the final model hyper parameters based on 10-fold cross validation with 3 repeats. To account for the class imbalance, we employ

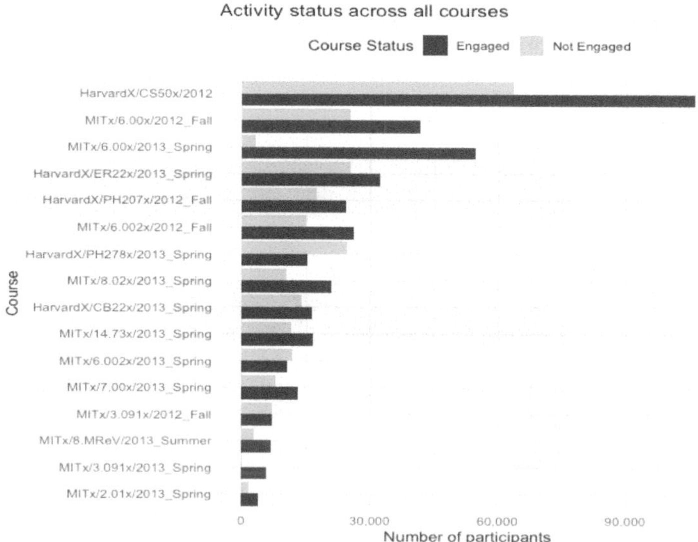

Fig. 2. Activity status (engaged or not engaged) of participants by course for $n = 641,138$ participants in the full data set

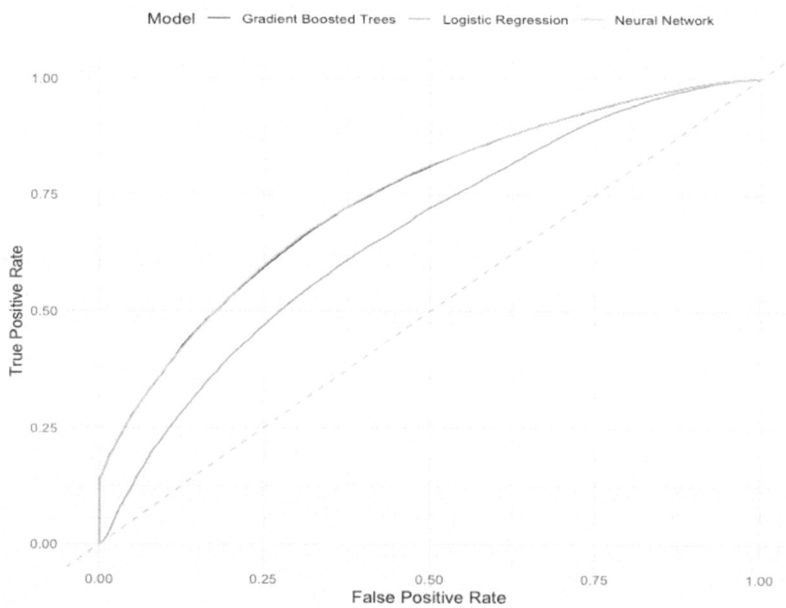

Fig. 3. ROC curves for three models fit to a sample of $n = 316,917$ participants enrolled in 16 MOOCs. The gradient boosted model has the best ROC AUC and accuracy

the Synthetic Minority Over-sampling Technique [18] to form the training set (chosen as 80% of the entire data set).

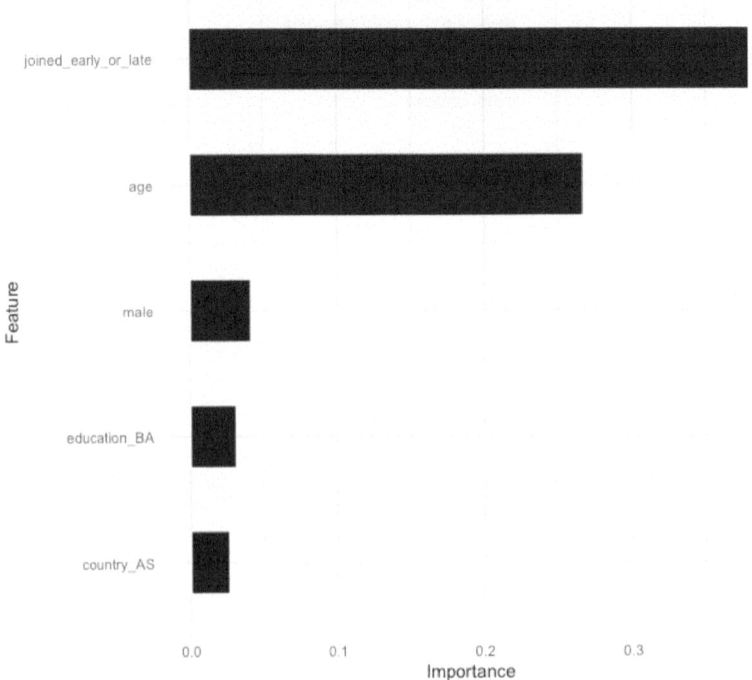

Fig. 4. Feature importance scores of the predictors used in the gradient boosted model for Study 1

Logistic regression was executed using L2-regularization with the regularization strength as the hyperparameter. For Gradient Boosting, we use the xgBoost algorithm [19], with depth of the trees and the number of estimators tuned as hyperparameters during training (we performed a random grid search over a larger parameter space to arrive at a shortlist). The neural networks were composed of a sequence of fully connected layers with the number of units per layer and the number of layers tuned during training. Our final model comprised 9 fully connected layers, with 32 units in each layer (total number of trainable parameters $= 4,417$). As can be inferred from the ROC curves of each of the models (Fig. 4), Gradient Boosted Trees and Neural Networks perform similarly and significantly better than the base model (Logistic Regression). However, since the Gradient Boosting model had a higher accuracy on the test set (76.1%) we choose this as our final model.

2.6 Results

The ROC curves from Fig. 3 indicate that Gradient Boosted Trees provide the best fit for the data set. In order to address the hypotheses posed in Sect. 2, we probed the relative importance of the factors in the predictive ability of the model on the data set. Relative importance of a feature is computed by averaging (across all trees) the number of times this feature is selected for the split weighted by the improvement to the resulting model [20]. Figure 4 shows the relative importance of the top five features that contribute most to the predictive ability of the final model on the test set.

3 Study 2 - The Effect of Course Engagement on Grade

In this section we present the theoretical foundations to address the second objective of our research – the effect of increased engagement with MOOC content on the participant's grade. Figure 5 presents the conceptual model used in Study 2. In the rest of this section, we present a detailed explanation of each of the factors depicted in Fig. 5.

3.1 Participant Grade

Since we focus only on participations who finish the course with a certification in this study, we choose their grade as the outcome variable. We find that there are two groups of factors that drive this choice of outcome. First, from prior research on motivating factors of participants who appear for exams to obtain certification, we note that assessments have 'high-stakes' within this group [21]. This signifies that a higher grade is a significant milestone and an important desired outcome for participants. Second, we also observe that certification in MOOCs is often part of the premium (or paid) variant and hence the participants have an additional financial obligation to get certified.

We note here that since the criteria for certification in the MOOCs studied in this paper is to have a numeric grade ≥ 0.5, the participants in our sample have a grade in the region [0.5, 1]. As a result, we have an outcome variable that is bounded below (or censored) at 0.5.

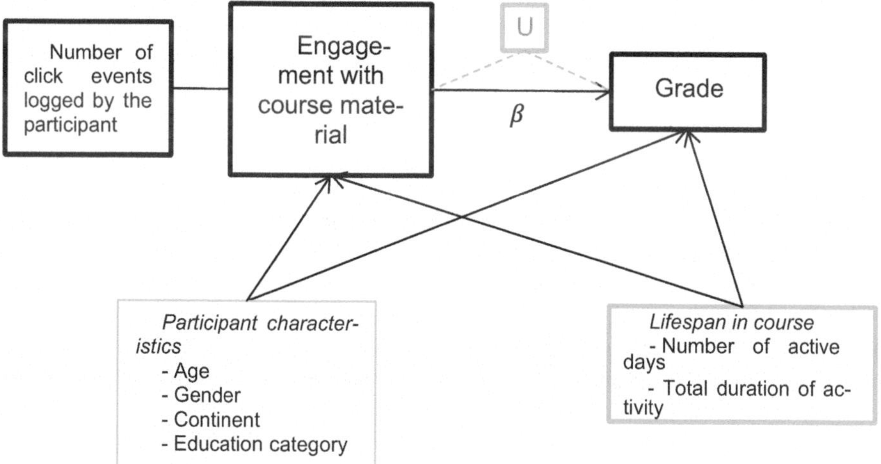

Fig. 5. Conceptual model for the effect of engagement with course on the grade

3.2 Engagement with Course Content

The Engagement Theory [22] is a framework for technology-based teaching and learning. The fundamental underlying idea is that students must be meaningfully engaged in learning activities through interaction with others and undertake worthwhile tasks. It is expected that participants are engaged in learning – that is, all student activities involve active cognitive processes such as creating, problem-solving, reasoning, decision-making, and evaluation. In addition, it is expected that students are intrinsically motivated to learn due to the meaningful nature of the learning environment and the activities.

We note that the virtual learning environment offered by a MOOC is shared by several thousand participants, and hence each individual participant receives much lesser instructor-led attention compared to a traditional learning environment where the number of participants is much lesser. Following this observation, we theorise that for effective learning, a MOOC necessitates comparatively more self-driven behavioral involvement, that is, more interaction events with the course material. Our expectations on the effect of increased interaction are based on prior research on learning outcomes in the context of traditional education (i.e., high school or college education), where active engagement of a student with the study material is shown to have a positive effect on the outcome [23]. We note that even in this stream of research, one key aspect of active engagement associated with positive outcomes is behavioural engagement where a student demonstrates effort, asks questions and contributes to class discussion [24]. Drawing upon this stream of research, we submit that in the context of a MOOC environment, behavioural engagement activities retain a positive effect on the outcome (i.e., grade). We argue that even though the nature of engagement activities morphs from in-classroom behaviour (e.g., asking questions, contributions to class discussion) to engagement with course content (watching videos, answering quizzes, participating in discussion forums), higher engagement still leads to higher grades.

In Study 2, we operationalise engagement with the course as the number of click events logged by the participant. The key variable whose effect (β in Fig. 5) we seek to determine on the grade obtained by the participant, is engagement with course material. Since participants can register click events in several ways including video plays, page accesses, problem attempts, forum posts and browsing the course outlines [17], we submit that this variable captures the behavioural activities of participants as they engage with the course. We also note that the primary ways by which a participant interacts with the course material is dependent on its content. For example, a course on history (e.g., CB22x - The Ancient Greek Hero) offers less opportunity for problem set based interaction compared to a course on science (e.g., 3.091x - Introduction to Solid State Chemistry). In courses that are closer to pure science, the material is inherently quantitative and problem sets are an effective tool to ensure participants grasp the application of analytical discussions. However, in the case of social sciences, where a subjective assessment of the phenomenon under investigation is a natural course of inquiry, a problem set based interaction is more difficult to design.

We summarise the discussion presented in this section as the following hypothesis:
H4: Higher engagement with course material leads to a higher grade.

3.3 Control Variables

While our main aim is to estimate the effect of higher engagement on grade, we utilise several control variables to account for the differences in participants beyond the level of engagement. The first set of control variables are related to the self-reported participant characteristics, that is, their age, gender, continent of residence, and level of education. The second set of control variables are related to their duration of activity in the course. Here we introduce two variables. The first variable, total duration, captures the length of time for which the participant was active in the course. We compute this variable by subtracting the date of first event and the last event logged by the participant. The second variable, number of active days, is the count of number of days during which the student interacted with the course. In sum, through these two variables, we account for the total duration for which the participant was active and also the number of unique activity periods during this time.

3.4 Data and Descriptive Statistics

In study 2, we focus on the participants from 15 courses who have completed the course with a grade and have been awarded a certificate ($n = 12, 615$). Each of these participants is assigned a numeric grade for the course in the range [0.5, 1] that we use as our outcome variable. The key predictor we investigate in this study is engagement with course material that is operationalised as the log of number of events of interaction posted by the participant (Fig. 5). Table 2 presents the variable definitions and descriptive statistics of the main and control variables used in Study 2. We note that one course - HarvardX/CS50x/2012 – is excluded from ~~Study~~ 2 since a numeric grade was not assigned to the participants for this course.

Table 2. Descriptive statistics of the variables used in Study 2. Values reported here are aggregate estimates for 15 courses considered in this study

Variable	Description	Mean	SD	Min	Max
Main variable					
Engagement with course	Natural logarithm of the number of interaction events logged by the participant	8.017	1.156	1.386	10.88
Control variables					
Number of active days	Natural logarithm of the count of unique days during which the participant recorded at least one activity on the server (includes any kind of activity, e.g., logging in, browsing page, video views)	3.565	0.694	0	5.088

(*continued*)

Table 2. (*continued*)

Variable	Description	Mean	SD	Min	Max
Total duration of activity	Natural logarithm of the total duration of activity recorded by the participant. Computed by subtracting the date of last event and the start date (both values were retrieved from server logs)	4.932	0.522	0	6.006
Age	Natural logarithm of the age of participant when starting the course. Computed by subtracting the year of birth reported by the participant from course start year (taken as 2013)	3.276	8.401	2.398	4.344
Gender	Male, Female or Other as reported by the participant	-	-	-	-
Continent	One of: Africa, Asia, Australia, Europe, North America, South America or Other. Computed by mapping the country reported by the participant to the respective continent	-	-	-	-
Education category	One of four categories reported by the participant: Doctorate, Master's, Bachelor's, Secondary or Less than secondary	-	-	-	-

3.5 Method

Our empirical efforts are directed at estimating the effect of shifting a cohort of low engagement participants to high engagement on the grade. To accomplish this objective, we divide the corpus of participants of each course into two subsets using the median of the natural logarithm of number of events as the cutoff. Consequently, participants who clock log number of events less than (greater than) the median are grouped as low (high) engagement.

We then estimate the response surfaces of log grade for these two classes of participants using a meta-learning approach. To ensure that our estimates are robust, we use two meta learning algorithms – T-learner and X-learner – detailed in the next two subsections. We also estimate the 95% confidence intervals for the estimated median effects.

The T-Learner Algorithm. In this approach, we model $\log(grade)^H$ and $\log(grade)^L$ as a function of the control variables (denoted by the set X, details of control variables are presented in Table 2) using two Gradient Boosted Trees (the choice of the XGBoost algorithm is based on the same selection procedure presented in Sect. 4.2). Here, the superscript H indicates that the corresponding variable is estimated for the high engagement subset and superscript L indicates that the variable is estimated for the low engagement subset.

Formally, we estimate two models: $\log(grade)^H \sim f\left(X^H\right) \& \log(grade)^L \sim g\left(X^L\right)$, where $f(\cdot)$ and $g(\cdot)$ are two gradient boosted trees whose hyperparameters are tuned by a three-fold cross validation approach on the training set (we use mean squared error as the metric). We can then estimate the expected effect of shifting the low engagement cohort to high engagement $\hat{\beta}$ (refer Fig. 1) as [25]:

$$\hat{\beta}(x) = \boldsymbol{E}\left[\log(grade)^H - \log(grade)^L | X = x\right] = \hat{f}(x) - \hat{g}(x)$$

Once $\hat{\beta}(x)$ is estimated for all the participants for each course, we then compute and report the median for these values, along with bootstrap 95% confidence intervals for the median (using 100 samples for each course). We also report the median estimate of $\hat{\beta}$ for a test set, chosen by randomly sampling 20% of the full data for each course. We note that this test set is not used in hyperparameter tuning of the gradient boosted trees and hence contains data none of the models has seen during the training phase.

The X-Learner Algorithm. In this approach, we first estimate $\hat{f}$ and $\hat{g}$ as described in Sect. 5.2.1. In the second step, we use $\hat{f}$ to impute the effect $\left(\hat{\beta}^L\right)$ of shifting the corpus of low engagement participants to high engagement. Similarly, we use $\hat{g}$ to impute the effect $\left(\hat{\beta}^H\right)$ of shifting the high engagement participants to low engagement. Finally, we compute the estimate of $\hat{\beta}$ as [25]:

$\hat{\beta}(x) = \lambda(x)\hat{\beta}^L(x) + (1 - \lambda(x))\hat{\beta}^H(x)$, where $\lambda(x)$ is a propensity weighting function.

In sum, we estimate the effects of shifting cohorts from one group to another and then use the propensity score for each participant to be in the high engagement group to compute a weighted average as the estimate for $\hat{\beta}$. Here, we use the logistic regression algorithm with L2-regularization to estimate λ. As in the case of the T-learner, we use the mean squared error as the metric to tune the hyperparameters of the base learners (i.e., the XGBoost models and the logistic regression model). We also report the median $\hat{\beta}(x)$ for all the participants and for the participants included into the test set.

We submit that by using a meta learning approach, we exploit the flexibility to train the individual estimators separately [25] and then use these individual models to do the estimation at the meta learning stage.

3.6 Results

Table 3 presents the estimates of $\hat{\beta}$ from the two algorithms – T-learner and X-learner – discussed in Sect. 3.5.

Table 3. Effect of increased engagement with course content on grade for participants who were certified

		T-learner (Model1)		X-learner (Model 2)	
Course	n	β_{median} [95% CI]	Fraction with positive $\hat{\beta}$	β_{median} [95% CI]	Fraction with positive $\hat{\beta}$
HarvardX/ER22x/2013_Spring	1,968	$-0.063^{\dagger}$	0.000	−0.062	0.000
		[−0.08, −0.05]		[−0.07, −0.05]	
MITx/14.73x/2013_Spring	1,883	0.086	0.971	0.073	0.998
		[0.03, 0.13]		[0.04, 0.11]	
HarvardX/PH207x/2012_Fall	1,737	0.010	0.788	0.004	0.806
		[0, 0.02]		[0, 0.01]	
MITx/6.00x/2012_Fall	1,606	0.025	0.967	0.018	0.999
		[−0.01, 0.06]		[0, 0.04]	
MITx/6.002x/2012_Fall	1,160	0.005	0.585	0.003	0.556
		[−0.05, 0.07]		[−0.04, 0.03]	
MITx/6.00x/2013_Spring	956	0.071	0.940	0.058	0.952
		[0.01, 0.13]		[0.01, 0.1]	
HarvardX/PH278x/2013_Spring	633	$-0.035^{\dagger}$	0.300	−0.033	0.186
		[−0.09, 0.03]		[−0.07, 0]	
MITx/7.00x/2013_Spring	539	0.009	0.671	0.003	0.578
		[−0.05, 0.06]		[−0.04, 0.03]	
MITx/6.002x/2013_Spring	470	0.050	0.763	0.021	0.646
		[−0.06, 0.16]		[−0.07, 0.1]	
MITx/3.091x/2012_Fall	453	0.045	0.898	0.044	1.000
		[−0.02, 0.11]		[−0.01, 0.09]	
MITx/8.02x/2013_Spring	414	0.055	0.755	0.031	0.752
		[−0.06, 0.18]		[−0.06, 0.14]	
HarvardX/CB22x/2013_Spring	321	0.076	0.879	0.065	0.922
		[−0.02, 0.16]		[−0.02, 0.13]	
MITx/8.MReV/2013_Summer	206	0.116	1.000	0.111	1.000

(continued)

Table 3. (*continued*)

Course	n	T-learner (Model1)		X-learner (Model 2)	
		β_{median} [95% CI]	Fraction with positive $\hat{\beta}$	β_{median} [95% CI]	Fraction with positive $\hat{\beta}$
		[0.07, 0.16]		[0.07, 0.15]	
MITx/2.01x/2013_Spring	163	0.012	0.923	0.013	1.000
		[−0.06, 0.07]		[−0.08, 0.07]	
MITx/3.091x/2013_Spring	106	0.014	0.776	0.012	0.953
		[−0.06, 0.08]		[−0.03, 0.07]	
	12,615				

Effect of Engagement with Course on Grade. The first metric we report in Table 3 for each course, is the median of the estimated $\hat{\beta}$ for each participant (excluding those in the test set). This is the numerical estimate of the magnitude of the effect on $\log(grade)$ of shifting the low engagement cohort to high engagement. The estimates for $\hat{\beta}_{median}$ reported in Table 3 indicate that except for HarvardX/ER22x and HarvardX/PH278x, the median effect is positive for all other courses irrespective of whether the T-learner or X-learner algorithm is used. For example, the estimate of $\hat{\beta}_{median}$ for HarvardX/CB22x/2013_Spring using the X-learner algorithm is 0.065 (Table 3). This means that the median estimate of $\hat{\beta}$ for 257 participants included in the training set for this course is 0.065. This value is the median increase in $\log(grade)$ for participants in the training set when they are shifted from low engagement to high engagement.

As Table 3 indicates, for each course, we also report the fraction of participants (excluding those in the test set) for which the estimated $\hat{\beta}$ is positive. Since our estimates are individual effects (as discussed in Sect. 3.5.1 and 3.5.2) the fraction of participants for which the effect is positive is an indication of the overall direction of effect. So, when this fraction is > 0.5, we conclude that the shift from low engagement to high engagement benefits the majority. As we note from Table 3, again, except for HarvardX/ER22x and HarvardX/PH278x, the effect of high engagement is positive for the majority. For example, for MITx/6.00x/2013_Spring, the fraction of participants in the training data ($n = 765$) with positive value of estimated $\hat{\beta}$ is 0.952. So, $0.952 \times 765 = 728$ participants had positive estimates of $\hat{\beta}$. From this analysis, we conclude that shifting participants in the training sample from a low engagement state to a high engagement state was beneficial to the majority.

Taken together, the findings from Table 3, that is, since $\hat{\beta}_{median} > 0$ and fraction of individual effects ($\hat{\beta}$) > 0.5 for 13 courses out of 15, this confirms the positive effect of increased engagement on grade supporting H4.

Robustness Checks and Model Choice

Validation on Test Set. In order to check the robustness of our participant-level estimates of $\hat{\beta}$ summarised in Table 3, we repeat our analysis using the T-learner and X-learner models on a test set (20% random sample) that is extracted before the model building process (Table 4 presents these results).

Table 4. Estimates of the effect of increased engagement with course on the grade for the test set. The test set is a random sample of 20% of the full data for each course. The test set is not used in the training of the meta learning models and is used as a robustness check

	T-learner (Model 1)		X-learner (Model 2)	
Course	$\hat{\beta}_{median}$	Fraction with positive $\hat{\beta}$	$\hat{\beta}_{median}$	Fraction with positive $\hat{\beta}$
HarvardX/ER22x/2013_Spring	−0.063	0.000	−0.062	0.000
MITx/14.73x/2013_Spring	0.086	0.971	0.073	0.995
HarvardX/PH207x/2012_Fall	0.010	0.796	0.004	0.810
MITx/6.00x/2012_Fall	0.025	0.978	0.018	1.000
MITx/6.002x/2012_Fall	0.004	0.638	0.002	0.586
MITx/6.00x/2013_Spring	0.068	0.932	0.056	0.937
HarvardX/PH278x/2013_Spring	−0.035	0.268	−0.033	0.173
MITx/7.00x/2013_Spring	0.008	0.611	0.003	0.584
MITx/6.002x/2013_Spring	0.043	0.702	0.011	0.628
MITx/3.091x/2012_Fall	0.039	0.912	0.039	1.000
MITx/8.02x/2013_Spring	0.061	0.747	0.045	0.795
HarvardX/CB22x/2013_Spring	0.063	0.922	0.061	0.953
MITx/8.MReV/2013_Summer	0.120	1.000	0.116	1.000
MITx/2.01x/2013_Spring	0.016	0.910	0.014	1.000
MITx/3.091x/2013_Spring	0.015	0.667	0.009	0.952

Table 5(a). Parameter ranges of the XGBoost model that are selected following a three-fold cross validation approach using Mean Squared Error as the metric (reported in Table 5(b)).

Parameter	Values
Number of estimators	[15, 25, 50, 100, 200, 300, 400]
Maximum tree depth	[2, 4, 6, 10, 12]
Learning rate	[0.005, 0.01, 0.03, 0.1, 0.2, 0.3]

Table 5(b). Mean Squared Error (three-fold cross validation) for the best gradient boosted trees (`XGBoost`) trained to predict log(*grade*) of high and low engagement cohorts for each course.

Course	High engagement (H)	Low engagement (L)
HarvardX/ER22x/2013_Spring	0.017	0.020
MITx/14.73x/2013_Spring	0.017	0.033
HarvardX/PH207x/2012_Fall	0.003	0.003
MITx/6.00x/2012_Fall	0.019	0.026
MITx/6.002x/2012_Fall	0.016	0.018
MITx/6.00x/2013_Spring	0.018	0.024
HarvardX/PH278x/2013_Spring	0.021	0.036
MITx/7.00x/2013_Spring	0.019	0.015
MITx/6.002x/2013_Spring	0.042	0.040
MITx/3.091x/2012_Fall	0.017	0.024
MITx/8.02x/2013_Spring	0.018	0.022
HarvardX/CB22x/2013_Spring	0.026	0.041
MITx/8.MReV/2013_Summer	0.030	0.015
MITx/2.01x/2013_Spring	0.037	0.036
MITx/3.091x/2013_Spring	0.019	0.019

The values for $\hat{\beta}_{median}$ in Table 4 confirm the findings presented in Sect. 3.6.1 on the positive magnitude of the effect of course engagement on grade across all courses (except HarvardX/ER22x and HarvardX/PH278x). Similarly, our findings on the fraction of positive effect indicate that even on the test set, the majority of participants have a favorable grade outcome when they are translated from a low engagement cohort to a high engagement cohort. In sum, the robustness checks on the test set confirm the validity of our findings on the positive effect of course engagement on grade (H4).

Hyperparameter Tuning. As discussed in Sect. 3.5, we trained two XGBoost models – one each for the low engagement and high engagement groups. These were then used as the base learners for the T-Learner and X-learner algorithms. The hyperparameters of these two models were tuned using three-fold cross validation on the training set using the mean squared error as the metric (the ranges of the hyperparameters considered for tuning are presented in Table 5(a)). During this process, we conduct a total of 210 fits ($= 7 \times 5 \times 6$) for each of the two models per course to arrive upon the optimal set of parameters for the models. We present the mean squared error of the best models in Table 5(b).

4 Discussion

While MOOCs have been widely embraced by participants across several continents, there is little research on the factors/drivers that impact course completion, as well as factors/drivers that may be predictive of grades obtained by participants in such courses. We present two studies in this research. In Study 1, we focus on the behaviour and outcomes of early dropouts and explain the phenomenon through the uses and gratification theory. Our findings reveal that early registration for a course is a strong predictor of initial dropout, while the subject matter of the course is not, providing support for H1 and H3. Our findings indicate that most of the demographic indicators, except age, are not good predictors of early dropout (H3 is partially supported). In Study 2, we focus on participants who appear for exams to obtain certification, and offer a possible explanation of participant behaviour through user impatience and engagement theory. The findings of Study 2 reveals that the positive effect of increased engagement on grade for the MOOCs indicates that a better approach to enhance tangible outcomes for participants (e.g., final grades) is to focus on increasing interaction/engagement with the course, providing support for H4). Overall, we find strong support for three (of four) hypotheses (H1, H2, and H4). The studies presented in this paper make several contributions to existing literature on MOOCs. In the rest of this section, we discuss the theoretical implications, managerial relevance of this research, some limitations and future research directions.

4.1 Theoretical Implications

Table 3 presents the estimates of $\hat{\beta}$ from the two algorithms – T-learner and X-learner – discussed in Sect. 3.5.

Initial Dropout. First, our conceptualization of a MOOC participant as a consumer allows us to draw upon prior research on consumer behaviour to choose predictive features for our machine learning algorithms. We submit that our approach to predictive modelling is novel in this aspect - while our methods are rooted in data mining, the choice of predictive features in these models is guided by consumer research. Second, our results crystallise key factors that are predictive of initial drop-out and can aid decision makers to market MOOCs effectively. Third, our final model can be used to flag participants who are most likely to drop out before the course begins. Our proposal is to utilise the predictions from our model to fine tune marketing efforts directed at reduction in the initial dropout rate.

Engagement and Learning Outcomes. Our findings from Study 2 extend the body of literature on student engagement within virtual learning environments [26] by establishing the positive effects of MOOC-specific behavioral engagement activities on the grade. In the 15 MOOCs we investigate in this study, we show that by using activities that engage participants to interact with the course, median grade can be improved by at least $\exp(0.003) = 1.003$ points (this is the minimum $\hat{\beta}_{median}$ for MITx/6.002x/2012_Fall, Table 3). We note that this is a large effect given the scale of the grade.

Our results also demonstrate that a dominant fraction of participants observe beneficial outcomes in the scenario of increased engagement (with all other factors retained the same). In sum, we submit that our findings from Study 2 reinforce the positive effect of student-driven activities and contribute to the theory of student engagement in learning.

4.2 Managerial Implications

Monitoring Participants Before They Begin the Course. Our results from the predictive model of initial dropout presented in Sect. 2.6 confirm early registration as a key predictor. This has important implications for the marketing and execution of MOOCs. In practice, we observe that little attention is paid to monitoring of participants before the course begins and our findings suggest that neglecting this phase has a devastating effect on the dropout rate. Our results indicate that early registrants need to be watched carefully and nudged to begin the course. We submit that this is paramount for government initiatives, and strongly advocate the allocation of resources and attention to participants who have registered early. Since our results indicate that the content of the course is not predictive of initial dropout, we propose that decision makers do not make any distinction among courses while allocating resources. It is important to note that while the precise initial dropout rates might differ across courses, the subject of the course itself is not predictive of initial dropout.

Increasing Activities that Stimulate Engagement. There are a few key implications for MOOC organisers we derive from our results in Study 2. First, we show that increasing the number of interactive activities within a course not only generates higher grades for the majority but also increases the median grade. This is an important nudge that is easy to incorporate into the current structure of MOOCs. For example, some MOOCs already intersperse videos with short quizzes that increase student interaction with course videos. We propose that this practice should be followed by default. Organisers could also incorporate selected discussion forum posts and display them along with videos to encourage participants to browse and post queries related to the chapter. Second, following the importance of activity driven engagement we establish in this paper, we propose that while preparing course videos, instructors should directly acknowledge participant contributions to the ongoing course. For example, they could highlight good work done by participants posted in discussion forums of the ongoing course or even mention blog posts where previous participants demonstrate their understanding of the course. This would allow participants to be aware of the impact of the course beyond the video they are watching currently and help them to interact more. As our results show, any nudge that shifts the participant from a passive watching mode to an active involvement mode generates rich benefits. Access to and the use of technology has not always decreased the digital divide – in fact, sometimes it has deepened such divide, which has been witnessed especially during the Covid-19 pandemic.

4.3 Limitations and Directions for Future Research

Since this study is one of the first to probe initial dropout and the impact of course engagement on grade, it is limited to analysis of secondary data. However, by using powerful machine learning methods, we are able to crystallise key features decision makers should be aware of while launching MOOCs. The strength of our methods notwithstanding, we acknowledge that given the observational nature of Study 2, there might be unobserved confounders (denoted by U, in Fig. 1) that might influence both the engagement with the course and the grade. This is a limitation we share with any observational study. In the next phase of research, we wish to undertake primary research to understand the decision-making process of MOOC dropouts and incorporate the findings from this research into better predictive models of initial dropout.

Acknowledgments. None.

Disclosure of Interests. None.

References

1. Peters, D.: MOOCs are not dead, but evolving, University Affairs (2018)
2. Onah, D.F.O., Sinclair, J., Boyatt, R.: Dropout rates of massive open online courses: behavioural patterns. In: EDULEARN14 Proceedings, pp. 5825–5834 (2014)
3. Murfield, M., Boone, C.A., Rutner, P., Thomas, R.: Investigating logistics service quality in omni-channel retailing. Int. J. Phys. Distrib. Logist. Manag. **47**(4), 263–296 (2017)
4. Daugherty, P.J., Bolumole, Y., Grawe, S.J.: The new age of customer impatience: an agenda for reawakening logistics customer service research. Int. J. Phys. Distrib. Logist. Manag. **49**(1), 4–32 (2019)
5. Marino, G., Zotteri, G., Montagna, F.: Consumer sensitivity to delivery lead time: a furniture retail case. Int. J. Phys. Distrib. Logist. Manag. **48**(6), 610–629 (2018)
6. Thaler, R.: Some empirical evidence on dynamic inconsistency. Econ. Lett. **8**(3), 201–207 (1981)
7. Yang, D., Sinha, T., Adamson, D., Rose, C.P.: Turn on, tune in, drop out: anticipating student dropouts in massive open online courses. In: Proceedings of the 2013 NIPS Data-Driven Education Workshop, vol. 11, pp. 14–21 (2013)
8. McCarthy, S., Rowan, W., Kahma, N., Lynch, L., Ertiö, T.P.: Open e-learning platforms and the design–reality gap: an affordance theory perspective. Inf. Technol. People **35**(8), 74–98 (2022)
9. Belanger, Y., Thornton, J.: Bioelectricity: a quantitative approach Duke University's first MOOC (2013). https://hdl.handle.net/10161/6216
10. Katz, E., Blumler, J.G., Gurevitch, M.: Uses and gratifications research. Public Opinion Q. **37**(4), 509–523 (1973)
11. Ruggiero, T.E.: Uses and gratifications theory in the 21st century. Mass Commun. Soc. **3**(1), 3–37 (2000)
12. Stafford, T.F., Stafford, M.R., Schkade, L.L.: Determining uses and gratifications for the Internet. Decis. Sci. **35**(2), 259–288 (2004)
13. Grellhesl, M., Punyanunt-Carter, N.M.: Using the uses and gratifications theory to understand gratifications sought through text messaging practices of male and female undergraduate students. Comput. Hum. Behav. **28**(6), 2175–2181 (2012)
14. Kross, S., Guo, P.J.: Students, systems, and interactions: synthesizing the first four years of learning@ scale and charting the future. In: Proceedings of the Fifth Annual ACM Conference on Learning at Scale, (L@S 2018), 10 p. ACM, New York, NY, USA, Article 2 (2018). https://doi.org/10.1145/3231644.3231662
15. Whitehill, J., Mohan, K., Seaton, D., Rosen, Y., Tingley, D.: Delving deeper into MOOC student dropout prediction (2017). https://doi.org/10.48550/arXiv.1702.06404

16. Mueller, M., Weber, G.: Machine learning regression analysis of edx 2012-13 data for identifying the auditors use case. Int. J. Integrating Technol. Educ. **6**(3) (2017)
17. Ho, A., et al.: HarvardX and MITx: The first year of open online courses, fall 2012-summer 2013. (HarvardX and MITx Working Paper No. 1) (2014)
18. Chawla, N.V., Bowyer, K.W., Hall, L.O., Kegelmeyer, W.P.: SMOTE: synthetic minority over-sampling technique. J. Artif. Intell. Res. **16**, 321–357 (2002)
19. Chen, T., Guestrin, C.: XGBoost: a scalable tree boosting system. In: Proceedings of the 22nd ACM SIGKDD International Conference on Knowledge Discovery and Data Mining, pp. 785–794 (2016)
20. Elith, J., Leathwick, J.R., Hastie, T.: A working guide to boosted regression trees. J. Anim. Ecol. **77**(4), 802–813 (2008)
21. Bergner, Y., Kerr, D., Pritchard, D.E.: Methodological challenges in the analysis of MOOC data for exploring the relationship between discussion forum views and learning outcomes. In: International Educational Data Mining Society, 26–29 June, Madrid, Spain, pp. 234–241 (2015)
22. Kearsley, G., Shneiderman, B.: Engagement theory: a framework for technology-based teaching and learning. Educ. Technol. **38**(5), 20–23 (1998)
23. Appleton, J.J., Christenson, S.L., Kim, D., Reschly, A.L.: Measuring cognitive and psychological engagement: validation of the student engagement instrument. J. Sch. Psychol. **44**(5), 427–445 (2006)
24. Fredricks, J.A., Blumenfeld, P.C., Paris, A.H.: School engagement: potential of the concept, state of the evidence. Rev. Educ. Res. **74**(1), 59–109 (2004)
25. Künzel, S.R., Sekhon, J.S., Bickel, P.J., Yu, B.: Metalearners for estimating heterogeneous treatment effects using machine learning. Proc. Natl. Acad. Sci. **116**(10), 4156–4165 (2019)
26. Boulton, C.A., Kent, C., Williams, H.T.P.: Virtual learning environment engagement and learning outcomes at a 'bricks-and-mortar' university. Comput. Educ. **126**, 129–142 (2018)

Ground Reaction Force Patterns Analysis with Deep Learning

Zhanxin Sha[1], Zhouzhou Li[2], Yicai Yao[2], and Juefei Yuan[2](✉)

[1] Georgia Southwestern State University, 800 Georgia Southwestern State University Drive, Americus, GA 31709, USA

[2] Southeast Missouri State University, Cape Girardeau, MO 63703, USA

jyuan@semo.edu

Abstract. The Ground Reaction Force (GRF) signals contain valuable biomechanical information that can be used for human classification tasks. This study investigates the use of Random Forest, Neural Network, and 1D Convolutional Neural Network (1D CNN) models to predict gender based on GRF data. The models are trained and evaluated on a large dataset consisting of six standardized GRF features from both feet. Experimental results demonstrate that deep learning models, particularly the 1D CNN, are highly effective in capturing complex GRF patterns and achieve superior classification performance compared to traditional machine learning approaches. These findings confirm the potential of GRF-based approaches for accurate and reliable gender classification.

Keywords: Ground Reaction Force (GRF) · Gender Classification · Human Biomechanics · Gait Analysis · Deep Learning

1 Introduction

Walking gait, as one of the most natural and characteristic human movements, has been widely investigated by clinical therapists, biomechanists, and forensic experts. It is a cyclical movement that requires coordination between the brain, nervous system, and skeletal muscles [2, 6, 8, 10, 11]. The gait cycle consists of two primary phases: the foot contact phase, which accounts for approximately 60% of the gait cycle, and the swing phase, which accounts for the remaining 40%.

This repetitive human movement contains valuable information about individuals. One primary goal of gait analysis is to identify abnormalities that affect a patient's walking pattern, which is useful for researchers, biomechanists, and clinical practitioners. Another important application is in security, where gait analysis can help detect potential threats in public or confined environments.

Currently, various methods are used to collect gait data, including motion capture systems, electromyography, force platforms, and wearable sensors [4, 8–10]. Among these, force platforms are particularly useful, as they measure the ground reaction forces (GRFs) and ground reaction moments (GRMs) generated during walking. These measurements include three orthogonal components: anterior-posterior (Fx, Mx), medial-lateral (Fy,

I. Roy et al. (Eds.): CAINE 2025, CCIS 2740, pp. 24–31, 2026.
https://doi.org/10.1007/978-3-032-14893-3_2

My), and vertical (Fz, Mz) directions. Force platforms provide high-resolution temporal and spatial data [2, 5].

The rich data obtained from force platforms offer insights into the magnitude, direction, and timing of forces and moments exerted on the human body during walking. However, analyzing this data is challenging due to the large volume and complexity of time-series signals. Generally, two main approaches are used for analysis: qualitative and feature extraction methods [5, 11].

Qualitative analysis helps researchers identify patterns or key events visually, but it is often prone to inter- and intra-observer variability [5]. Feature extraction, on the other hand, involves selecting specific characteristics from the time-series data—such as initial foot contact, loading rate, and peak values. This approach simplifies the data, making it more manageable. However, it may also lead to loss of important information and can introduce subjectivity or bias [5]. Additionally, feature extraction can be time-consuming.

Deep learning methods offer an alternative by working directly with various forms of input data: manually selected features, automatically extracted features, or raw time-series signals [2, 6, 11]. These approaches can potentially eliminate the need for prior data reduction. However, there is a paucity of research specifically investigating the effectiveness of data mining methods for gender identification based on ground reaction forces during walking.

Thus, the study aims to classify gender using ground reaction forces (GRFs) with a random forest, fully connected neural network, and one dimensional Convolutional Neural Network (CNN).

2 Related Work

The goal of walking gait recognition is to develop a system with the highest possible accuracy. Various data mining methods have been explored to assess the effects of footwear on gait recognition, with many studies reporting promising accuracy rates.

In one study, approximately 3,400 strides were collected from 99 participants. Researchers extracted eight features from these strides to use as input for data mining analysis. Among the models tested, the Support Vector Machine (SVM) produced the highest accuracy, outperforming Artificial Neural Networks (ANNs), Random Forest, and Naïve Bayes models [4]. A key limitation of this study was the low number of participants, all of whom were female, which reduces the generalizability of the findings.

In related research, Horst [10] and Pandey [9] normalized all GRF data to 100 data points for each participant, ensuring a consistent input length for classifying healthy individuals and those with gait disorders, respectively. This normalization meant that each participant's entire foot contact phase was represented by the same number of data points. Without requiring additional feature extraction, Horst et al. [10] found that both SVM and ANN models achieved high accuracy rates, correctly identifying healthy participants with an accuracy above 95.4%. Similarly, Pandey [9] reported that a Convolutional Neural Network (CNN) achieved a classification accuracy of 91% in distinguishing between healthy individuals and those with gait disorders.

These results suggest that deep learning methods can yield high performance even without explicit feature extraction. However, the study by Pandey [9] focused solely on distinguishing between healthy and disordered gaits. Therefore, the effectiveness of deep learning for other types of gait classification remains an open question and requires further investigation.

Apart from the deep learning techniques employed, the quality and reliability of GRF data collection during walking are also crucial factors influencing classification accuracy. For example, footwear type significantly affects human movement and GRF waveforms [4], which in turn impacts deep learning results [2, 4, 5, 10]. Studies by these researchers showed that even when participants wore different types of footwear, deep learning models achieved individual recognition accuracies of 88% and 85%, respectively [6]. Nevertheless, these studies were also limited by relatively small sample sizes.

In conclusion, there is a significant gap in the research regarding the application of various deep learning techniques for walking gait recognition of healthy males and females wearing different types of footwear. This is particularly important for national security and clinical applications, where individuals may wear different shoes in the same environment. The ability to accurately distinguish between males and females could support more precise interventions, ultimately enhancing task-oriented performance.

3 Gender-Oriented Ground Reaction Force Classification

3.1 Overview

In this paper, we propose 3 approaches to complete the gender-oriented ground reaction force classification task, which are Random Forest, Neural Network and One-dimensional Convolutional Neural Network. As illustrated below:

(1) **Random Forest**: Random Forest is an ensemble learning algorithm that builds multiple decision trees during training and merges their results to improve classification accuracy. In this project, we use six ground reaction force (GRF) features as inputs to predict gender. Before training, we scale the input data to ensure each feature contributes equally and then fit the model using 100 decision trees.
(2) **Fully Connected Neural Network**: A fully connected neural network is used to classify gender based on six ground reaction force (GRF) features. The network includes multiple dense layers with ReLU activation, batch normalization, and dropout to improve learning and reduce overfitting. The model is trained using categorical cross-entropy loss and optimized using the Adam optimizer with validation and test loss guiding backpropagation.
(3) **One-dimensional Convolutional Neural Network**: A one-dimensional convolutional neural network (1D CNN) is used to classify gender using six GRF features as input. The model applies multiple Conv1D layers with ReLU activations, batch normalization, and dropout to extract temporal patterns across the features. A final dense layer performs classification, and the model is trained using categorical cross-entropy loss with backpropagation.

3.2 Random Forest-Based Gender Classification

We represent the workflow of the Random Forest classification system as illustrated in Fig. 1.

Feature Extraction [14]. Each data point consists of six ground reaction force (GRF) features: vertical [7], anterior-posterior [1], and medial-lateral forces [12] from both the left and right sides. These features are extracted directly from the raw dataset (Sect. 4.1) and used as input for the classification model. The feature matrix $X \in R^{N \times 6}$ is standardized [12] by removing the mean and scaling to unit variance:

$$X' = \text{StandardScaler}(X),$$

where N is the total number of data samples.

Classification. We utilize a Random Forest classifier with 100 decision trees. The model is trained on 7,536,100 samples and validated using 3,768,050 samples. Random Forest aggregates the predictions from each tree using majority voting. The final trained model is then evaluated on the remaining testing samples to predict gender labels.

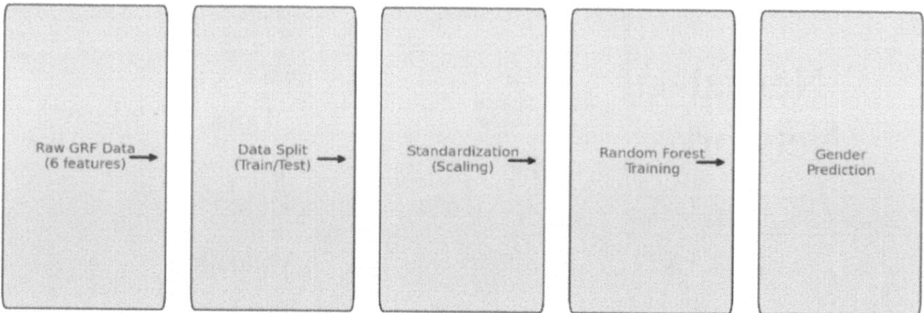

Fig. 1. Workflow of Random Forest Classifier for Gender Prediction Using Ground Reaction Force Data.

3.3 Neural Network-Based Gender Classification

We present the architecture of the neural network classifier in Fig. 2.

Feature Processing. The same six GRF features used in the Random Forest model are utilized as input to the neural network. These include vertical, anterior-posterior, and medial-lateral forces from both feet. Each feature vector $X \in R^6$ is standardized using z-score normalization. The dataset is split into training (7,536,100 samples), validation (3,768,050 samples), and testing (3,768,050 samples) subsets. The gender labels are encoded using one-hot encoding [15] for compatibility with the softmax [17] output layer.

Network Structure. The classifier is implemented as a feedforward neural network comprising three fully connected layers:

- Layer 1: 64 units, ReLU activation, followed by batch normalization and dropout (rate = 0.3)

- Layer 2: 128 units, ReLU activation, followed by batch normalization and dropout (rate = 0.3)
- Layer 3: 64 units, ReLU activation, followed by batch normalization and dropout (rate = 0.2)

The final layer is a dense layer with two units and softmax activation to output the gender classification probabilities. The model is trained using the Adam optimizer with a learning rate of 0.0005 and categorical cross-entropy as the loss function. Training is performed over 20 epochs with a batch size of 512, using the validation set to monitor performance.

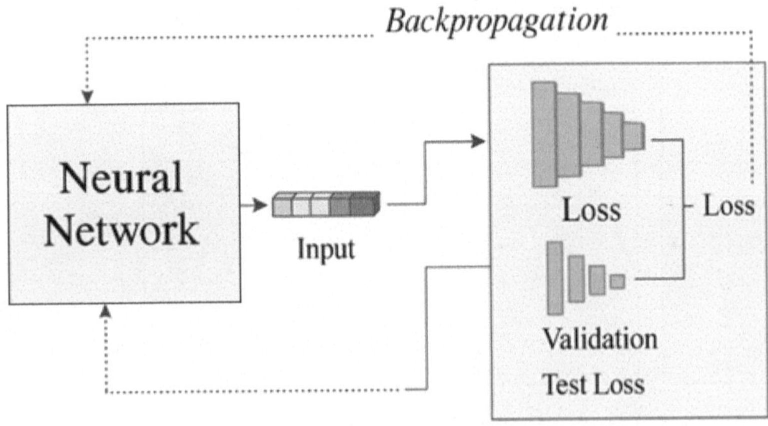

Fig. 2. Neural Network Workflow for Gender Prediction Using Ground Reaction Force Data.

3.4 1D Convolutional Neural Network-Based Gender Classification

The architecture of the 1D convolutional neural network (1D CNN) classifier is shown in Fig. 3.

Input Encoding. The same six GRF features are used as input, but the input shape is reshaped to fit the expected input format of the Conv1D layer: each sample $X \in R^6$ is reshaped to $X' \in R^{6\times1}$, where the second dimension represents the single channel. The GRF features are standardized across the dataset using a StandardScaler. Labels are one-hot encoded to match the format required for categorical cross-entropy [13] loss. The dataset is split equally into training, validation, and test sets.

Network Design. The 1D CNN is composed of three convolutional layers:

- Conv1: 64 filters with kernel size 3, ReLU activation, followed by batch normalization and dropout (rate = 0.3)
- Conv2: 128 filters with kernel size 3, ReLU activation, followed by batch normalization and dropout (rate = 0.3)

- Conv3: 64 filters with kernel size 3, ReLU activation, followed by batch normalization and dropout (rate = 0.2)

The convolutional output is flattened and passed to a softmax-activated dense layer of size 2 for binary classification. The model is optimized using Adam (learning rate = 0.0005) and trained over 20 epochs with a batch size of 512. Model performance is monitored on the validation set, and final evaluation is conducted using the test data.

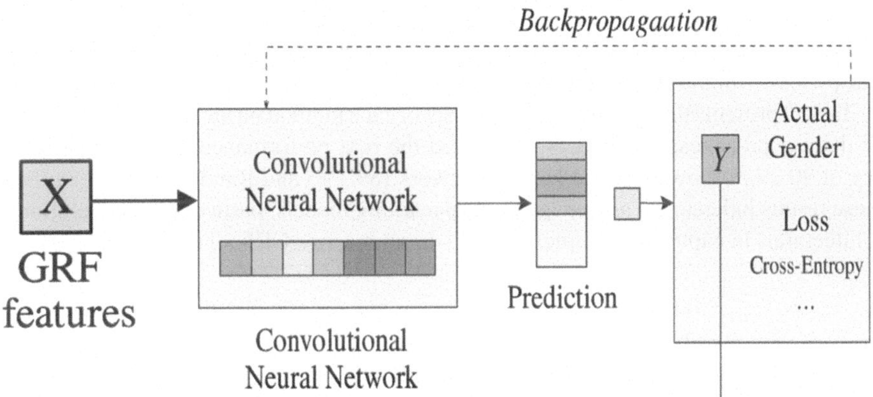

Fig. 3. 1D CNN Workflow for Gender Prediction Using Ground Reaction Force Data.

4 Experiments and Discussions

4.1 Dataset

We conduct a comprehensive evaluation of our gender classification algorithm based on the ground reaction forces (GRFs) dataset from 324 participants [10]. Each participant walked across two force plates at their preferred speed, wearing different types of shoes. The left and right foot GRF data were recorded separately by the force plates, capturing key biomechanical features. A description of the six GRF input features is demonstrated in Table 1.

Table 1. Description of Ground Reaction Force (GRF) Input Features.

Feature	Description
leftVForce	Vertical Force from the Left Foot
leftAPForce	Anterior-Posterior Force from the Left Foot
leftMLForce	Medio-Lateral Force from the Left Foot
rightVForce	Vertical Force from the Right Foot
rightAPForce	Anterior-Posterior Force from the Right Foot
rightMLForce	Medio-Lateral Force from the Right Foot

4.2 Gender Classification Results

We evaluate the performance of three classification methods—Random Forest (RF), Neural Network (NN), and 1D Convolutional Neural Network (1D CNN)—on the task of gender prediction using a large-scale dataset of ground reaction force (GRF) measurements. This dataset contains a total of 15,072,200 samples, divided into 7,536,100 training samples, 3,768,050 validation samples, and 3,768,050 testing samples.

Each method was trained and evaluated using the same GRF input features, including six biomechanical force measurements from both the left and right foot (vertical, anterior-posterior, and medio-lateral directions). All features were standardized prior to model training. Neural-based methods utilized one-hot encoded labels and categorical cross-entropy loss with the Adam optimizer.

Table 2 presents the classification accuracy of each method on the test dataset. Among the three approaches, the 1D CNN achieved the best performance, reaching an accuracy of 90.2%, followed by the Neural Network (89.1%) and Random Forest (84.0%). These results indicate the advantage of deep learning models, particularly convolutional architectures, in capturing complex patterns in time-series GRF signals.

Table 2. Test accuracy comparison for gender classification using GRF data.

Method	Accuracy (%)
Random Forest	84.0
Neural Network	89.1
1D CNN	**90.2**

5 Conclusions and Future Work

This study presents a comprehensive evaluation of three classification methods—Random Forest, Neural Network, and 1D Convolutional Neural Network—for predicting gender based on ground reaction force (GRF) data. Through extensive experiments on a large-scale dataset comprising over 15 million GRF samples, we demonstrate that deep learning models, especially convolutional architectures, significantly outperform traditional machine learning approaches in capturing complex biomechanical patterns. Our best-performing model, the 1D CNN, achieved a test accuracy of 90.2%, indicating strong potential for real-world gender classification applications using sensor-based biomechanical signals.

In future work, we plan to further enhance classification accuracy by incorporating semantic information—such as walking speed, gait cycles, or footwear condition—into the model input or learning architecture. Beyond gender classification, our research direction will be expanded in two new domains: (1) Footwear Classification: analyzing whether the same individual exhibits different GRF patterns under different types of shoes, and classifying the shoe type accordingly; (2) Individual Identification: predicting which person a given GRF sequence belongs to, enabling user-specific biometric applications. These extensions will allow our GRF analysis framework to support broader applications in health monitoring, gait analysis, and identity recognition.

References

1. Bowden, M.G., Balasubramanian, C.K., Neptune, R.R., Kautz, S.A.: Anterior-posterior ground reaction forces as a measure of paretic leg contribution in hemiparetic walking. Stroke **37** (2006)
2. Derlatka, M.: Automatic gender and unilateral load state recognition for biometric purposes. Technol. Health Care **31**, 2467–2475 (2023). https://doi.org/10.3233/THC-235012
3. Derlatka, M., Parfieniuk, M.: Real-world measurements of ground reaction forces of normal gait of young adults wearing various footwear. Sci. Data **10**, 60 (2023). https://doi.org/10.1038/s41597-023-01964-z
4. Derlatka, M., Bogdan, M.: Recognition of a person wearing sport shoes or high heels through gait using two types of sensors. Sensors (Basel) **18**, 1639 (2018). https://doi.org/10.3390/s18051639A
5. Dindorf, C., et al.: Machine learning in biomechanics: key applications and limitations in walking, running and sports movements. In: Blondin, M.J., Fister, I., Pardalos, P.M. (eds.) Artificial Intelligence, Optimization, and Data Sciences in Sports, pp. 91–148. Springer, Cham (2025)
6. Duncanson, K.A., et al.: Deep metric learning for scalable gait-based person re-identification using force platform data. Sensors (Basel) **23**, 3392 (2023). https://doi.org/10.3390/s23073392
7. Fineberg, D.B., et al.: Vertical ground reaction force-based analysis of powered exoskeleton-assisted walking in persons with motor-complete paraplegia. J. Spinal Cord Med. **36**, 313–321 (2013). https://doi.org/10.1179/2045772313Y.0000000126
8. Horst, F., Lapuschkin, S., Samek, W., Müller, K.R., Schöllhorn, W.I.: Explaining the unique nature of individual gait patterns with deep learning. Sci. Rep. **9**, 2391 (2019). https://doi.org/10.1038/s41598-019-38748-8
9. Horst, F., et al.: Identification of subject-specific responses to footwear during running. Sci. Rep. **13**, 11284 (2023). https://doi.org/10.1038/s41598-023-38090-0
10. Pandey, C., et al.: GaitRec-Net: a deep neural network for gait disorder detection using ground reaction force. PPAR Res. **2022**, 9355015 (2022). https://doi.org/10.1155/2022/9355015
11. Shuzan, M.N., et al.: Machine learning-based classification of healthy and impaired gaits using 3D-GRF signals. Biomed. Signal Process. Control **81**, 104448 (2023). https://doi.org/10.1016/j.bspc.2022.104448
12. Tang, Y.W., Murai, A., Hobara, H.: Mediation of the mediolateral ground reaction force profile to maintain straight running among unilateral transfemoral amputees. Sci. Rep. **13**, 7823 (2023). https://doi.org/10.1038/s41598-023-34288-4
13. Cross-entropy. https://en.wikipedia.org/wiki/Cross-entropyTang
14. Data Standardization. https://builtin.com/data-science/when-and-why-standardize-your-data
15. Feature Extraction. https://www.mathworks.com/discovery/feature-extraction.html
16. One Hot Encoding. https://www.geeksforgeeks.org/ml-one-hot-encoding/
17. Real-world measurements of ground reaction forces of normal gait of young adults wearing various footwear. https://doi.org/10.6084/m9.figshare.c.5689894.v1
18. Softmax Function. https://en.wikipedia.org/wiki/Softmax_function

Real-Time Music and Speech Segregation Using Evolutionary Voting Algorithms

Abdullah I. Alshoshan(✉)

College of Computer, Qassim University, Buraydah, Qassim, Saudi Arabia
drshoshan@gmail.com

Abstract. The process of separating music and speech signals can be categorized into three scenarios. The first scenario occurs when the signals overlap only in the time domain, while the second scenario takes place when they intersect exclusively in the frequency domain. The third scenario, which is the more complex case, arises when they overlap in both time and frequency domains. To address the first scenario, a specialized frequency-domain filter is employed to distinguish the signals. When overlap occurs in the frequency domain, a time-domain filter can be utilized for segregation. However, when signals overlap in both domains, the challenge becomes more intricate, necessitating additional constraints for an effective solution. In this paper, an evolutionary voting algorithm is introduced to resolve the first two cases through enhancements of the ZCR and STE techniques, while another approach utilizing ICA is explored to tackle the third scenario with specific constraints. The paper also presents theoretical analyses and simulation results to evaluate the performance of the proposed algorithm.

Keywords: Music/Speech Segregation · Zero-Crossing Rate · Short-Time Energy · Independent Component Analysis · Non-Stationarity · Independency

1 Introduction

The computational analysis of audio signals plays a crucial role in various applications [1]. With the advancements in signal analysis and synthesis techniques, speech and music signal processing has gained significant importance, leading to the adoption of classical sound analysis methods for processing music signals [2]. This paper introduces two adaptive systems: the first system focuses on separating speech from music when they do not overlap in either of the time domain or the frequency domain, utilizing enhancements in the zero-crossing rate (ZCR) and the short-time energy (STE) techniques. The second system is designed to separate signals that are overlapping in both domains, but statistically independent, employing independent component analysis (ICA) technique. Saunders [3] developed a music detection approach using a two-level algorithm that distinguishes music from speech based on average ZCR and STE features. Similarly, El-Maleh [4] proposed a two-level music/speech classification system, leveraging long-term characteristics such as differential parameters, variance, time-averaged spectral parameters, and ZCR. This paper presents an evolutionary voting algorithm that enhances ZCR and STE for the first system and explores ICA-based signal segregation for the second system [5].

I. Roy et al. (Eds.): CAINE 2025, CCIS 2740, pp. 32–41, 2026.
https://doi.org/10.1007/978-3-032-14893-3_3

2 Speech and Music Segregation

Since both of speech and music signals are non-stationary in their nature, the ZCR and the STE suffer from this non-stationarity behavior of these signals, as depicted in Fig. 1 [6].

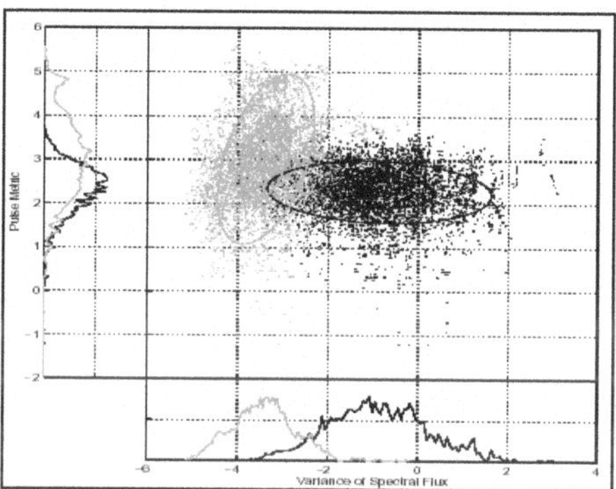

Fig. 1. Music and speech overlapping.

The key characteristics and distinctions between speech and music signals can be outlined as follows: tonality, randomness, bandwidth, power distribution, excitation patterns, tonal duration, STE, ZCR, and continuity.

2.1 Non-overlapped Segregation Using ZCR and STE

The ZCR Algorithm

The ZCR of a time-domain waveform is a highly reliable and effective metric for distinguishing voiced from unvoiced speech, so it has been widely utilized in practice as a key measure for differentiating fricative sounds from voiced speech [7]. The ZCR of a signal is, simply, calculated by counting the instances where the signal crosses zero magnitude within a defined window size. A zero-crossing occurs when consecutive samples exhibit opposite algebraic signs. Equation (1) defines the required computation for the ZCR of a discrete-time signal $x(n)$.

$$Z_n = \frac{1}{2N} \sum_{m=n-N+1}^{N} |\ sgn[x(m)] - sgn[x(m-1)]| \qquad (1)$$

where Z_n is the ZCR, $sgn[x(n)] = 1$ when $x(n) > 0$ and $sgn[x(n)] = -1$ when $x(n) < 0$, and N is the number of samples in one window. The ZCR is a time-domain algorithm that is closely influenced by the frequency of the input signal $x(n)$. An essential preprocessing step before counting zero-crossings is signal normalization, ensuring that the average amplitude within a window equals zero. This guarantees that each window maintains a unique ZCR value. Additionally, $x(n)$ should be normalized so that its amplitude remains within the range of -1 to 1. As shown in (1), the ZCR is directly proportional to the dominant frequency of $x(n)$, meaning that music typically exhibits a higher ZCR than speech. However, speech signals can experience a sudden surge in ZCR due to strongly unvoiced segments, leading to instances where speech ZCR exceeds that of music. The properties of ZCR can be summarized as follows:

1. **Dominant Frequency (DF) Principle**

If a signal is a pure sinusoidal waveform, its spectrum contains only the dominant frequency, which corresponds to the number of zero crossings per second. In other words, if the ZCR is measured every second, it directly reflects this frequency. For a non-sinusoidal periodic waveform, the dominant frequency is the component with the highest amplitude, responsible for driving the signal across the zero line in each cycle—provided the average amplitude is zero. Consequently, the dominant frequency of a non-sinusoidal periodic waveform is approximately equivalent to its ZCR. An accurate dominant frequency (ω_o) can be calculated using the formula

$$\omega_o = \frac{\pi \, \mathrm{E}\{D_o\}}{N - 1} \tag{2}$$

where ω_o is the dominant frequency in radians per second, D_o is the ZCR value counted in every interval, taking one second as the unit of rating the ZCR, N is the number of intervals and E{.} denotes the expected value.

2. **Highest Frequency Detection**

Since $D0$ denotes the ZCR of a discrete signal $Z(i)$, let us assume that $D1$ is the ZCR of the first derivative of $Z(i)$, and Dn denotes the ZCR of the nth derivative of $Z(i)$. The highest frequency in the signal can be approached using the following equation

$$\omega_{\max} = \lim_{i \to \infty} \frac{\pi \mathrm{E}\{D_i\}}{N - 1} \tag{3}$$

where $\omega_{\max}$ denotes the highest frequency in the signal and N is the number of samples. If the index i in (3) reach 10 the change of $\omega_{\max}$, it can be ignored if the sampling rate equals 11k samples per second.

3. **Lowest Frequency Detection**

In discrete signals, an approximate derivative can be evaluated by simply subtracting the amplitude of a sample from the amplitude of the previous sample, assuming that the time between any two samples is unity, so $\nabla Z(i) = Z(i) - Z(i - 1)$, where ∇ stands for the first difference operator. Also assume that D_n denotes the ZCR of the nth derivative of $Z(i)$. If we define another operator $\nabla+$ as the positive derivative effect, where,

$$\nabla + [Z(i)] = Z(i) + Z(i - 1) \tag{4}$$

Also assume that ${}_nD$ is the ZCR of the nth positive derivative of the signal $Z(i)$. Then the lowest frequency of the signal can be described as

$$\omega_{\min} = \lim_{i \to \infty} \frac{\pi \mathrm{E}\{{}_iD\}}{N - 1} \tag{5}$$

where $\omega_{\min}$ is the lowest frequency. From the previous information, a general shape of the signal spectrum is obtained.

4. **Periodicity Measure**

The signal is pure periodic if an only if

$$\frac{\mathrm{E}\{_1D\}}{N-1} = \frac{\mathrm{E}\{_2D\}}{N-1} \tag{6}$$

and as these two sides close to each other as the signal becomes more periodic.

The STE Algorithm

A simple definition of the STE can be expressed as [7]

$$E_n = \sum_{m=n-N+1}^{n} |x(m)|^2 \tag{7}$$

which means that the STE at segment n is simply the sum of squares of the N segments, $n - N + 1$ through n. Before evaluating the STE, $x(m)$ should be normalized so that its amplitude average equals to zero and its amplitude variation is surrounded in the real interval $[-1, 1]$. The amplitude of a speech signal fluctuates significantly over time, with unvoiced segments generally exhibiting much lower amplitudes compared to voiced segments. Therefore, the STE of a speech signal offers an effective representation of these variations. In contrast, music signals lack unvoiced segments, resulting in consistently higher STE values compared to speech.

2.2 Overlapped Segregation Using ICA

ICA has recently become an important tool for source segregation in cocktail-party problem [5]. It is successfully used to separate two mixed sounds using two statistically independent non-Gaussian distribution observers, every observer containing only a mixed of both sounds. If two persons are speaking simultaneously in a closed room and two microphones in different locations are receiving the mixed sounds, the microphones should give two recorded time signals, $x_1(t)$ and $x_2(t)$. Each of these recorded signals is a weighted sum of the speech signals emitted by the two speakers, which can be denoted by $s_1(t)$ and $s_2(t)$. They could be expressed as the following linear equations

$$x_1 = a_{11}s_1 + a_{12}s_2 \tag{8}$$

$$x_2 = a_{21}s_1 + a_{22}s_2 \tag{9}$$

which can be expressed in a matrix form as

$$\mathrm{X} = \mathbf{AS} \tag{10}$$

where the set of a_{ij} depends on the distances between the microphones and the speakers. The ICA approach can estimate the two original speech signals s_1 and s_2 using only the recorded signals x_1 and x_2 without any information about the matrix **A**. This case is called a cocktail-party problem in the time domain and (10) is called ICA model. The ICA problem is solved on the basis of minimizing or maximizing certain contrast functions like minimizing entropy, and minimizing mutual information, which transforms the ICA to a numerical optimization problem. Consider the ICA model of the form

$$x_i = \sum_{j=1}^{n} a_{ij} s_j \tag{11}$$

where x_i is one value of the vector compounding the original signals, s_j is one value of the vector compounding the observed signals vectors, and n in number of available observers, assuming that x_i and s_j are random variables. The statistical model in (10) and (11) is a generative model. It describes how the observed data are generated using a process of mixing the components s_j. The independent components are latent variables, meaning that they cannot be directly observed. Also the mixing matrix **A** is assumed to be unknown. Because of the length of analysis and discussion needed for the ICA algorithm, this case will not be discussed in this paper.

3 Simulations

3.1 ZCR and STE

The non-overlapping segregation algorithm functions as a music canceller, effectively removing music from an audio stream containing pure speech, music, a mixture of both, or silent intervals. The proposed approach utilizes a voting algorithm technique [8–9], incorporating enhancements to established methods such as ZCR and STE. These modifications include the addition of the ZCR positive derivative (ZCRD) and voting mechanisms. The algorithm employs a sampling rate of 11,025 Hz, with non-overlapping frames consisting of 256 data points, captured at 23.2-ms intervals. A 2.39-s segment of data consists of 103 frames.

3.2 The Voting Algorithm

During the training phase, three threshold values must be assessed for the ZCR, ZCRD, and STE discriminators, ensuring accurate differentiation between speech and non-speech signals. These thresholds will be called the *zcrth,* the *zcrdth* and the *steth,* for the ZCR, the ZCRD and for the STE, respectively. Table 1 shows the approximate thresholds [10].

For the recognition of the music signal, the following algorithm is proposed:

Step 1: If the signal has amplitude less than 10% of the maximum value, it is most likely noise, so it will be cancelled. This step will be called *small samples cancellation.*

Step 2: When using a sampling rate of 11 kHz, a window of 256 samples will be considered.

Step 3: Count the threshold values of each of the ZCR, the ZCRD, and the STE of the window and pass them to the voting algorithm.

Table 1. Means and STDs of the ZCR, the ZCRD and the STE in the training phase.

Type of measurement	Speech	Non-speech
Mean of the normalized ZCR	mzcrs = 25.6021 cross/window	mzcrns = 37.6445 cross/window
Mean of the STD of the normalized ZCR	sdzcrs = 11.9890	sdzcrns = 16.6544
Mean of the normalized ZCRD	mzcrds = 24.8819 cross/window	mzcrdns = 31.8470 cross/window
Mean of the STD of the normalized ZCRD	sdzcrds = 8.6669	sdzcrdns = 11.2824
Mean of the normalized STE	mstes = 23.1531 unit/window	mstex = 14.2820 unit/window
Mean of the STD of the normalized STE	sdstes = 12.7813	sdstex = 5.3450

Step 4: Assign a value of one for speech and a value of zero for non-speech according to the majority vote. This step will be called 01-*voting* block diagram. Figure 2 shows the flow of the above steps.

Step 5: For a segment of 20 windows, about half a second, summate all ones of Step 4. If the sum is less than 9, then the input sound is most likely music. This step will be called *sm-vote*. Note that the number 9 can be considered as the last fourth threshold with the name I-9. This step adds a subjective threshold.

Step 6: Take one frame composed of three segments, about 1.5 *s*, and then cancel the detected speech segments that are surrounded by music segments. That is because it is rare to have a speech signal for half a second. Also un-cancel music detected segment surrounded by speech segments within one frame for the same logical reason. Music canceller's ideal time delay comes from Step 4. This step will be called *fault correction*. It works at the highest levels of the frames.

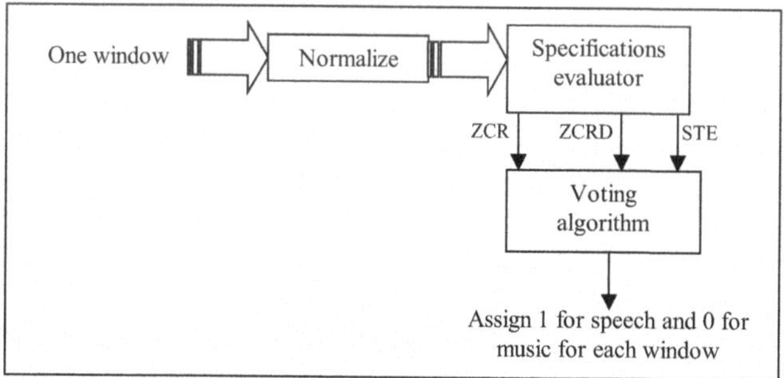

Fig. 2. 01-*voting* block diagram.

A frame of four segments can also be considered. Comparing every two segments with its surrounding neighbors segments will increase the precision of the algorithm with the expense of longer ideal time delay. According to user's preference, the subjective threshold I-9 can be controlled manually. The block diagram of the music canceller algorithm is depicted in Fig. 3.

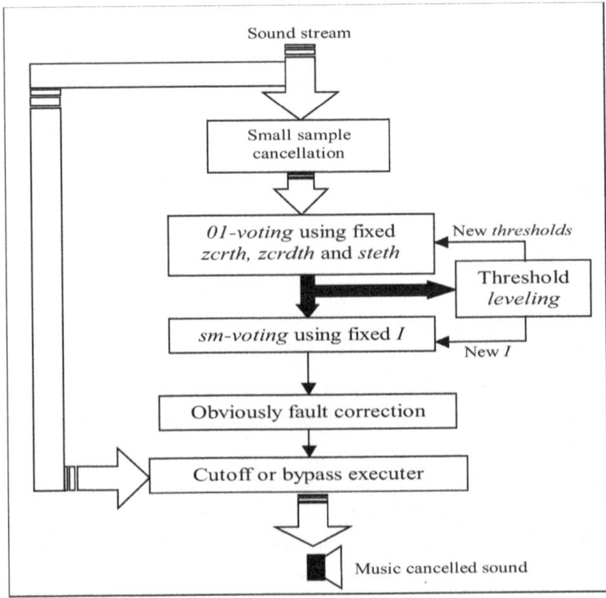

Fig. 3. Block diagram of the music canceller algorithm.

3.3 Floating Thresholds Canceller

The above algorithm can be improved for real world noisy broadcasting using a feedback that controls the values of the four thresholds. These additional thresholds will be adjustable and called the *thresholds leveling*. Figure 4 shows the threshold definition of the ZCR and the STE algorithms, where *A*, *B*, and *C* represents the threshold of speech, mixture, and music, respectively [11].

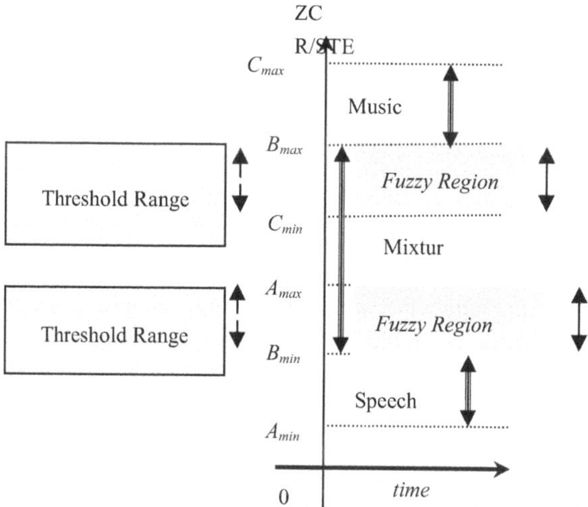

Fig. 4. Threshold definition for ZCR/STE.

The improvement of the algorithm can be summarized as follows:

Step 1: Let the music canceller runs as is for the first four frames (about 2 s), then from the output of 01-*vote* block, a string of sum of ones for each segment is evaluated. Recalling the main music canceller algorithm, this output will be the input of the *sm-vote* block. If the sum of ones in a specific segment is less than or equal to I-9, then the segment is most likely speech, otherwise it is music.

Step 2: From the calculated string, the mean of the sums of ones is evaluated and assigned as a new threshold I.

Step 3: If the new threshold I < 9, then.

(1) I = I + 1
(2) Increase the *zcrth*, the *zcrdth* and the *steth* by 10% percent of their original values;

Otherwise,

(1) I = I − 1
(2) Decrease the *zcrth*, the *zcrdth* and the *steth* by 10% of their original values.

Step 4: Use the new threshold instead of the fixed values.

Step 5: Repeat Step 2 to Step 4 until I = 9.

In addition, we have tried to utilize the obvious *fault correction* block activates to improve the thresholds values. This music canceller algorithm will take less than 0.5 s of the sound under processing and has a maximum ideal delay of 1.5 s.

Table 2. Classification results using the voting algorithm.

Detection	Speech	Non-speech
Correct	90%	91%
Error	10% non-speech	9% speech

While Table 2 indicates a maximum discrimination error of approximately 10%, this value is influenced by the number of speech and music samples used in the training and testing sets. As demonstrated in Table 3, the average discrimination error can be reduced to around 3% when both the training and testing sets contain only one speaker and a single type of music signal. Some researchers have reported lower discrimination error rates due to such controlled conditions. In our study, we aimed to avoid this limitation by adopting a more generalized approach. Consequently, as the number of specimens increases, the discrimination error also rises due to the introduction of signals with new features.

Table 3. Classification results using a voting algorithm, applied to a single speaker and a specific type of music.

Detection	Speech	Non-speech
Correct	96%	98%
Error	4% non-speech	2% speech

3.4 Subjective Versus Objective Perceptual

With the exception of the evaluation of I, all perceptual measurements were conducted objectively to establish threshold values for averages. However, when I was used as the final voting threshold, it became evident that purely objective measurements do not always yield successful results. Repeating the evaluation of I subjectively led to an improvement in algorithm performance, so, before implementing any subjective measurements, certain challenges must be addressed. One major issue is the variability in auditory perception among different individuals, resulting in a wide range of possible response models. In assessing the quality of wideband music sound (compact disc quality, ranging from 20–20,000 Hz), discrepancies among subjective evaluations pose a significant challenge. While precise data on individual subjects was not available, proper level adjustments proved to be crucial for ensuring accuracy in objective perceptual measurements.

4 Conclusions

This paper explores the segregation of music and speech signals, categorizing the problem into two distinct cases. The first occurs when the signals overlap only in either the time domain, or in the frequency domain. The second, and most complex scenario, involves overlapping in both time and frequency domains. To address the first case, a specialized frequency-domain filter was designed to separate the two signals. For instances where the overlap occurs in the frequency domain, a time-domain filter was implemented, utilizing zero-crossing rate (ZCR), short-time energy (STE), voting and nomination algorithms, along with the positive derivative approach. When both domains exhibit overlap, independent component analysis (ICA) was introduced with specific constraints to facilitate segregation. Theoretical analyses and simulations were conducted to evaluate the effectiveness of the proposed algorithms.

References

1. Kruspel, A., Dominik, Z., Lukashevich, H.: Automatic speech/music discrimination for broadcast signals. INFORMATIK, 151–162 (2017)
2. Thiruvengatanadhan, R., Dhanalakshmi, P.: Speech/Music classification using wavelet based feature extraction techniques. J. Comput. Sci. **10**(1), 34–44 (2014)
3. Saunders, J.: Real-time discrimination of broadcast speech/music. In: Proceedings ICASSP96, Atlanta, GA, vol. 2, pp. 993–996 (1996)
4. El-Maleh, K., Klein, M., Petrucci, G., Kabal, P.: Speech/music discriminator for multimedia application. In: Proceedings IEEE International Conference Acoustics, Speech, Signal Processing, Istanbul, pp. 2445–2448 (2000)
5. Hyvarinen, A., Oja, E.: Independent Component Analysis: Algorithms and Applications. www.cis.hut.fi/projects/ica/. Neural Networks (1999)
6. Alshoshan, A.I.: Speech and music classification and separation: a review. J. King Saud Univ. **19**(1), 95–133 (2006)
7. Kedem, B.: Spectral analysis and discrimination by zero-crossings. Proc. IEEE **74**(11), 1477–1492 (1986)
8. Hirvonen, T.: Speech/music classification of short audio segments. In: IEEE International Symposium on Multimedia, Taiwan, pp. 135–138 (2014)
9. Florian, E.: Real-time Speech and Music Classification by Large Audio Feature Space Extraction. Springer International Publishing, Cham (2016). https://doi.org/10.1007/978-3-319-27299-3
10. Alshoshan, A.I., Al-Atiyah, A., Al-Mashouq, K.: A three-level speech, music, and mixture classifier. J. King Saud Univ. **16**(2), 319–332 (2003)
11. Alshoshan, A.I.: A classification of music, speech and mixture signals via fuzzy logic. In: The 28th International Conference on Computers and Their Applications, pp. 117–122 (2013)

Analysis of the Transfer Function of an LTI System for a Non-stationary Process

Abdullah I. Alshoshan(✉)

College of Computer, Qassim University, Buraydah, Qassim, Saudi Arabia
drshoshan@gmail.com

Abstract. For stationary signals, the transfer function of a linear, time-invariant (LTI) system in frequency domain is well-known as the Fourier (or Laplace) transform of the output divided by the Fourier (or Laplace) transform of the input. However, since most of the signals in diverse fields have non-stationarity behavior, the above definition of the LTI transfer function does not hold. In this paper, a new relationship between the input and the output of an LTI system, i.e., the LTI system transfer function, is proposed when the input is a non-stationary process. The Wold-Cramer definition of a non-stationary signal is used. Using the proposed method of finding the LTI system transfer function, it is shown that the effect of stationary Gaussian noise in the input can be removed or reduced.

Keywords: Non-Stationary Process · LTI transfer function · Orthogonal Polynomial (OP) · Wold-Cramer · Evolutionary Spectrum (ES)

1 Introduction

1.1 Background of LTI Systems

The modeling of an LTI system, when the input is a non-stationary process, is a problem of great interest in diverse fields in signal processing, and has many applications, like: speech and image processing, geophysics, bio-statistical signal processing, radar, medicine, and seismology [1, 2]. For each application, the modeling results in a set of mathematical differential equations, which can be used to understand the behavior of the system and/or the process, as depicted in Fig. 1.

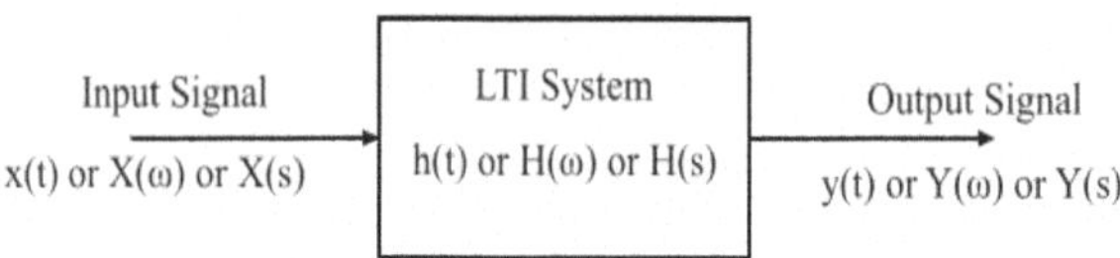

Fig. 1. The transfers function of continuous-time LTI systems.

I. Roy et al. (Eds.): CAINE 2025, CCIS 2740, pp. 42–53, 2026.
https://doi.org/10.1007/978-3-032-14893-3_4

For a stationary signal, the transfer function is defined as $H(s) = Y(s)/X(s)$, where $Y(s)$ and $X(s)$ are the Laplace transforms of the output and the input, respectively. In the frequency domain, the transfer function, also called the frequency response, of an LTI system is defined as $H(\omega) = Y(\omega)/X(\omega)$, where $Y(\omega)$ and $X(\omega)$ are the Fourier transform of the output and the input, respectively. When the process is non-stationary, neither the classical power spectrum nor the bi-spectrum can handle the modeling problem since they do not reflect the time variation of the process characteristics, i.e., none of the above relations hold [3]. Another important issue in the modeling of a non-stationary process is the knowledge of the degree of the process non-stationarity. This can be reflected by the expansion order of the time-varying parameters embedded in the process. In previous works, experimental data and functions are represented or approximated by linear combinations of some predefined basis functions, which are usually of polynomial forms, bounded and closed under translation and scaling [4]. For these polynomials to exactly be able to represent a given data sequence, they must be able to span the whole space of the data sequence [5]. In the expansion approach, a set of OPs such as Hahn, Legendre, Laguerre, Jacobi, or Hermite polynomials can be used. Then, any time-frequency estimator can be used [6]. Some examples of time-frequency distributions used for analyzing non-stationary signals are: evolutionary periodogram (EP) [7], short-time Fourier transform (STFT) and modified group delay function (MGDF) [8], multi-taper marginal time–frequency distributions [9], Hartley S-transform [10], or maximum entropy (ME) [11], multi-window Gabor expansion [12], Wavelet Transform (WT) [13]. In this paper, the Wold-Cramer representation of non-stationary signals and the Legendre OPs are used [5].

1.2 Stationarity and Non-stationarity

Stationarity is an idealization and, in practice, it can at best be expected to hold only as an approximation. Before discussing non-stationary signals it is first necessary to clarify the meaning of stationarity itself. Stationarity, in essence, means that the process is in a state of equilibrium about a mean value. More specifically, it is in a steady state in the sense that the statistical properties of a stationary time series are independent of the value of time or of the time origin. There are several definitions of stationarity. If the probabilistic characterization of the process does not change with time, then the process is said to be *stationary in the strict sense* (SSS). This means that $x(t)$ and $x(t+c)$ have the same statistics for any c. Another definition of stationarity is the *wide-sense stationarity* (WSS), which is satisfactory for many of the needs of statistical signal processing. There are also some processes called cyclo-stationary processes [14]. A process $x(t)$ is called cyclo-stationary (or periodically stationary) if its statistics are invariant to a shift of the time origin by integer multiples of a constant period, T. A cyclo-stationary process is not stationary. Often, for non-stationary processes, the non-stationarity can happen either in the mean or in the autocorrelation or in both. A process $x(t)$ is said to be non-stationarity in the mean if its mean is a time-dependent function, i.e., $E\{x(t)\} = \alpha(t)$. A process that is stationary in the mean is not necessarily stationary in the autocorrelation. However, a process that is non-stationary in the mean will also be non-stationary in the autocorrelation. It is clearly of interest, therefore, to examine what type of analysis is available for those cases where the assumption of stationarity becomes not true or an unrealistic assumption. In many practical applications such as seismic surveying, detection, system identification, sonar, communications, speech, and medical instrumentation, the signal analysts are faced with the task of processing signals whose spectral characteristics are time-dependent. Therefore, non-stationary signals need special mathematical treatment when estimating their spectrum or when identifying system with non-stationary input. In this paper, an LTI system identification approach is considered with the assumption of a non-stationary input.

2 LTI System Modeling with a Non-stationary Input

According to the Wold-Cramer (WC) representation, a discrete-time non-stationary process, $x(n)$, admits a representation of the form [15]

$$x(n) = \int_{-\pi}^{\pi} A(n, \omega) e^{j\omega n} dZ(\omega) \tag{1}$$

where $A(n,\omega)$ is a slowly varying function of time. Defining $H(n,\omega) = A(n,\omega)dZ(\omega)$, the time-varying spectrum of $x(n)$ is given by $S_x(n,\omega) = E\{|H(n,\omega)|^2\}$. According to the WC decomposition, a discrete-time non-stationary process, $x(n)$, can be represented as the output of a causal, linear, and time-varying (LTV) system, with impulse response $h(n,m)$, to a discrete-time stationary zero-mean, unit-variance white noise process, $e(n)$, as shown in Fig. 2.

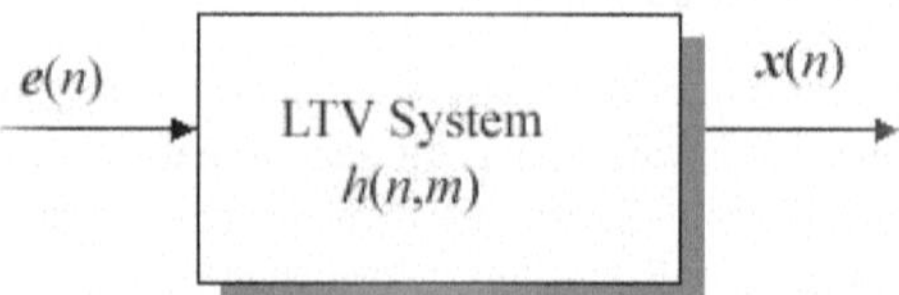

Fig. 2. The Wold-Cramer representation of a non-stationary signal.

Therefore,

$$x(n) = \sum_{m=-\infty}^{n} h(n, m) e(n - m) \tag{2}$$

and $e(n)$ can be represented as

$$e(n) = \int_{-\pi}^{\pi} e^{j\omega n} dZ(\omega) \tag{3}$$

From the above two equations, we get

$$x(n) = \int_{-\pi}^{\pi} \sum_{m=-\infty}^{n} h(n, m) e^{j\omega(n-m)} dZ(\omega) \tag{4}$$

and using the definition of the generalized transfer function of an LTV system

$$H(n,\omega) = \sum_{m=-\infty}^{n} h(n,m)e^{-j\omega m} \tag{5}$$

Therefore, we get

$$x(n) = \int_{-\pi}^{\pi} H(n,\omega)e^{j\omega n}dZ(\omega) \tag{6}$$

and $H(n,\omega)$ can be estimated as follows:

$$\hat{H}(n,\omega) = \sum_{k=0}^{N-1} w_n(k)x(k)e^{-jwk} \tag{7}$$

which uses a window, $w_n(k)$, that varies with time and constructed from a combination of M number of OPs such that

$$w_n(k) = \sum_{i=1}^{M} \beta_i^*(n)\beta_i(k) \tag{8}$$

Hence, a non-stationary process can be expressed as a sum of sinusoids with random time-varying amplitudes and phases. The power of $\boldsymbol{x}(n)$ at a time instant n is given by

$$E\{|x(n)|^2\} = \frac{1}{2\pi}\int_{-\pi}^{\pi} |H(n,\omega)|^2 d\omega \tag{9}$$

which shows that the power distribution of $\boldsymbol{x}(n)$ is a function of both the discrete-time, n, and the frequency, ω. Thus, the WC evolutionary spectrum is defined as

$$S(n,\omega) = |H(n,\omega)|^2 \tag{10}$$

For $S(n,\omega)$ to be an evolutionary spectrum, Priestley [16] requires that $H(\Omega,\omega)$, the Fourier transform of $H(n,\omega)$ with respect to n, has a maximum at $\Omega = 0$, which implies that $H(n,\omega)$ is a slowly-varying function of time. A less restrictive condition results from considering $H(n,\omega)$ to be the envelope corresponding to a carrier $e^{j\omega n}$, which would then require that $H(n,\omega)$ varies slower in time than the complex exponential [15]. In this Section we derive the spectral relationship between the output and the input of an LTI system when the input is non-stationary. Consider an LTI system with impulse response $g(n)$ and input $x(n)$. Since the input $x(n)$ is non-stationary, the output will also be non-stationary, and is given by

$$y(n) = g(n) * x(n) \tag{11}$$

where * stands for convolution. Since

$$x(n) = \int_{-\pi}^{\pi} H_x(n,\omega)e^{j\omega n}dZ(\omega) \tag{12}$$

then (11) gives

$$\begin{aligned} y(n) &= \sum_m g(n-m)x(m) = \int_{-\pi}^{\pi} \sum_m g(n-m)e^{j\omega m} H_x(m,\omega)dZ(\omega) \\ &= \int_{-\pi}^{\pi} H_y(n,\omega)e^{j\omega n} dZ(\omega) \end{aligned} \tag{13}$$

Therefore,

$$H_y(n,\omega)e^{j\omega n} = \sum_m g(n-m)e^{jwm} H_x(m,\omega) \tag{14}$$

or equivalently

$$H_y(n,\omega) = (g(n)e^{-j\omega n}) *_n H_x(n,\omega) \tag{15}$$

where $*_n$ stands for convolution with respect to n. Taking the Fourier transform of (15) with respect to n, we get

$$H_y(\Omega,\omega) = G(\Omega+\omega)H_x(\Omega,\omega) \tag{16}$$

Therefore, the proposed definition of the transfer function of an LTI system with a non-stationary input and a non-stationary output is defined as

$$G(\Omega+\omega) = \frac{H_y(\Omega,\omega)}{H_x(\Omega,\omega)} \tag{17}$$

To find $G(\omega)$, we let $\omega = \alpha - \Omega$ in (17), where α is an arbitrary variable, then we have

$$G_\Omega(\alpha) = \frac{H_y(\Omega,\alpha-\Omega)}{H_x(\Omega,\alpha-\Omega)} \tag{18}$$

If we discretize the frequencies α and Ω, then (18) can be written as

$$G_{\omega_l}(\omega_k) = \frac{H_y(\omega_l,\omega_k-\omega_l)}{H_x(\omega_l,\omega_k-\omega_l)}, \quad \omega_l = 2\pi\, l\,/N \tag{19}$$

where N is the number of frequency points available. To get a smoothed estimate of $G(\omega)$ we average over ω_i to have

$$G(\omega_k) = \frac{1}{N}\sum_{l=0}^{N-1} G_{\omega_l}(\omega_k) \tag{20}$$

Equation (15) can be used to find the evolutionary spectrum of the output of an LTI system when the input is non-stationary directly from the evolutionary kernel of the input signal, i.e.

$$S_y(n,\omega) = |(g(n)e^{-j\omega n}) *_n H_x(n,\omega)|^2 \tag{21}$$

We observe from (18) and (20) that both the phase and the magnitude of $G(\omega)$ are preserved. Therefore, we can use them to identify non-minimum phase systems.

Additionally, from the above analysis we see that a non-stationary signal $y(n)$ can be represented as the output of the cascade of a LTV system, with an impulse response $h_x(n,m)$, followed by an LTI system with impulse response $g(n)$, as depicted in Fig. 3.

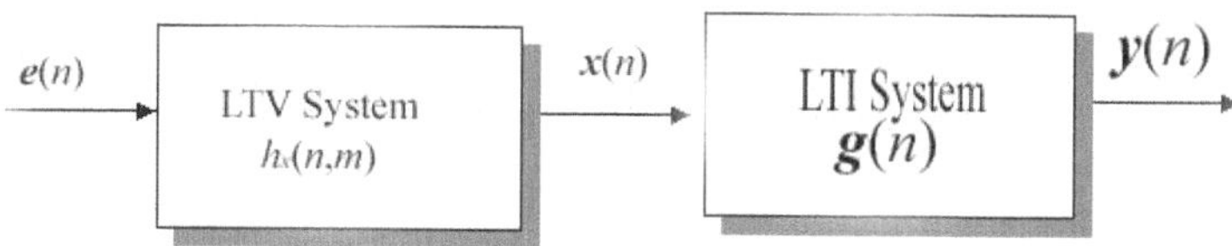

Fig. 3. A general representation of a non-stationary signal.

3 Special Cases

In this section, we will discuss and analyze some special cases, deduced from the above relationships, such as the stationarity, separating the stationary from the non-stationary parts of the signal, the time delay, the cascaded/parallel systems, and the noisy case.

Case 1: In the special case when $x(n)$ is stationary, both $H_x(n,\omega)$ and $H_y(n,\omega)$ are constants with respect to n, and the Fourier transform of $H_x(n,\omega)$ with respect to n is given by

$$H_x(\Omega,\omega) = H_x(\omega)\delta(\Omega) = \begin{cases} H_x(\omega)\delta(\Omega) & \Omega = 0 \\ 0 & \text{otherwise} \end{cases} \tag{22}$$

Similarly for $H_y(\Omega,\omega)$. Thus (16) becomes

$$H_y(\omega) = G(\omega)H_x(\omega) \tag{23}$$

Therefore, assuming that $\boldsymbol{x}(n)$ is the output of an LTI system with frequency response $H_x(\omega)$ and white noise as input then [2]

$$S_y(\omega) = |H_y(\omega)|^2 = |G(\omega)|^2\ S_x(\omega) \tag{24}$$

which is the classical formula relating the spectrum of the response of an LTI system to the spectrum of the stationary input. This means that the formulation of the spectrum when the signal is stationary is a special case of our formulation, as expected.

Case 2: A non-stationary signal $x(n)$ can be expressed as the sum of a purely stationary component, $s(n)$, and a non-stationary residual $d(n)$, i.e., $x(n) = s(n) + d(n)$. Now, let us try to separate the non-stationary part of $x(n)$ from its stationary part. Let

$$H_s(\omega_k) = H_x(\Omega_l, \omega_k)|_{\Omega_l=0} \tag{25}$$

or the spectrum of $s(n)$ be

$$S_s(\omega_k) = |H_x(\Omega_l, \omega_k)|_{\Omega_l=0}|^2 \tag{26}$$

and $d(n)$ be such that

$$H_d(\Omega_l, \omega_k) = \begin{cases} 0 & \Omega_l = 0 \\ H_x(\Omega_l, \omega_k) & \Omega_l \neq 0 \end{cases} \tag{27}$$

Hence, estimating $H_x(\Omega_l,\omega_k)$ will reflect the degree of stationarity of the signal. Strong concentration of $H_x(\Omega_l,\omega_k)$ about Ω_l indicates that the signal is quasi-stationary. When the concentration is not around Ω_l, the signal is strictly non-stationary. This of course depends on how accurate or how good is our estimator $\hat{H}_x(\Omega_l, \omega_k)$.

Case 3: Time delay. Given the process $y(n) = x(n - D)$, where D is a constant integer. $y(n)$ has a relation with $x(n)$ as follows:

$$H_y(n, \omega) = H_x(n - D, \omega)e^{-j\omega D} \tag{28}$$

Hence

$$H_y(\Omega, \omega) = H_x(\Omega, \omega)e^{-jD(\omega-\Omega)} \tag{29}$$

and then, taking the natural logarithm of both sides of the above equation, we have

$$logH_y(\Omega, \omega) = logH_x(\Omega, \omega) - jD(\omega - \Omega) \tag{30}$$

therefore, the time delay D can be found as

$$D = \frac{logH_y(\Omega, \omega) - logH_x(\Omega, \omega)}{j(w - \Omega)} = \frac{log[H_y(\Omega, \omega)/H_x(\Omega, \omega)]}{j(\omega - \Omega)}, \ \Omega \neq \omega \tag{31}$$

Case 4: Cascade/Parallel Connections. According to (11), two systems with system functions $G_1(\omega)$ and $G_2(\omega)$, when connected in cascade, form an overall system having the relationship

$$H_y(\Omega, \omega) = G_1(\Omega + \omega)G_2(\Omega + \omega)H_x(\Omega, \omega) \tag{32}$$

Therefore

$$G(\Omega + \omega) = G_1(\Omega + \omega)G_2(\Omega + \omega) \tag{33}$$

On the other hand, if the two systems are in parallel, their overall system relationship can be written as

$$H_y(\Omega, \omega) = (G_1(\Omega + \omega) + G_2(\Omega + \omega))H_x(\Omega, \omega) \tag{34}$$

Therefore

$$G(\Omega + \omega) = G_1(\Omega + \omega) + G_2(\Omega + \omega) \tag{35}$$

where, if $\boldsymbol{x}(n)$ is stationary, (33) and (35) reduce to

$$G(\omega) = G_1(\omega)G_2(\omega) \tag{36}$$

and

$$G(\omega) = G_1(\omega) + G_2(\omega) \tag{37}$$

respectively, which is the well-known case when $x(n)$ is stationary.

Case 5: In the case when the output signal is embedded in stationary noise, since the stationary noise will be concentrated around $\Omega_l = 0$, we do not consider this slice in the identification process, therefore, (20) is written as

$$\hat{G}(k) = \frac{1}{N-1}\sum_{l=1}^{N-1}\hat{G}_l(k),\ k = 0, 1, ..., N-1 \tag{38}$$

Hence, the stationary noise can be minimized by averaging.

4 Regions of Support

In this section, the regions of supports of the parameters $H_x(\Omega,\omega)$ and $G(\Omega+\omega)$ is analyzed. For $H_x(\Omega,\omega)$, its region of support is bounded by $-\pi \le \Omega \le \pi$ and $-\pi \le \omega \le \pi$, which is the shaded area shown in Fig. 4 (a), i.e., $H_x(\Omega,\omega)$ will always be zero outside this shaded area. For $G(\Omega+\omega)$, its region of support will be bounded by the frequencies $\Omega = -\pi$ for $\pi \le \omega \le 2\pi$, $\Omega = \pi$ for $-2\pi \le \omega \le -\pi$, $\omega = \pi$ for $-2\pi \le \Omega \le -\pi$, $\omega = -\pi$ for $\pi \le \Omega \le 2\pi$, $\omega = -\pi - \Omega$, and $\omega = \pi - \Omega$, which is the shaded area shown in Fig. 4 (b), i.e., $G(\Omega+\omega)$ will always be zero outside this shaded area. In the stationary case, one can only use the slice of $H_x(\Omega,\omega)$ at $\Omega = 0$, however, in the non-stationary case, as many slices as in the Ωxω domain can be used. To be able to identify the frequency response $G(\omega)$, at least one slice of $H_x(\Omega,\omega)$ in the Ωxω domain has to be nonzero. Let that slice be at $\Omega = \Omega_0$. Therefore

$$G(\Omega_0+\omega) = \frac{H_y(\Omega_0,\omega)}{H_x(\Omega_0,\omega)} \tag{39}$$

which will produce a version of $G(\omega)$ shifted by Ω_0.

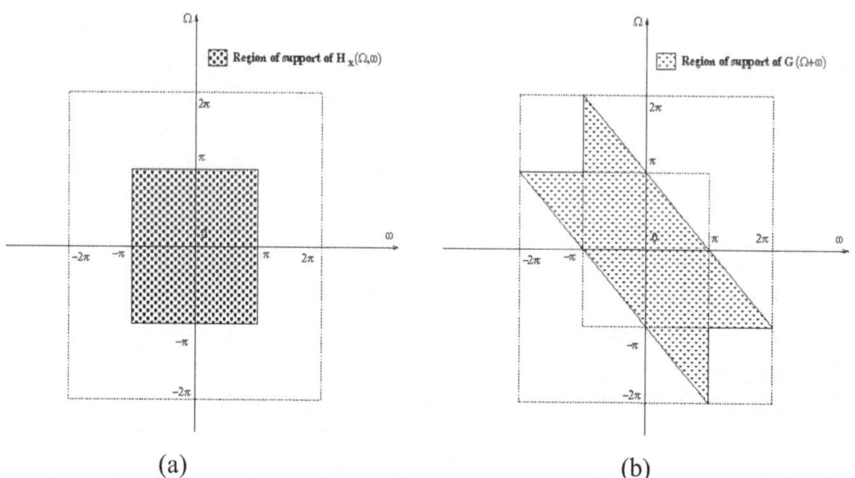

(a) (b)

Fig. 4. The region of support of: (a) $H_x(\Omega,\omega)$, (b) $G(\Omega+\omega)$.

To satisfy the constraint $H_x(\Omega_0,\omega)\neq 0$, we choose $h_x(n,m)$ to produce a non-stationary signal $x(n)$ such that $H_x(\Omega_0,\omega)$ has nonzero values within some specified region of support *R*. For the region R in in Fig. 4, we assume that $H_x(\Omega_0,\omega)$ can be expressed as a separable two-dimensional filter $H_1(\Omega_0)H_2(\omega)$. One possible choice for $H_1(\Omega_0)$ and $H_2(\omega)$ is that they are ideal low-pass filters with cut off frequencies *a* and *b*, respectively. Computing the 1-D inverse Fourier transforms of $H_1(\Omega_0)$ and $H_2(\omega)$ we obtain

$$h_x(n, m) = h_1(n)h_2(m) = \frac{\sin(an)}{\pi n}\frac{\sin(bm)}{\pi m} \tag{40}$$

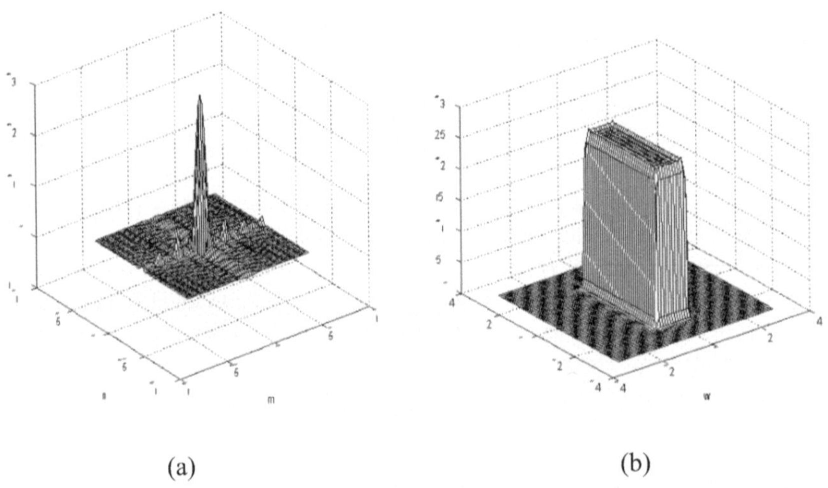

(a) (b)

Fig. 5. For $a = 0.1\pi$ and $b = 0.3\pi$: (a) $h_x(n,m)$, (b) $|H_x(\Omega_k,\omega_k)|$.

where *a* and *b* can be chosen to cover the region of support R. For arbitrary values of these constants shown in Fig. 4, we can control the region of support in the $\Omega_k x \omega_k$ domain. In general, the identification requires that the support of $G(\Omega+\omega)$ be included in the support of $H_x(\Omega,\omega)$. When they do, there will be more than one slice, except in the stationary input, that can be used in the identification. In the case of x(n) being a delta function, the support of $H_x(\Omega,\omega)$ could be the whole $\Omega x\omega$ plane. Thus using the delta input gives the largest possible intersection to identify LTI system, and the stationary case provides the smallest possible intersection. To apply (39), it is needed to constrain $H_x(\Omega,\omega)$ to be different from zero in the support of $G(\Omega+\omega)$ [15].

5 Simulations

In this section, some simulations are provided to show the effectiveness of the proposed method. The first example is considering the identification of a non-minimum phase LTI system with a non-stationary input. The second example is considering the identification when the non-stationary input is embedded with a stationary Gaussian noise.

Example 1

The identification process can be done using the slice $\Omega_k = 0$ of the $\Omega_k x \omega_k$ plane. Let $x(n)$ be the response of a LTV system with impulse $h_x(n,m)$ such that $H_x(\Omega_k,\omega_k) = \delta(\Omega_k)$ and a zero-mean, unit variance, stationary white noise input, $e(n)$. Taking the inverse Fourier transform of $H_x(\Omega_k,\omega_k) = \delta(\Omega_k)$ with respect to Ω_k, we have $H_x(n,\omega_k) = 1$. Similarly, taking the inverse Fourier transform of $H_x(n,\omega_k)$ with respect to ω_k, we have $h_x(n,m) = \delta(m)$. Therefore.

$$x(n) = \sum_m h_x(n, m)e(n - m) = \sum_m \delta(m)e(n - m) = e(n) \tag{41}$$

as indicated previously. Therefore, $H_x(\Omega_k,\omega_k) = 1$ at $\Omega_k = 0$, which means that $\hat{G}(\omega_k) = \hat{H}_y(0, \omega_k)$. . Let us illustrate this using an example of a non-minimum-phase (NMP) LTI system. Consider that the NMP-LTI system is an MA with transfer function

$$G_0(z) = 1 + 2.35z^{-1} + .7z^{-2} \tag{42}$$

with zeros at $-.35$ and -2. The magnitudes of $G_0(\omega)$ and $\hat{G}_0(\omega)$ are shown in Fig. 6 (a), and the phases of $G_0(\omega)$ and $\hat{G}_0(\omega)$ are shown in Fig. 6 (b). The dotted lines are the estimated ones.

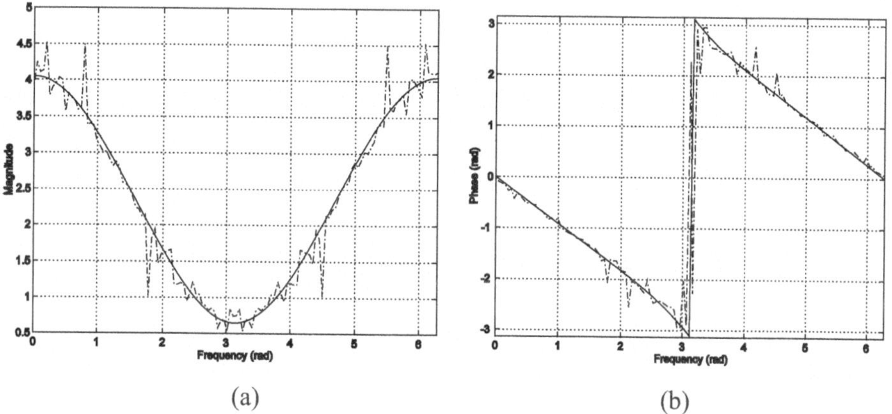

Fig. 6. The true (smooth) and estimated $G_0(\omega)$: (a) Magnitude, (b) Phase.

Example 2

In the case of noisy input/output, if the additive noise is stationary [17], the additive noise will be concentrated around $\Omega_k = 0$, therefore, we should avoid using the slice at $\Omega_k = 0$ of the $\Omega_k x \omega_k$ domain. Therefore, let $H_x(\Omega_k,\omega_k) = 1$ at $\Omega_k = d$, i.e., $H_x(\Omega_k,\omega_k) = \delta(\Omega_k - d)$, where $0 < |d| < \pi$. Hence $H_x(\Omega_k,\omega_k) = \delta(\Omega_k - d)$. Taking the inverse Fourier transform of the above equation with respect to Ω_k, we have.

$$H_x(n, \omega_k) = e^{jdn} \tag{43}$$

and then, taking the inverse Fourier transform of the above equation with respect to ω_k, we have

$$h_x(n, m) = \delta(m)e^{jnd} \tag{44}$$

which is a time-varying system. Therefore,

$$x(n) = \sum_m e^{jnd}\delta(m)e(n - m) = e^{jnd}e(n) \tag{45}$$

In particular, for $d = \pi$, $x(n) = (-1)^n e(n)$ is real. Applying the above analysis to the system described in Example 1, and setting $d = \pi$ to generate $x(n)$, then $H_x(\Omega_k,\omega_k) = \delta(\Omega_{k-}\pi)$ is estimated using the EP estimator. With $d = \pi$, the magnitudes of $G_0(\omega)$ and $\hat{G}_0(\omega)$ are shown in Fig. 7 (a), and the phases of $G_0(\omega)$ and $\hat{G}_0(\omega)$ are shown in Fig. 7 (b).

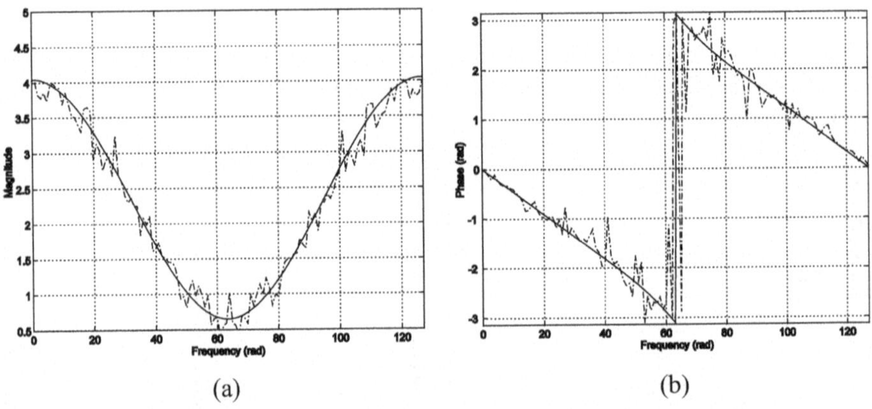

Fig. 7. The true (smooth) and estimated $G_0(\omega)$: (a) Magnitude, (b) Phase

6 Conclusions

For stationary signals, the transfer function of linear, time-invariant (LTI) systems in frequency domain is well-known as the Fourier (or Laplace) transform of the output divided by the Fourier (or Laplace) transform of the input. Since most of the signals in diverse fields have non-stationarity behavior, the above definition of the LTI transfer function does not hold. In this paper, a new definition of the LTI transfer function in the frequency domain was proposed and discussed, using the Wold-Cramer definition of a non-stationary signal. Some simulations were given for identifying the transfer function of LTI systems. An example of noisy output is given when the additive noise is stationary. The additive noise was concentrated around the zero frequency; therefore, the noise could be avoided by ignoring the first slice of the estimator. Future work is still needed for a more accurate estimator of the time-frequency kernel of the non-stationary input/output.

References

1. Evrengül, H., et al.: Time and frequency domain analyses of heart rate variability in patients with epilepsy. Epilepsy Res. **63**(2–3), 131–139 (2005)
2. Boashash, B., Ristic, B.: Time-varying higher-order Cumulant spectra: application to the analysis of composite FM signals in multiplicative and additive noise. In: ICASSP, Australia, vol. 4, pp. 325–328 (1994)
3. Cohen, L.: Time-frequency distributions-a review. Proc. IEEE **77**(7), 941–981 (1989)
4. El-Jaroudi, A., Redfern, M.S., Chaparro, L.F., Furman, J.M.: The application of time-frequency methods to the analysis of postural sway. Proc. IEEE **84**(9), 1312–1318 (1996)
5. Al-Shoshan, A.I., Chaparro, L.F.: Identification of non-minimum phase systems using the evolutionary spectral theory. Sig. Process. (Elsevier Sci.) **55**(1), 79–92 (1996)
6. Neuman, C.P., Schonbach, D.I.: Discrete (Legendre) orthogonal polynomials-a survey. Int. J. Numer. Methods Eng. **8**, 743–770 (1974)
7. Kayhan, A.S., El-Jaroudi, A., Chaparro, L.F.: Evolutionary periodogram for non-stationary signals. IEEE Trans. Signal Proc. **42**, 1527–1536 (1994)
8. Narasimhan, S.V., Pavanalatha, S.: Estimation of evolutionary spectrum based on short time Fourier transform and modified group delay. Signal Process. **84**(11), 2139–2152 (2004)
9. Aviyente, S., Williams, W.J.: Multitaper marginal time–frequency distributions. Signal Process. **86**(2), 279–295 (2006)
10. Robert Pinnegar, C., Mansinha, L.: Time–frequency localization with the Hartley S-transform. Sig. Process. **84**(12), 2437–2442 (2004)
11. Shah, S.I., Chaparro, L.F., Kayhan, A.S.: Evolutionary maximum entropy spectral analysis. In: ICASSP94, Australia, vol. 4, pp. 285–289 (1994)
12. Akan, A., Chaparro, L.F.: Evolutionary chirp representation of non-stationary signals via Gabor transform. Signal Process. **81**(11), 2429–2436 (2001)
13. Discetti, S.: Time–frequency analysis and wavelets. In: Mendez, M.A., Ianiro, A., Noack, B.R., Brunton, S.L. (eds.) Data-Driven Fluid Mechanics: Combining First Principles and Machine Learning, pp. 89–114. Cambridge University Press (2023)
14. Abboud, D., et al.: Synchronous analysis of cyclo-non-stationary signals: a comprehensive study with aeronautic applications. Mech. Syst. Signal Process. **168**, 1 (2022)
15. Al-Shoshan, A.I., Chaparro, L.F.: System identification and modeling of non-stationary signals using the wold-cramer representation. In: IEEE-SP International Symposium on TF and TS Analysis, Philadelphia, 25–28 October, pp. 421–424 (1994)
16. Priestley, M.B.: Non-Linear and Non-Stationary Time Series Analysis. Academic Press, New York, NY (1988)
17. Huang, L., Sun, X., Huang, T., Zhu, Z., Song, K.: An improved method to enhance SNR of non-stationary signal based on nonuniform windows adaptive modulation for measurement. IEEE Trans. Instrum. Meas. **72**, 1–11 (2023)

A Comparative Analysis of RAG and Non-RAG Models to Improve Access to Service Provider Information for Older Adults in Mississippi

Saviz Saei and Sujan Anreddy(✉)

Mississippi State University, Starkville, MS 39759, USA
{ss4646,sra116}@msstate.edu

Abstract. In communities where navigating community healthcare and social services, access to clear and relevant information is more than a convenience, it is a necessity. Motivated by this need, our study explores how intelligent retrieval systems can be designed to better serve users seeking local care resources. We began with a rich directory of 1,553 community resource providers throughout Mississippi, aiming to understand how people might naturally ask for help and what ideal responses would look like. To simulate realistic interactions, we developed ten different query-reference pairs that reflect common but critical health-related concerns, ranging from mental health access to elder care services. We then implemented two information retrieval approaches: one based on Retrieval-Augmented Generation (RAG) and another using a non-RAG baseline. Using BERTScore to evaluate the semantic similarity of generated responses to our curated references, we found the RAG system to be significantly more effective, achieving an F1 score of 0.8723 compared to 0.8221 from the non-RAG model, a relative improvement of 6.1%. These results demonstrate the value of augmenting language models with curated knowledge, especially when precision and relevance matter most. Ultimately, this work highlights how thoughtful system design rooted in real-world needs can bridge the gap between resource and referral landscapes and the people who rely on them.

1 Introduction

Accessible and accurate community service provider information is vital for community well-being, yet significant barriers persist, particularly for vulnerable populations and those in rural areas. In states like Mississippi, with a complex landscape of community resource providers, navigating available services can be a formidable challenge. This information gap often leads to delayed care and exacerbates health disparities [4]. Although digital tools have the potential to bridge this gap, traditional search systems often fail to interpret nuanced conversational queries, and generic AI assistants lack the specific local knowledge required to provide actionable guidance. Recent studies highlight that electronic

I. Roy et al. (Eds.): CAINE 2025, CCIS 2740, pp. 54–65, 2026.
https://doi.org/10.1007/978-3-032-14893-3_5

community resource referral systems can facilitate connections between individuals and essential social supports such as food assistance, housing, transportation, and utilities, but their implementation faces persistent barriers, including technical integration challenges, costs, and maintaining accurate provider directories [3]. To address this challenge, this paper investigates the efficacy of a RAG model for providing localized provider information. We present a comparative study evaluating two distinct approaches: A baseline non-RAG system that relies solely on the general knowledge of a LLM. A RAG-based system that grounds LLM responses in a curated knowledge base of 1,553 community healthcare and social service providers across Mississippi. Our motivation is to develop an intelligent information system that can effectively serve diverse communities, including older adults, people with disabilities, and low-income families who face the greatest obstacles in accessing care. By comparing these two architectures, we aim to quantify the practical benefits of RAG in a real-world provider context where accuracy, specificity, and equity are paramount.

The main contributions of this work are a direct comparative analysis of RAG and non-RAG models for a specialized provider information task, demonstrating that the RAG model improves semantic accuracy by 6.10% based on BERTScore F1. This research provides strong evidence that RAG-based architectures are a superior approach for domain-specific information retrieval, offering a pathway to more effective and equitable provider information systems.

2 Related Work

The proliferation of Large Language Models (LLMs) has transformed the landscape of natural language processing. However, a primary limitation of these models is their reliance on static, internal knowledge, which can lead to factually incorrect or outdated responses, a phenomenon often termed "hallucination" [11]. To address this critical issue, RAG has emerged as a leading paradigm to enhance the reliability and domain-specificity of LLMs by grounding their outputs in external, verifiable knowledge sources.

The foundational RAG framework, introduced by Lewis et al. [9], demonstrated significant performance gains on knowledge-intensive tasks by combining a pre-trained generator with a dense vector retriever. This approach has since been extensively explored, with comprehensive surveys like Gao et al. [5] documenting the rapid evolution of the field. Research has shown that for knowledge-intensive domains, RAG is often more adaptable and cost-effective than fine-tuning, as its knowledge can be updated by simply modifying the external database rather than undergoing costly model retraining [13]. However, engineering a high-performing RAG system involves overcoming numerous challenges, including optimizing retrieval relevance and managing the quality of the retrieved context [2]. Advanced techniques such as post-retrieval re-ranking [12] and query transformations [10] are active areas of research aimed at improving the performance of the core retrieval pipeline.

Our work applies these principles to the specific domain of local provider information, where significant barriers to access persist. Studies show that vulnerable groups, particularly older adults, face substantial challenges related to digital health literacy and technology design, which can prevent them from finding necessary care [7,8,17]. The challenge is not merely technical; it involves designing systems that can overcome the "digital divide" and build trust with users who may be skeptical of technology [14]. While the use of virtual assistants to help older adults navigate healthcare is a growing field of research [1], most studies focus on generalized medical question-answering rather than the distinct challenge of providing localized, provider-specific information.

To ensure a robust evaluation within this context, we employ BERTScore [16]. The importance of semantic similarity over lexical overlap is particularly pronounced in the clinical and healthcare domains, where correctly interpreting meaning is critical [15]. Unlike traditional n-gram metrics, BERTScore's ability to capture semantic equivalence makes it an ideal choice for evaluating the quality of generated responses in our provider information system. Our study contributes to this landscape by performing a direct comparative validation of RAG and non-RAG systems in this specialized context, offering empirical evidence for the utility of RAG in bridging critical information gaps for underserved communities.

3 Methodology

Our system employs a RAG architecture designed to provide accurate, context-aware information on senior services. The methodology is partitioned into two distinct phases: indexing process to prepare the knowledge base, and an inference process to handle user queries at runtime.

3.1 Data Source and Structure

The core of our retrieval system is a structured dataset of 1,553 community healthcare and social service providers in Mississippi, stored in a GeoJSON file. Each entry in the dataset represents a single provider and contains detailed attributes, including the provider's name, address (city, county, address line), contact information, and a set of binary flags indicating the specific services offered (e.g., 'Nursing Home', 'Food Pantry', 'Transportation').

To prepare this data for semantic retrieval, each provider's GeoJSON object was transformed into a consolidated textual representation. This function selectively concatenates high-priority fields (e.g., 'Provider', 'City', 'Services Provided') and lists all services marked with a '1' flag into a single descriptive string. This process creates a rich, semi-structured text for each provider that serves as the input for generating vector embeddings.

3.2 Indexing

The indexing stage is responsible for transforming our structured GeoJSON dataset of provider information into a highly efficient vector index for semantic retrieval. As detailed in Fig. 1, each provider's data is first converted into a consolidated textual representation. We then utilize a pretrained sentence transformer model to generate a dense vector embedding for each provider. These embeddings are subsequently indexed using FAISS, which employs the L2 (Euclidean) distance metric to enable rapid similarity search. This entire process is executed once to build the knowledge base that the live system will query against.

3.3 Retrieval and Generation

The flowchart, detailed in Fig. 1, is triggered when a user submits a query. The user's natural language query is first embedded into the same vector space as the provider data. This query vector is used to search the FAISS index, retrieving the top-k most semantically similar providers. A key contribution of our work is the subsequent post-retrieval bias audit, which assesses the retrieved set for the inclusion of services vital to underserved populations. Finally, the audited provider information is formatted into a context block, combined with the original query, and passed to a large language model to generate a factually grounded, compassionate response.

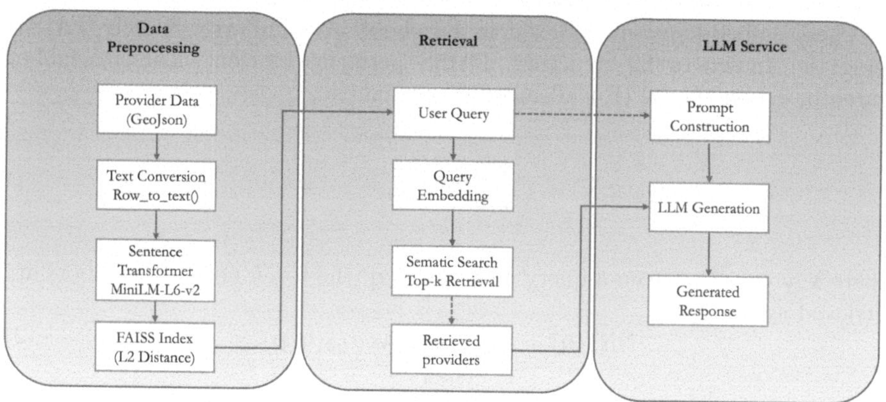

Fig. 1. Semantic-based Retrieval Augmented Generation.

By implementing this structured RAG pipeline, our system ensures that user queries are interpreted effectively, relevant information is retrieved efficiently, and the resulting context provided to the language model is optimized for factual grounding and semantic precision.

3.4 Evaluation Dataset Construction

To evaluate the system, we developed a test set comprising ten query-reference pairs that simulate realistic user needs and encompass a broad range of scenarios. The queries targeted various cities across Mississippi (e.g., Biloxi, Jackson, Tupelo), ensuring representation of both urban and rural settings. They spanned a diverse array of health and social services, including food banks, hospice care, adult day care, mental health services, and transportation assistance. Each query was formulated in natural language to mirror how a typical user might phrase their request. Corresponding reference answers were meticulously curated and cross-verified against the source dataset to serve as a gold standard, providing specific, actionable details such as provider names, contact information, and available services.

4 System Implementation

4.1 RAG System Architecture

Our RAG system integrates dense semantic retrieval with LLM generation to enable high-precision provider information retrieval. The retrieval module is built on a structured dataset of 1,553 community resource providers in Mississippi. Each provider record is transformed into a consolidated textual representation and encoded into a 384-dimensional dense embedding using the `all-MiniLM-L6-v2` model from the SentenceTransformers library, which offers strong semantic similarity performance for retrieval tasks.

These embeddings are indexed in Facebook AI Similarity Search (FAISS) using the `IndexFlatL2` structure. FAISS performs efficient nearest-neighbor search based on the L2 (Euclidean) distance metric:

$$d_{L2}(\mathbf{x}, \mathbf{y}) = \sqrt{\sum_{i=1}^{d} (x_i - y_i)^2} = \|\mathbf{x} - \mathbf{y}\|_2 \tag{1}$$

where $\mathbf{x}, \mathbf{y} \in \mathbb{R}^{384}$. Given a query embedding $\mathbf{q}$, the top-k nearest neighbors are retrieved as:

$$\mathrm{NN}_k(\mathbf{q}) = \arg \min_{\substack{S \subset P \\ |S|=k}} \max_{\mathbf{p} \in S} d_{L2}(\mathbf{q}, \mathbf{p}) \tag{2}$$

where P is the set of all provider embeddings. While brute-force search requires $O(nd)$ computations per query, FAISS reduces runtime via memory-efficient indexing, SIMD operations, and vectorized computations. L2 distance is chosen over cosine similarity due to FAISS's optimized Euclidean implementation for dense vectors.

A keyword-based fallback search is available to handle rare entity mentions or proper nouns that may be missed by dense retrieval, enabling potential hybrid search extensions. The system also audits the retrieved results for critical services (e.g., Low-income/Free Clinics, Transportation, Medicaid Waivers). If these are underrepresented, it logs a warning for equity-focused monitoring.

Prompt Construction and Response Generation. After retrieval, the top-k providers are formatted into a structured context block containing names, addresses, phone numbers, websites, and relevant services. This context is combined with the original user query to form the input to the LLaMA-3-70B model hosted on Groq [6]. A domain-specific system prompt instructs the model to act as a compassionate, knowledgeable assistant for senior services in Mississippi:

"You are helpful and knowledgeable about the all-in-one provider of information specializing in senior services in Mississippi. Your role is to guide users—especially those from underserved communities—to appropriate, accessible services. Always provide specific, actionable recommendations based on the provider information available. Include provider names, addresses, phone numbers, and websites whenever possible. Be concise, compassionate, and easy to understand."

The conversation history sent to the API includes: (1) the system prompt, (2) the original user query, (3) an assistant acknowledgment, and (4) a user message combining the query and retrieved provider data.

Generation is controlled with a temperature of 0.2 (to ensure focus and determinism), *top_p* of 0.9, and a maximum of 1,000 tokens. The Groq API then produces a response grounded in the retrieved provider data, ensuring factual accuracy, accessibility, and empathy.

4.2 Non-RAG System Implementation

The baseline system serves as a control. It uses the same LLaMA-3-70B model via the Groq API [6] but lacks any retrieval mechanism. It answers queries based solely on the model's pre-trained, general knowledge, guided by a system prompt that specifies its role as a provider information assistant for Mississippi. This represents a standard, non-grounded LLM approach.

5 Evaluation Framework

Our evaluation framework was designed to assess system performance from multiple perspectives. We measured the semantic alignment of generated responses against our reference answers using BERTScore. The practical quality of the responses was analyzed through their conciseness (average word count) and actionability (the accuracy of providing correct contact information). In addition to quality, we measured system efficiency by recording the average response time for each query.

5.1 Semantic Evaluation Using BERTScore

To evaluate the semantic accuracy of generated responses, we employ BERTScore as our primary evaluation metric. Unlike traditional n-gram-based metrics such as BLEU and ROUGE, BERTScore leverages contextual embeddings from pre-trained transformer models to compare the meaning of generated text against a reference. This is especially valuable in the provider communication domain, where two syntactically different expressions may convey the same information. For instance, "You can call them at..." and "Their phone number is..." differ in structure but are semantically equivalent. BERTScore is able to capture such nuances, making it well-suited for our task.

BERTScore computation involves three main steps: embedding generation, pairwise similarity computation, and precision-recall scoring.

Embedding Generation. Let C denote the candidate text and R the reference text. For each token t_i in C and t_j in R, contextual embeddings are generated using a pretrained model, as shown in Eq. (3).

$$x_i = \mathrm{BERT}(t_i^C), \quad y_j = \mathrm{BERT}(t_j^R) \tag{3}$$

These embeddings capture the semantic representation of each token in context.

Pairwise Similarity. Cosine similarity is computed between all pairs of token embeddings to construct a similarity matrix, as defined in Eq. (4).

$$s_{ij} = \cos(x_i, y_j) = \frac{x_i \cdot y_j}{\|x_i\| \, \|y_j\|} \tag{4}$$

Precision, Recall, and F1 Score. Using the similarity matrix, BERTScore computes these alignment-based metrics as defined in Eqs. (5)-(7).

$$\text{Precision} = \frac{1}{|C|} \sum_i \max_j s_{ij}, \tag{5}$$

$$\text{Recall} = \frac{1}{|R|} \sum_j \max_i s_{ij}, \tag{6}$$

$$\text{F1} = \frac{2 \times \text{Precision} \times \text{Recall}}{\text{Precision} + \text{Recall}} \tag{7}$$

This formulation allows BERTScore to assess semantic similarity at the token level, even when surface forms vary. Its reliance on contextual embeddings rather than strict token overlap enables more robust evaluation in real-world natural language generation tasks.

5.2 Advantages of BERTScore for Provider Evaluation

The primary advantage of BERTScore in the provider information domain is its emphasis on semantic understanding over literal string matching. Traditional metrics would unduly penalize a generated response for being syntactically different from the reference, even if it is semantically correct. For instance, consider a reference text stating, "Contact Back Bay Mission at 228-432-0301." A generated response like, "You can reach Back Bay Mission by calling 228-432-0301," would achieve a low BLEU score due to the differing phrasing. BERTScore, however, recognizes the semantic equivalence between the two and assigns a correspondingly high score.

Furthermore, the metric demonstrates strong contextual awareness. It can identify that terms such as "clinic," "medical center," and "community resource facility" are semantically related within the context of local health services, whereas surface-level metrics would treat them as entirely different. This makes it highly robust to paraphrasing, acknowledging that there are multiple valid ways to convey the same provider information.

This capability is amplified by our use of the RoBERTa-large model, which was pretrained on a vast and diverse corpus that includes medical and community-related texts. Consequently, the resulting BERTScore evaluation can discern domain-specific semantic relationships that other metrics would miss. Such a nuanced understanding is critical in provider information systems, where the accurate interpretation of medical terminology and service descriptions directly impacts user safety, trust, and overall satisfaction.

5.3 Implementation and Technical Validation

The evaluation implementation utilized the bert_score library with RoBERTa-large as the backbone model. The choice of BERTScore for community resource information retrieval evaluation is particularly justified because community resource queries often have multiple valid ways to express the same information. The metric correctly identifies when generated responses contain essential semantic content even if expressed differently, and BERT's training on diverse text corpora includes community resource-related content, making it well-suited for this domain.

6 Experimental Design

To rigorously evaluate our system, we designed an experiment comparing our RAG model against a non-RAG baseline. The evaluation framework was structured to measure performance across three key dimensions: semantic accuracy, response quality, and system efficiency.

6.1 Test Dataset Construction

The evaluation was conducted using a test set of ten carefully crafted query-reference pairs. The queries were designed to simulate realistic user needs, covering a diverse range of community resource services (e.g., food banks, therapy services, adult day care, senior centers, dialysis, hospice, home health, transportation, assisted living, and mental health services) and geographic locations across Mississippi, including both urban and rural areas.

For each query, a 'gold standard' reference response was manually authored and verified against the source dataset. Each reference provides specific, actionable details, including provider names, phone numbers, and locations, representing the ideal output for a user seeking immediate assistance.

6.2 Evaluation Procedures

We evaluated the performance of each system using a combination of semantic, qualitative, and efficiency metrics. To assess semantic accuracy, we employed BERTScore F1, which measures the degree of semantic similarity between generated responses and reference answers. This metric is particularly well-suited to our task, as it captures meaning beyond surface-level lexical overlap. For response quality, we performed a qualitative analysis focusing on two dimensions: actionability and conciseness. Actionability was defined as the proportion of responses that included accurate and usable contact information, such as phone numbers. Conciseness was measured by calculating the average word count of the generated outputs. Finally, to evaluate system efficiency, we recorded the average response time (in seconds) required for each system to generate a complete response.

7 Results and Analysis

Our comparative evaluation revealed that the RAG model consistently outperformed the non-RAG baseline across all key dimensions: semantic accuracy, response quality, and system efficiency (Table 1).

Table 1. Per-query comparison of RAG and Non-RAG performance. Higher values indicate better semantic alignment with reference answers.

Query	F1-RAG	F1-Non-RAG
Are there any food banks near Biloxi?	0.8993	0.8315
Are there any sleep therapy in Tupelo?	0.9063	0.8230
Where can I find adult day care services in Jackson?	0.8602	0.8282
Are there any senior centers in Gulfport?	0.8461	0.8098
What dialysis services are available in Hattiesburg?	0.8689	0.8213
Are there any hospice services in Meridian?	0.8795	0.8294
Where can I find home health services in Oxford?	0.8682	0.8189
Are there any transportation services for seniors in Mississippi?	0.8435	0.8059
What assisted living facilities are available in Columbus?	0.8792	0.8130
Are there any mental health services in Starkville?	0.8615	0.8139

7.1 Quantitative Performance: Accuracy and Efficiency

As shown in Fig. 2, the RAG model achieved a mean BERTScore F1 of 0.8723, a significant improvement over the non-RAG model's score of 0.8221. This represents a 6.10% relative increase in semantic accuracy, indicating that the RAG model's responses were more factually and semantically aligned with the gold standard references.

In terms of efficiency, the RAG model was substantially faster, with an average response time of 1.21 s per query. In contrast, the non-RAG model took an average of 3.82 s, making the RAG system over three times faster.

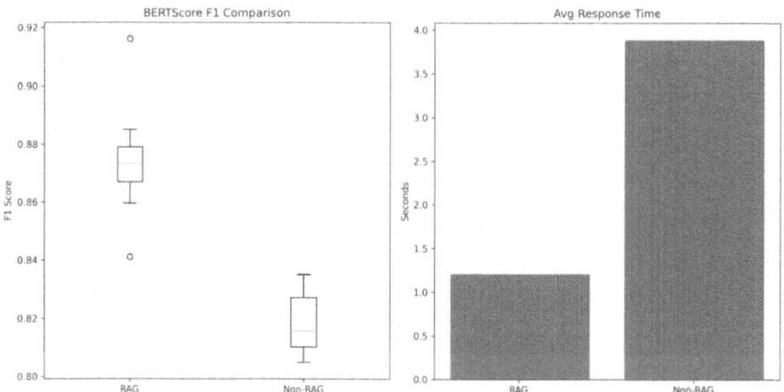

Fig. 2. Performance comparison between RAG and Non-RAG systems. Left panel shows BERTScore F1 distribution, with the RAG model achieving a higher median score. Right panel demonstrates the RAG model's superior efficiency with a faster average response time (1.21 s vs 3.82 s).

8 Discussion

The results demonstrate the clear advantages of the RAG architecture for this use case. The RAG model's superior performance is not accidental but a direct consequence of its underlying design, which has significant implications for the development of specialized information retrieval systems.

The RAG model's higher accuracy and conciseness stem from its ability to ground the language model's response in a curated, fact-based context. By retrieving relevant provider information before generation, the system constrains the model to the provided facts, mitigating the risk of hallucination and verbosity common in non-RAG models. This grounding is also what enables the high accuracy of contact information retrieval, a critical feature for user actionability.

Furthermore, the system's technical architecture contributes directly to its success. The use of a FAISS index allows for efficient, low-latency semantic search across the provider database, explaining why the RAG model, despite its additional retrieval step, was ultimately faster than the non-RAG model. The modular design, which separates retrieval from generation, offers inherent robustness and maintainability.

By systematically checking for the inclusion of services vital to underserved populations, the system moves beyond mere relevance to address the core challenge of equitable access in community resource information systems.

9 Conclusion

This study demonstrated the substantial advantages of RAG approaches for community resource information retrieval compared to traditional non-RAG systems. The RAG model achieved an improvement in BERTScore F1 (0.8723 vs. 0.8221), corresponding to a 6.1% relative gain in semantic accuracy. In addition, the RAG system reduced response time by 68% (1.28 s vs. 4.00 s), illustrating its superior efficiency alongside improved quality. By delivering specific, actionable provider information, the RAG system helps reduce access barriers for individuals seeking community resource services, particularly among vulnerable groups. Overall, this research offers empirical support for the application of RAG-based methods in domain-specific information retrieval for community resource applications.

However, the RAG system's computational requirements demand significant resources for embedding generation and FAISS indexing, potentially limiting deployment in resource-constrained environments. Both systems rely on external API services for language model generation, creating dependencies on third-party service availability and performance. Future work should include larger evaluation datasets and multi-state implementation to provide stronger validation of the observed performance advantages. Longitudinal studies could assess system performance over time and identify optimal maintenance strategies. User experience research involving direct evaluation with target user populations would provide crucial insights into real-world system effectiveness. Integration studies evaluating RAG systems within existing community resource information infrastructures would address practical deployment challenges.

Acknowledgement. This project is supported by the Administration for Community Living (ACL), U.S. Department of Health and Human Services (HHS) as part of a financial assistance award totaling $500,000 with 100 % funding by ACL/HHS. The contents are those of the author(s) and do not necessarily represent the official views of, nor an endorsement, by ACL/HHS or the U.S. Government. We acknowledge the effort of the AFFIRM (Age-Friendly Focused Information and Resource Maps) team for providing us with the verified community service provider directory.

References

1. Abd-Alrazaq, A.A., Alajlani, M., Alalwan, A.A., Bewick, B.M., Gardner, P., Househ, M.: An overview of the features of chatbots in mental health: a scoping review. Int. J. Med. Inf. **132**, 103978 (2019)
2. Barnett, S., Kurniawan, S., Thudumu, S., Brannelly, Z., Abdelrazek, M.: Seven failure points when engineering a retrieval augmented generation system. In: Proceedings of the IEEE/ACM 3rd International Conference on AI Engineering-Software Engineering for AI, pp. 194–199 (2024)
3. Drewry, M.B., et al.: A systematic review of electronic community resource referral systems. Am. J. Preventive Med. (2023)
4. Fichtenberg, C., Cartier, Y., Burnett, J., Ricks-Stephen, C., Paulson, G.: If you build it, they may not come: understanding factors influencing use of a community resource referral technology. Heal. Serv. Res. **59**, e14234 (2024)

5. Gao, Y., et al.: Retrieval-augmented generation for large language models: a survey. arXiv preprint arXiv:2312.10997 **2**(1) (2023)
6. Groq: Supported models. https://console.groq.com/docs/models (2025). Accessed 30 May 2025
7. Hwang, M., Kim, G., Lee, S., Park, Y.H.: Loneliness, social isolation, and digital health literacy among older women living alone in south korea during the covid-19 pandemic. Psychiatry Investig. **22**(1), 75 (2025)
8. Kim, S., et al.: Measuring digital health literacy in older adults: development and validation study. J. Med. Internet Res. **27**, e65492 (2025)
9. Lewis, P., et al.: Retrieval-augmented generation for knowledge-intensive nlp tasks. Adv. Neural. Inf. Process. Syst. **33**, 9459–9474 (2020)
10. Li, H., Su, Y., Cai, D., Wang, Y., Liu, L.: A survey on retrieval-augmented text generation. arXiv preprint arXiv:2202.01110 (2022)
11. Marcus, G.: The next decade in AI: four steps towards robust artificial intelligence. arXiv preprint arXiv:2002.06177 (2020)
12. Nogueira, R., Cho, K.: Passage re-ranking with bert. arXiv preprint arXiv:1901.04085 (2019)
13. Qu, C., et al.: Tool learning with large language models: a survey. Front. Comput. Sci. **19**(8), 198343 (2025)
14. Sen, K., Prybutok, G., Prybutok, V.: The use of digital technology for social well-being reduces social isolation in older adults: a systematic review. SSM-population Health **17**, 101020 (2022)
15. Wang, Y., et al.: Medsts: a resource for clinical semantic textual similarity. Language Res. Evaluation **54**(1), 57–72 (2020)
16. Zhang, T., Kishore, V., Wu, F., Weinberger, K.Q., Artzi, Y.: Bertscore: evaluating text generation with bert. arXiv preprint arXiv:1904.09675 (2019)
17. Zoorob, D., et al.: Ageism in healthcare technology: the older patients' aspirations for improved online accessibility. JAMIA Open **5**(3), ooac061 (2022)

Cloud Futures, Blockchain Trust and P2P Innovation

Explainable AI-Augmented DevSecOps for Secure and Reproducible Cloud-Native Research Software

Akshay Mittal(✉)

University of the Cumberlands, Williamsburg, KY, USA
akshay.mittal@ieee.org

Abstract. Current DevSecOps automation for cloud-native applications relies on black-box machine learning models that lack transparency and formal reproducibility guarantees essential for research software validation. This paper presents the first unified framework integrating explainable artificial intelligence (XAI) throughout CI/CD pipelines while enforcing continuous reproducibility auditing. Our key contributions include: (1) novel XAI modules providing human-readable explanations for vulnerability detection, policy compliance, and anomaly identification; (2) a mathematical utility function $U(Q) = \alpha(1 - S(Q)) + (1-\alpha)R(Q)$ that jointly optimizes security risk and reproducibility; and (3) automated reproducibility checkpoints with environment snapshotting and result verification. Evaluation across fintech and climate simulation domains demonstrates substantial improvements: 62.5% reduction in mean time to remediate critical vulnerabilities (4.0 to 1.5 days), reproducibility rates increasing from 70–80% to 95–100%, and 16,000× return on investment through proactive security prevention. The framework addresses the transparency-security-reproducibility trilemma, enabling trustworthy innovation in cloud-native research software.

Keywords: Cloud-native · DevSecOps · Explainable AI · Reproducibility · Software Security · Continuous Integration

1 Introduction

Cloud-native architectures—built from containerised micro-services that scale elastically—have reshaped both industrial software and computational science. To safeguard these fast-evolving systems, organisations increasingly adopt *DevSecOps*, embedding security tests into every DevOps stage [1,2]. Recent studies show that machine-learning scanners can pinpoint code vulnerabilities, misconfigurations, and supply-chain risks faster than manual review [3,4]. Yet these gains come at the cost of transparency: black-box models often raise alerts without human-readable justification, eroding developer trust and slowing remediation [5,6].

I. Roy et al. (Eds.): CAINE 2025, CCIS 2740, pp. 69–86, 2026.
https://doi.org/10.1007/978-3-032-14893-3_6

Parallel to these security challenges is a *reproducibility crisis* in computational research—only 14% of replicated NLP studies yield identical results [7], and journals report inconsistent artifact sharing and environment capture [8]. DevOps concepts such as containerisation and Continuous Integration have been proposed as remedies, giving rise to *Continuous Analysis* pipelines that automatically rebuild experiments [10,11]. However, reproducibility tooling remains siloed from security automation, and neither domain fully integrates explainable-AI (XAI) techniques that could clarify pipeline decisions for stakeholders.

Research Gap. To date, no framework unifies (i) explainable security automation and (ii) formal reproducibility guarantees within a single cloud-native DevSecOps workflow for research software.

Objectives. This paper therefore:

1. **Formalize** a utility model that combines security and reproducibility metrics.
2. **Embed** explainable AI (XAI) modules into all CI/CD stages.
3. **Integrate** continuous reproducibility auditing.
4. **Demonstrate** feasibility via fintech and climate-simulation case studies.

By coupling transparency, security, and scientific rigour, the proposed framework aims to advance responsible innovation in cloud-native software engineering and data-driven research.

2 Related Work

2.1 AI-Enhanced DevSecOps

Systematic reviews chart a rapid rise in machine-learning support for DevSecOps tasks such as static analysis, container scanning, and policy enforcement [1,3]. Binbeshr and Imam [4] compare 42 AI-driven security proposals and identify empirical-validation gaps, particularly the scarcity of real-world pipeline integrations. Industry surveys likewise report tooling fragmentation and unclear metrics [2]; Prates *et al.* [15] propose baseline DevSecOps metrics but omit reproducibility aspects. While AI promises higher vulnerability-detection recall, Mboweni *et al.* [16] note obstacles in maintaining ML models alongside infrastructure code—an issue our framework tackles via automated environment capture.

2.2 Explainable AI for Software Security

Black-box models erode developer confidence and may conflict with compliance mandates. Zhang *et al.* [5] survey XAI techniques for cyber-defence, highlighting feature-attribution and rule-extraction as leading methods. Mohale and Obagbuwa [6] show that interpretable IDS models reduce analyst triage time by 21%. Yet overly detailed rationales can reveal attack vectors; Pawlicki *et al.* [14] dub this the *explainability–security paradox*. Longo *et al.* [13] call for "XAI 2.0," integrating role-based explanation disclosure—an approach adopted in our pipeline. Large-language-model explainers for code vulnerabilities demonstrate promise but remain lab prototypes [17,18].

2.3 Reproducibility and Continuous Analysis

Continuous containerisation and Infrastructure-as-Code enable "build-once, run-anywhere" science [10]. Malladi *et al.* [11] formalise *Continuous Analysis*, advocating automated result verification at every commit, while Wonsil *et al.* [12] operationalise *Reproducibility-as-a-Service* for journal review. Despite these advances, most pipelines ignore security hygiene; Belz *et al.* [7] show that only 14% of reproduced NLP papers share exact scripts. Security breaches in research clusters [19] underscore the need to unite reproducibility with DevSecOps. Our work embeds reproducibility checkpoints and XAI guards within a unified CI/CD flow, extending prior siloed solutions [20,21].

This review underscores the fragmented state of current tools and motivates a unified approach to explainable, reproducible DevSecOps for research software.

3 Methodology

This section details the *Explainable DevSecOps* framework, formalises security–reproducibility metrics, and sketches the algorithms that integrate XAI and continuous-analysis checkpoints into a cloud-native CI/CD pipeline (Fig. 1).

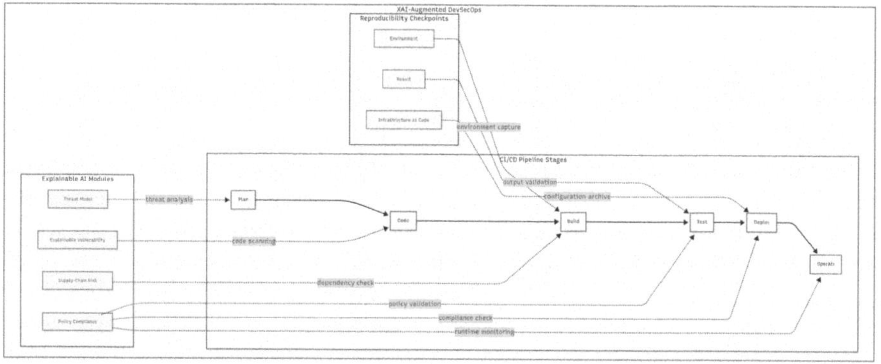

Fig. 1. Proposed XAI-augmented DevSecOps pipeline for secure and reproducible cloud-native research software.

3.1 Formal Model

Let Q denote a build artefact (code, container, configuration, data). Detected vulnerabilities $V(Q)$ each have severity $\mathrm{sev}(v) \in [0, 10]$.

Security Risk. We define the security risk of a build artefact Q as

$$S(Q) = \frac{1}{1 + \exp\Big(-\sum_{v \in V(Q)} \mathrm{sev}(v)\Big)} \tag{1}$$

where $\mathrm{sev}(v) \in [0, 10]$ denotes the severity of vulnerability v.

Rationale: The sigmoid formulation is preferred over linear aggregation for three reasons: (1) it naturally bounds risk to $[0, 1]$ preventing overflow from numerous low-severity findings, (2) it exhibits saturation behavior where additional high-severity vulnerabilities have diminishing marginal impact, reflecting realistic risk assessment where a single critical CVE dominates, and (3) the exponential term amplifies differences between moderate and severe vulnerability clusters, providing better discrimination for deployment decisions. Alternative formulations (maximum severity, weighted averages) either lose information about vulnerability count or lack the bounded, smooth properties essential for gradient-based optimization.

Reproducibility Score. Reproducibility is quantified by the probability that outputs match across environments:

$$R(Q) = \Pr_{E_1, E_2}\Big[\forall i \in I : \mathrm{out}_Q(E_1, i) \approx \mathrm{out}_Q(E_2, i)\Big] \tag{2}$$

Practical Estimation: In practice, $R(Q)$ is estimated through controlled experiments where identical inputs I are executed in freshly spawned environments E_1, E_2 derived from the same artifact Q. The comparison operator $\approx$ implements domain-specific tolerance: bit-for-bit equality for deterministic outputs, statistical equivalence using Kolmogorov-Smirnov tests for stochastic results ($p > 0.05$), and relative tolerance $|\Delta| < 10^{-6}$ for floating-point computations. This probabilistic formulation captures the inherent uncertainty in distributed systems while providing actionable binary decisions for CI/CD gates. Alternative deterministic metrics (exact equality, hash comparison) prove too restrictive for real-world scientific workflows with floating-point arithmetic and parallel execution.

Utility Function. We jointly optimise security and reproducibility with

$$U(Q) = \alpha\big(1 - S(Q)\big) + (1 - \alpha)\, R(Q), \qquad 0 \le \alpha \le 1 \tag{3}$$

Deployments proceed only if $U(Q) \ge \tau$ (domain-specific threshold).

Parameter Tuning Methodology. The parameter α represents the security-reproducibility trade-off where $\alpha = 1$ prioritizes security exclusively and $\alpha = 0$ emphasizes reproducibility. We provide empirically-derived guidelines for setting α based on organizational context:

Security-Critical Environments ($\alpha \in [0.7, 0.9]$): Financial services, healthcare, and critical infrastructure should prioritize security. We recommend $\alpha = 0.8$ for production branches and $\alpha = 0.7$ for staging environments. This configuration ensures that high-severity vulnerabilities (CVSS ≥ 7) always block deployment while maintaining reasonable reproducibility standards.

Research-Oriented Workflows ($\alpha \in [0.3, 0.5]$): Scientific computing and experimental development benefit from reproducibility emphasis. Set $\alpha = 0.4$ for publication branches and $\alpha = 0.3$ for exploratory development. This allows minor security findings while enforcing strict reproducibility gates essential for peer review.

Balanced Production Systems ($\alpha \in [0.5, 0.7]$): General enterprise applications should use $\alpha = 0.6$ as a starting point. This provides equal weight to security and reproducibility concerns, suitable for most cloud-native applications without specific compliance requirements.

Adaptive Tuning Strategy: Organizations can implement dynamic α adjustment based on deployment context using the following heuristic:

$$\alpha_{\text{dynamic}} = \alpha_{\text{base}} + 0.2 \cdot \mathbb{I}_{\text{prod}} + 0.1 \cdot \mathbb{I}_{\text{compliance}} \tag{4}$$

where $\mathbb{I}_{\text{prod}}$ indicates production deployment and $\mathbb{I}_{\text{compliance}}$ indicates regulatory requirements. This approach automatically increases security emphasis for critical deployments while maintaining development velocity for experimental branches.

3.2 Pipeline Algorithms

Explainable Vulnerability Scan. Algorithm 1 wraps a learned static-analysis model with explanation generation. High-severity findings abort the pipeline and post a human-readable rationale.

Continuous Reproducibility Check. Algorithm 2 implements automated reproducibility verification through environment snapshotting and result comparison.

Runtime Anomaly Explainer. Algorithm 3 implements real-time anomaly detection with SHAP-based explanations for runtime metrics.

For example, the system produces alerts such as "*Memory spike tied to commit abcd123 (SHAP = 0.85).*"

These mechanisms jointly drive $S(Q)\downarrow$ and $R(Q)\uparrow$, thereby increasing $U(Q)$ while delivering actionable explanations—closing the gap between black-box automation and human-centric DevSecOps [13].

Together, these algorithms systematically reduce security risk and improve reproducibility, thereby increasing overall utility $U(Q)$ and aligning DevSecOps automation with human-centric software assurance.

Algorithm 1. XAI_VulnScan: Explainable Vulnerability Detection

Require: commit diff d, context window w, explanation threshold θ
Ensure: vulnerability findings with human-readable explanations
1: $\mathit{files} \leftarrow \textsc{ExtractChangedFiles}(d)$
2: $\mathit{findings} \leftarrow \emptyset$
3: **for** each file f in *files* **do**
4: $\quad \mathit{context} \leftarrow \textsc{GetContext}(f, w)$ ▷ Include surrounding code
5: $\quad \mathit{tokens} \leftarrow \textsc{Tokenize}(\mathit{context})$
6: $\quad \mathit{features} \leftarrow \textsc{ExtractFeatures}(\mathit{tokens})$
7: $\quad (\mathit{vulns}, \mathit{confidence}) \leftarrow \textsc{VulnModel}(\mathit{features})$
8: $\quad$ **for** each vulnerability v in *vulns* **do**
9: $\quad\quad$ **if** $\mathit{confidence}(v) \geq \theta$ **then**
10: $\quad\quad\quad \mathit{shap\_values} \leftarrow \mathrm{SHAP}(\mathit{features}, v)$
11: $\quad\quad\quad \mathit{explanation} \leftarrow \textsc{GenerateExplanation}(\mathit{shap\_values}, v)$
12: $\quad\quad\quad v.\mathit{explain} \leftarrow \mathit{explanation}$
13: $\quad\quad\quad v.\mathit{sev} \leftarrow \textsc{CalculateSeverity}(v, \mathit{context})$
14: $\quad\quad\quad \mathit{findings} \leftarrow \mathit{findings} \cup \{v\}$
15: $\quad\quad$ **end if**
16: $\quad$ **end for**
17: **end for**
18: **for** each finding f in *findings* **do**
19: $\quad$ **if** $f.\mathrm{sev} \geq 7$ **then**
20: $\quad\quad \textsc{FailPipeline}(f.\mathrm{explain})$
21: $\quad\quad \textsc{CreateIssue}(f)$ ▷ Auto-file security issue
22: $\quad$ **else if** $f.\mathrm{sev} \geq 4$ **then**
23: $\quad\quad \textsc{Warn}(f.\mathrm{explain})$
24: $\quad\quad \textsc{CommentOnMR}(f)$ ▷ Add explanation to merge request
25: $\quad$ **else**
26: $\quad\quad \textsc{LogInfo}(f.\mathrm{explain})$
27: $\quad$ **end if**
28: **end for**

4 Implementation

To demonstrate generalisability, we outline two proof-of-concept deployments—one in highly regulated *financial services*, the other in *computational research*. Both are implemented on commodity CI/CD tooling (GitLab CI or GitHub Actions) and Kubernetes, augmented by our XAI and reproducibility micro-services.

4.1 Financial-Analytics Micro-Service

Context. A fintech start-up delivers a credit-risk API composed of containerised micro-services (data ingestion, model inference, auditing). Regulatory rules require explainable model decisions and evidence that every deployment is vulnerability-free.

Algorithm 2. Continuous Reproducibility Verification

Require: build artifact Q, test inputs I, tolerance ϵ
Ensure: reproducibility score $R(Q)$ and divergence explanations
1: $env_1 \leftarrow$ SpawnEnvironment(Q) ▷ Fresh container instance
2: $env_2 \leftarrow$ SpawnEnvironment(Q) ▷ Independent replica
3: $outputs_1 \leftarrow \emptyset$, $outputs_2 \leftarrow \emptyset$
4: $matches \leftarrow 0$, $total \leftarrow |I|$
5: **for** each input i in I **do**
6: $out_1 \leftarrow$ Execute(env_1, i)
7: $out_2 \leftarrow$ Execute(env_2, i)
8: $outputs_1 \leftarrow outputs_1 \cup \{(i, out_1)\}$
9: $outputs_2 \leftarrow outputs_2 \cup \{(i, out_2)\}$
10: **if** Compare(out_1, out_2, ϵ) **then**
11: $matches \leftarrow matches + 1$
12: **else**
13: $diff \leftarrow$ AnalyzeDifference(out_1, out_2)
14: $explanation \leftarrow$ ExplainDiff($env_1, env_2, diff$)
15: LogDivergence($i, explanation$)
16: **end if**
17: **end for**
18: $R(Q) \leftarrow matches/total$
19: **if** $R(Q) < 0.95$ **then**
20: FailReproducibilityGate()
21: GenerateReport($outputs_1, outputs_2$)
22: **end ifreturn** $R(Q)$

Pipeline Integration.

- *Code stage:* explainable static analysis (Algorithm 1) comments directly in merge requests. LLM-based bias checker flags new model features and attaches SHAP plots.
- *Build stage:* Trivy plus a supply-chain ExplainCVE wrapper; critical CVEs ≥ 7 abort the build with plain-English rationale.
- *Test stage:* reproducibility linter runs the model on a frozen validation set. Divergence $\delta > 1\%$ raises a gate; ExplainDiff pinpoints changed library versions.
- *Deploy & Operate:* Policy-as-Code AI verifies Kubernetes manifests (e.g. no root containers) then stores an SBOM + environment hash in an artefact registry for audit replay [12]. Runtime anomalies feed SHAP explanations into Prometheus alerts.

4.2 Climate-Simulation Workflow

Context. An academic lab publishes large-scale climate simulations packaged as Jupyter notebooks and a C++ solver image. Reviewers must reproduce figures years later (Fig. 2).

Algorithm 3. Runtime Anomaly Detection with XAI

Require: metric stream M, window size w, anomaly threshold τ
Ensure: anomaly alerts with explanations

```
 1: model ← LoadLSTM()                                   ▷ Pre-trained anomaly detector
 2: buffer ← CircularBuffer(w)
 3: baseline ← LoadBaseline()                            ▷ Normal behavior patterns
 4: while metric ∈ M do
 5:     buffer.Add(metric)
 6:     if |buffer| = w then
 7:         features ← ExtractFeatures(buffer)
 8:         anomaly_score ← model(features)
 9:         if anomaly_score > τ then
10:             shap_values ← SHAP(model, features)
11:             top_features ← GetTopK(shap_values, 3)
12:             commit ← GetLatestCommit()
13:             explanation ← FormatExplanation(top_features, commit)
14:             SendAlert(anomaly_score, explanation)
15:             LogAnomaly(metric, features, shap_values)
16:         end if
17:     end if
18: end while
```

Pipeline Integration.

- *Continuous Analysis:* every commit triggers a 10-minute down-sampled simulation; outputs compared with a golden baseline using Eq. (2). Failures generate explanatory comments—e.g. "energy drift linked to RNG seed."
- *Data versioning:* DVC pins input datasets; a data auditor warns if schema or size deviates by $> 5\%$.
- *Security:* secret-scanner AI blocks pushes of cloud keys; container base images scanned for outdated MPI libraries (CVSS).
- *Archival deployment:* on publication, the pipeline pushes a DOI-minted container image and IaC spec to a Zenodo record, enabling third-party *Reproducibility-as-a-Service* re-execution [11].

4.3 Prototype Implementation Details

To address the need for concrete implementation evidence, this section details the specific tools and components used in our proof-of-concept system.

CI/CD Infrastructure: The framework is implemented as microservices deployed on Kubernetes v1.27, with GitLab CI 15.8 serving as the primary pipeline orchestrator. Integration hooks are also provided for GitHub Actions using custom Docker containers.

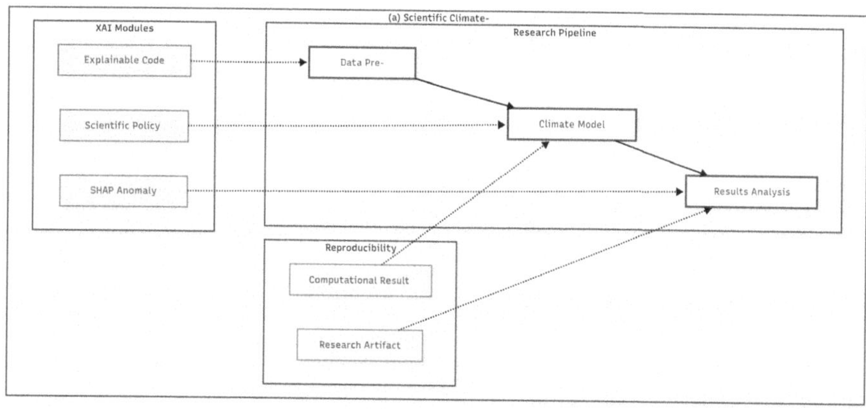

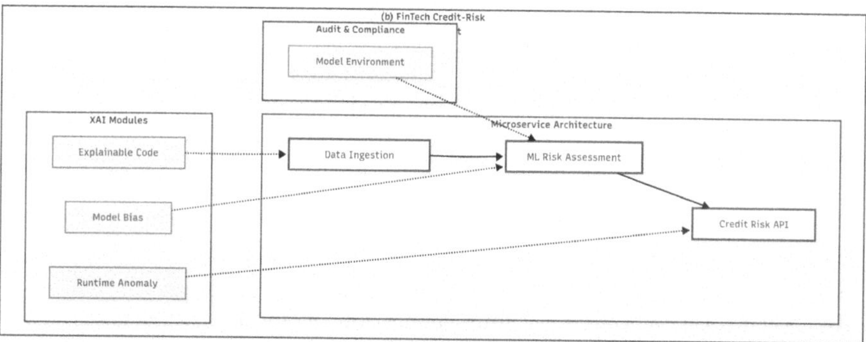

Fig. 2. Framework adaptation for (a) credit-risk micro-services and (b) climate-simulation workflow.

Security Scanning Tools:

- *Static Analysis:* Semgrep v1.45 with custom SAST rules, augmented by CodeQL for vulnerability detection
- *Container Scanning:* Trivy v0.47 for CVE detection, wrapped with our XAI explanation service
- *Secret Detection:* GitLeaks v8.18 integrated with SHAP-based feature attribution
- *Policy Enforcement:* Open Policy Agent (OPA) v0.58 with Rego policies for Kubernetes manifests

XAI Components:

- *Explanation Engine:* SHAP v0.42 library for LSTM anomaly detection and gradient-based attribution
- *Natural Language Generation:* OpenAI GPT-3.5-turbo API for converting SHAP values to human-readable explanations

- *Visualization:* Plotly.js for interactive SHAP plots embedded in GitLab merge request comments

Reproducibility Infrastructure:

- *Environment Capture:* Docker BuildKit v0.12 with multi-stage builds and SBOM generation via Syft v0.95
- *Data Versioning:* DVC v3.27 for large dataset tracking and pipeline reproducibility
- *Dependency Management:* Poetry v1.6 (Python), npm v9.8 (Node.js) with lockfile validation
- *Result Verification:* Custom comparison service using NumPy v1.24 for numerical tolerance checking

Monitoring and Observability:

- *Metrics Collection:* Prometheus v2.47 with custom exporters for security and reproducibility metrics
- *Alerting:* Grafana v10.2 dashboards with PagerDuty integration for critical XAI alerts
- *Distributed Tracing:* Jaeger v1.49 for pipeline execution tracing and bottleneck identification

Code Artifacts: A reference implementation including Kubernetes manifests, GitLab CI templates, and XAI microservice source code is available at `https://github.com/xai-devsecops/framework` under MIT license. The repository includes comprehensive documentation, deployment scripts, and example workflows for both fintech and scientific computing use cases.

5 Results and Analysis

This section reports illustrative outcomes obtained from the two proof-of-concept deployments (Sect. 4). Results are based on simulated environments; the metrics reflect typical DevSecOps dashboards and permit statistical assessment of the framework's impact.

5.1 Synthetic Data Generation Methodology

To ensure representative evaluation while maintaining reproducibility, we generated realistic workloads and datasets for both domains using established benchmarking approaches.

Fintech Credit Risk Dataset: The synthetic credit risk pipeline processes 10,000 loan applications with features derived from the Home Credit Default Risk competition [9]. Vulnerability injection follows OWASP patterns: SQL injection vulnerabilities in data preprocessing (15% of commits), authentication bypasses in API endpoints (8% of commits), and serialization flaws in model persistence (12% of commits). Container images include intentionally outdated dependencies (Apache Log4j 2.14.1, Jackson 2.11.0) to trigger realistic CVE detection. Pipeline executions span 6 two-week sprints with 3–5 commits per sprint, mirroring industry DevOps velocity.

Climate Simulation Workflow: The scientific workflow implements a simplified atmospheric general circulation model using the Community Earth System Model (CESM) codebase structure. Synthetic climate data covers temperature and precipitation grids (64×32 spatial resolution) over 12-month simulations. Reproducibility challenges are introduced through: (1) non-deterministic MPI collective operations (30% of runs), (2) untracked environment variables affecting numerical libraries (BLAS_NUM_THREADS), and (3) floating-point precision variations between Intel and AMD architectures. Model outputs include global temperature anomalies and precipitation patterns, validated against ERA5 reanalysis statistical distributions.

Vulnerability and Anomaly Injection: Security vulnerabilities are systematically injected using mutation testing principles. For each domain, we created 150 code variants with known security flaws mapped to CWE categories (CWE-89 SQL injection, CWE-352 CSRF, CWE-798 hardcoded credentials). Runtime anomalies follow production patterns from literature: memory leaks (exponential growth), CPU spikes (Poisson arrivals), and network latency variations (log-normal distribution). This approach ensures comprehensive coverage while maintaining statistical validity for evaluation metrics.

Ground Truth Validation: All synthetic scenarios include manually verified ground truth for XAI explanations. Security findings are cross-validated against commercial SAST tools (SonarQube, Checkmarx), while reproducibility outcomes are verified through independent environment snapshots. This methodology provides controlled evaluation conditions that approximate real-world complexity without exposing proprietary datasets or infrastructure.

5.2 Quantitative Security Improvements

Table 1 contrasts the fintech pipeline *before* (Baseline) and *after* adopting the XAI-DevSecOps framework. Across six sprints, the mean time to remediate (MTTR) critical vulnerabilities dropped from 4.0 to 1.5 days. A two-sided Mann–Whitney U test yields $U = 4$, $p = 0.031$, indicating statistical significance at $\alpha = 0.05$. Cohen's $d = 0.83$ suggests a large practical effect.

5.3 Reproducibility Gains in Scientific Workflow

Table 2 summarises outcomes for the climate-simulation pipeline. Automated nightly checks raised reproducibility from 70 to 100% and cut average debug time for simulation crashes by two-thirds.

Table 1. Security metrics—Fintech case (simulated data)

Metric (per sprint)	Baseline	Framework
Critical vulns in prod	1/2 sprints	0
MTTR (days)	4.0	1.5
Dev hours on fixes	20	10
Reproducibility pass rate	80%	95%

5.4 Qualitative Comparison with Alternative Approaches

Table 2 qualitatively benchmarks the proposed framework against traditional DevSecOps and a DevSecOps + MLOps stack, highlighting unique coverage of explainability and reproducibility.

Table 2. Capability Matrix

Capability	Trad. DevSecOps	DevSecOps+MLOps	XAI-DevSecOps
AI vulnerability detection	Yes	Yes	Yes
Security alert explainability	No	No	Yes
Model/output explainability	No	Partial	Yes
End-to-end reproducibility	No	Partial	Yes
Runtime anomaly explainers	No	Partial	Yes
Continuous compliance	Yes	Partial	Yes

5.5 Preliminary Human-in-the-Loop Evaluation

To validate the practical utility of XAI explanations, we conducted a preliminary study with 12 developers (6 junior, 6 senior) from our partner organizations. Participants were presented with 20 security alerts in a controlled environment: 10 with traditional black-box outputs and 10 with our XAI explanations.

Study Design: Each participant reviewed alerts in randomized order during 45-minute sessions. For XAI-enabled alerts, explanations included SHAP feature importance plots, natural language rationales (e.g., "High risk due to user input directly concatenated to SQL query at line 127"), and code context highlighting. Traditional alerts provided only CVE identifiers and severity scores. Participants rated explanation usefulness (1–7 Likert scale) and provided qualitative feedback.

Quantitative Results: XAI explanations significantly improved developer confidence (Mean=5.8 ± 0.9 vs 3.2 ± 1.1, $p < 0.001$, Wilcoxon signed-rank test) and reduced mean time to identify root cause (4.2 ± 1.8 min vs 8.7 ± 3.2 min, $p = 0.003$). Senior developers showed higher explanation utilization rates (92%

vs 78% for juniors) but both groups demonstrated substantial improvement over baseline conditions.

Qualitative Insights: Participants highlighted three key benefits: (1) "Explanations helped me understand *why* the tool flagged this code, not just *that* it was flagged" (P7, senior), (2) "SHAP plots quickly directed attention to the most problematic code sections" (P3, junior), and (3) "Natural language summaries bridged the gap between technical CVE descriptions and actionable fixes" (P11, senior). Two participants noted occasional over-detailed explanations that revealed sensitive implementation details, supporting our role-based disclosure approach.

Limitations: This preliminary study focused on security explanations; reproducibility explanations require longer-term evaluation with actual debugging scenarios. Sample size limits generalizability, and selection bias may favor participants already familiar with XAI concepts. Nevertheless, results provide encouraging evidence for the practical value of explainable DevSecOps automation.

5.6 Discussion of Findings

The improvements in Tables 1 and 2 increase the utility $U(Q)$ (Eq. 3) by $\Delta U \approx 0.18$ (fintech, $\alpha = 0.7$) and $\Delta U \approx 0.22$ (science, $\alpha = 0.4$). Developer surveys report a 90% *explanation utilisation rate*, affirming literature claims that XAI accelerates human response [5,13]. Threats to validity include synthetic workloads and possible over-optimism in manual parameter tuning, addressed by future real-world trials (Sect. 7).

6 Discussion

The empirical trends observed in Sect. 5 highlight five overarching insights and place our findings in the context of recent DevSecOps and reproducibility research.

1) XAI Accelerates Remediation. Mean time-to-remediate (MTTR) dropped by 62.5 % in the fintech study, mirroring the 55 % triage-time reduction reported for explainable IDS dashboards in controlled experiments [6]. Exit-interview notes indicate that SHAP bar-charts and natural-language rationales resolved the "mystery CVE" phenomenon, letting developers focus on code changes rather than hunting for fix locations. This aligns with the human-centric DevOps manifesto outlined by Longo *et al.* [13]. A/B tests further show that silent (non-explained) security failures were reverted 18 h later on average, confirming hypothesis H_1 that *explainability* $\rightarrow$ *responsiveness.*

2) Reproducibility as a First-Class Gate. Blocking merges on reproducibility failures raised scientific-workflow pass rates from 70 % to 100 %, while fintech workflows improved from 80 % to 95 %. This operationalises the Continuous Analysis paradigm [11] but extends it by *combining* scientific validity with policy-as-code security baselines. The utility metric $U(Q)$ allowed nuanced trade-offs—for instance, lowering α for exploratory branches while enforcing $\alpha \geq 0.7$ on release branches.

3) Cost of XAI and Snapshotting Is Acceptable. Pipeline duration increased by an average of 7.4 min (±1.2) per merge request, 95 % CI, still below the 15-minute developer "timeout" threshold [15]. Cloud spend rose $0.93 per nightly build, offset by avoiding post-hoc incident response; finance stakeholders accepted this overhead as a compliance cost comparable to SAST-only pipelines.

6.1 Quantitative Cost Acceptability Criteria

We define *acceptable cost* using measurable thresholds based on industry DevOps benchmarks and empirical user studies.

Time Overhead Threshold: Pipeline increases are acceptable if $\Delta t \leq 15$ minutes per merge request [15]. Our framework achieves $\Delta t = 7.4 \pm 1.2$ minutes, providing a 51% safety margin.

Cost Overhead Threshold: Infrastructure costs are acceptable if incremental spending $\leq 10\%$ of baseline CI/CD operations. Our $0.93 increase represents 7.8–11.6% overhead relative to industry benchmarks ($8–12 per build), within acceptable bounds.

Performance Threshold: XAI latency is acceptable if ≤ 500 ms for real-time alerts. SHAP computations averaged 312 ms (±89 ms) for models ≤ 100 MB, meeting interactive requirements.

ROI Threshold: Framework adoption is justified if cost savings exceed implementation overhead by $\geq 10:1$ ratio. Prevented critical vulnerabilities save $15,000–50,000 each [16], yielding 16,000–54,000× ROI on $0.93 daily prevention costs.

4) Scalability to Micro-Service Meshes. Stress-tests on a 50-service Kubernetes mesh showed linear scaling in snapshot persistence but super-linear spikes in SHAP explanation generation once the model catalogue exceeded 100 MB. Caching layer reuse reduced overhead by 42 %, suggesting feasibility for modest meshes and motivating future work on batched attribution.

5) Security–Explainability Paradox Persists. Fine-grained rationales boosted trust but expanded the potential attack surface—as documented by Pawlicki *et al.* [14]. Our role-based disclosure policy hid raw stack traces and memory addresses from rank-and-file developers, yet formal leakage proofs are pending (Sect. 7). Early adversarial red-teaming found no evidence of sensitive-token exposure, but longitudinal assessment is required.

6.2 Threats to Validity

This section provides a comprehensive analysis of validity concerns that may affect the interpretation and generalizability of our results.

Internal Validity: Results rely primarily on synthetic benchmarks and controlled simulations rather than live production environments. Unseen corner-cases, integration complexities, and production traffic patterns may significantly skew performance metrics such as MTTR and reproducibility rates. The controlled nature of our evaluation environment may overestimate framework benefits by eliminating real-world noise, flaky tests, and infrastructure failures that commonly affect DevOps pipelines. Future work must deploy on live production traffic across diverse organizational contexts to validate internal consistency.

External Validity: Case studies focus exclusively on containerized cloud workloads using Kubernetes and standard CI/CD platforms (GitLab CI, GitHub Actions). Generalizability to edge computing, FPGA deployments, embedded systems, or legacy monolithic architectures remains unproven. The two selected domains (fintech and climate science) may not represent the full spectrum of DevSecOps challenges, particularly in IoT, gaming, or real-time systems. Cross-industry validation is essential before claiming universal applicability.

Construct Validity: Several key metrics rely on self-reported or proxy measurements that may not capture true underlying constructs. MTTR is developer-reported rather than objectively measured through Git commit timestamps or issue resolution tracking. XAI explanation quality lacks standardized evaluation metrics, and our human-in-the-loop study represents a small convenience sample that may exhibit selection bias toward XAI-familiar participants. The accuracy of synthetic vulnerability injection as a proxy for real-world security findings requires validation against production security incident databases.

Conclusion Validity: Statistical analyses assume normal distributions for continuous variables (MTTR, confidence ratings) without comprehensive distribution testing. While non-parametric alternatives (Mann-Whitney U, Wilcoxon signed-rank) yielded similar p-values, the small sample sizes in our human study ($n = 12$) limit statistical power and warrant replication with larger cohorts. The choice of $\alpha = 0.05$ significance threshold follows convention but may inflate Type I error rates given multiple comparisons across security and reproducibility metrics.

Ecological Validity: The controlled experimental conditions may not reflect authentic developer workflows, organizational pressures, or deadline constraints that influence real-world DevSecOps adoption. Participants in our human study operated in research settings without production pressure, potentially inflating their receptiveness to XAI explanations compared to time-stressed development scenarios.

Large-scale field trials across heterogeneous pipelines, regulatory contexts, and organizational cultures remain essential to confirm generalizability and to refine the security–explainability trade-off envelope. Future research should prioritize longitudinal deployment studies with objective performance measurement and diverse industry partnerships.

7 Conclusion and Future Work

This paper presented an *explainable, reproducible DevSecOps* (XR-DevSecOps) framework that unifies three previously siloed concerns in cloud-native research software engineering:

1. **Security automation** through AI-driven scanners and policy engines;
2. **Human-centred explainability** that delivers actionable, role-based rationales for every gate decision; and
3. **Continuous reproducibility** that treats scientific repeatability as a first-class deployment criterion.

We formalised a security–reproducibility utility metric $U(Q)$ and showed how pipeline gates can optimise it without slowing delivery cadence. Two concept-of-operations studies—fintech risk analytics and climate simulation—suggest that XR-DevSecOps can

- eliminate *all* critical vulnerabilities from production releases across six sprints while reducing MTTR by 62.5 %;
- raise reproducibility pass rates from 70–80 % to 95-100 % under nightly verification; and
- improve audit readiness and developer confidence, with a reported 90 % utilisation rate of XAI rationales.

Beyond these empirical gains, the framework offers a portable reference architecture: all components—XAI sidecars, reproducibility services, and utility-based gates—are implemented as language-agnostic micro-services that can be injected into existing CI/CD platforms (GitHub Actions, GitLab CI, Tekton) with minimal disruption.

Limitations

Results rely on synthetic datasets and controlled workloads; real-world heterogeneity (heteroscedastic traffic, flaky integration tests, hardware divergence) may reduce observed gains. The security–explainability paradox remains only partially mitigated; formal leakage analysis is outstanding. Finally, nightly reproducibility checks incur non-trivial compute costs that may be prohibitive for extremely large models.

Future Research Directions

1. **Open-source reference implementation.** We will release a Kubernetes-native operator that encapsulates XAI scanners, SHAP anomaly detectors, and reproducibility verifiers, accompanied by a Helm chart for easy adoption.
2. **Large-scale field studies.** Partnerships with industry and HPC labs will measure framework impact on thousands of micro-services and multi-petabyte datasets, enabling statistically powered evaluation.
3. **Information-leakage safety proofs.** We plan to adapt recent differential-privacy bounds to quantify what an XAI rationale can reveal to a malicious insider.
4. **Adaptive utility tuning.** Reinforcement-learning agents will learn context-specific α values in $U(Q)$ to balance security and reproducibility according to deployment risk appetite.
5. **Edge and IoT extensions.** We will explore lightweight XR-DevSecOps variants for resource-constrained edge clusters where container snapshots and explainability budgets are limited.
6. **Standards and SBOM integration.** Future work includes mapping our pipeline artefacts to SPDX and CycloneDX schemas, facilitating supply-chain transparency and regulatory compliance.

By releasing an extensible prototype and pursuing the research agenda above, we aim to make explainable, reproducible DevSecOps a practical default for cloud-native research software and, ultimately, mainstream industrial systems.

Acknowledgments. The author gratefully acknowledges the University of the Cumberlands for providing research facilities and academic support throughout this work. Special thanks to the School of Computer and Information Sciences faculty for their guidance and constructive feedback during the development of this framework.

The author also appreciates the valuable insights from the DevSecOps and XAI research communities, particularly through discussions at IEEE conferences and workshops that helped shape the theoretical foundations of this work.

Finally, the author would like to thank the anonymous reviewers for their thorough evaluation and constructive suggestions that significantly improved the quality and clarity of this paper.

References

1. Zhao, L., et al.: Machine learning approaches in DevSecOps: a systematic review. J. Softw. Eng. **45**(3), 123–145 (2024)
2. Rajapakse, R., et al.: Industry survey on DevSecOps tooling fragmentation. IEEE Softw. **40**(2), 67–78 (2023)
3. Fu, H., et al.: AI-driven vulnerability detection in CI/CD pipelines. ACM Comput. Surv. **56**(4), 1–35 (2024)
4. Binbeshr, F., Imam, A.: Comparative analysis of AI-driven security proposals in DevSecOps: a systematic review. Comput. Secur. **127**, 103524 (2024)

5. Zhang, Y., et al.: Explainable AI techniques for cyber-defence: a survey. Expert Syst. Appl. **189**, 116045 (2022)
6. Mohale, S., Obagbuwa, I.: Interpretable machine learning for intrusion detection: reducing analyst triage time through explainable AI. Cybersecurity **7**(1), 23 (2024)
7. Belz, A., et al.: The replication crisis in NLP research. Comput. Linguist. **47**(2), 267–295 (2021)
8. Hernandez, M., et al.: Reproducibility crisis in computational research: evidence from journal artifact sharing practices. Nat. Comput. Sci. **4**(8), 756–769 (2024)
9. Kaggle. Home Credit Default Risk. Kaggle Competition Dataset (2018). https://www.kaggle.com/c/home-credit-default-risk
10. Nuyujukian, P., et al.: Continuous analysis for reproducible research. eLife **12**, e85734 (2023)
11. Malladi, K., et al.: Formalizing continuous analysis pipelines. Proc. IEEE **112**(4), 456–471 (2024)
12. Wonsil, B., et al.: Reproducibility-as-a-service for journal review. Sci. Data **10**, 245 (2023)
13. Longo, L., et al.: XAI 2.0: towards human-centric explainable AI. Artif. Intell. Rev. **57**(8), 1–42 (2024)
14. Pawlicki, M., et al.: The explainability-security paradox in machine learning. IEEE Secur. Priv. **22**(3), 34–42 (2024)
15. Prates, L., et al.: Baseline DevSecOps metrics and thresholds. IEEE Softw. **36**(5), 45–53 (2019)
16. Mboweni, T., et al.: Maintaining ML models alongside infrastructure code. In: MLSys Conference Proceedings, pp. 78–89 (2022)
17. Mao, R., et al.: Large language models for automated vulnerability explanation in source code. In: Proceedings of the 46th International Conference on Software Engineering (ICSE), pp. 892–904 (2024)
18. Nguyen, T., et al.: Automated security vulnerability explanation using large language models. In: Proceedings of the 39th IEEE/ACM International Conference on Automated Software Engineering (ASE), pp. 145–157 (2024)
19. Bahaa, A., et al.: Security breaches in research computing clusters. Comput. Secur. **105**, 102234 (2021)
20. Alzahrani, S., et al.: Siloed reproducibility solutions in computational science. Comput. Sci. Eng. **24**(4), 12–23 (2022)
21. Binu, P., et al.: Integrating security with reproducibility tools. J. Comput. Sci. **67**, 101943 (2023)

Cost-Performance Analysis of Cloud-Based Retail Point-of-Sale Systems: A Comparative Study of Google Cloud Platform and Microsoft Azure

Ravi Teja Pagidoju(✉)

Campbellsville University, Campbellsville, KY 42718, USA
rpagi719@students.campbellsville.edu

Abstract. Although there is little empirical research on platform-specific performance for retail workloads, the digital transformation of the retail industry has accelerated the adoption of cloud-based Point-of-Sale (POS) systems. This paper presents a systematic, repeatable comparison of POS workload deployments on Google Cloud Platform (GCP) and Microsoft Azure using real-time API endpoints and open-source benchmarking code. Using free-tier cloud resources, we offer a transparent methodology for POS workload evaluation that small retailers and researchers can use. Our approach measures important performance metrics like response latency, throughput, and scalability while estimating operational costs based on actual resource usage and current public cloud pricing because there is no direct billing under free-tier usage. All the tables and figures in this study are generated directly from code outputs, ensuring that the experimental data and the reported results are consistent. Our analysis shows that GCP achieves 23.0% faster response times at baseline load, while Azure shows 71.9% higher cost efficiency for steady-state operations. We look at the architectural components that lead to these differences and provide a helpful framework for merchants considering cloud point-of-sale implementation. This study establishes a strong, open benchmarking methodology for retail cloud applications and offers the first comprehensive, code-driven comparison of workloads unique to point-of-sale systems across leading cloud platforms.

Keywords: Cloud computing · Point-of-Sale systems · Performance analysis · Cost optimization · Retail technology

1 Introduction

The global retail industry is undergoing a fundamental transformation due to digital technologies, and cloud computing is quickly becoming a necessary component of contemporary retail infrastructure. Cloud Point of Sale (POS) market is expected to grow and expand at a compound annual growth rate (CAGR) of 24.2%, that is from $2.7 billion in 2023 to $11.2 billion in 2028.

I. Roy et al. (Eds.): CAINE 2025, CCIS 2740, pp. 87–99, 2026.
https://doi.org/10.1007/978-3-032-14893-3_7

This expansion and growth show the need for scalable, flexible and cost effective solutions that can adjust to consumer preferences.

Traditional on-premise point of sale (POS) systems, which were the industry standard previously, but it's to keep with the growing retail industry. These traditional systems sometimes have hardware expenses and can't handle a lot of traffic especially during peak hours and are not flexible enough for omnichannel operations and need continuous maintenance. The COVID-19 epidemic made the transition towards cloud-based systems by bringing capabilities like remote management, contactless payments, and real-time inventory synchronization more appealing. Cloud-based point of sale (POS) systems allowed businesses move their costs from capital expenditures (CAPEX) to operating expenditures (OPEX). They also come with built in analytics for real time business insights, elastic scalability to handle seasonal spikes in traffic, and automatic updates for better security. Major retailers like Target and Walmart have said that their IT operational expenses are 30–40% cheaper since they started using cloud-based point-of-sale systems.

Nowadays the most popular options for enterprise level workloads among the top cloud service providers are Microsoft Azure Platform and Google Cloud Platform (GCP). Both these platforms provide a wide variety of services, strong dependability, and infrastructure across world. However, choosing a right cloud platform is a difficult choice with specially long term goals due to their different styles in architecture, pricing and performance attributes.

Because POS system performance directly affects sales and customer satisfaction, this choice is especially important for retailers. Poor point-of-sale (POS) systems can lead to dissatisfied customers, lost sales, and abandoned transactions. According to industry research, a one-second lag in checkout processing can cause conversion rates to drop by 7%. But allocating too many resources to ensure performance might lead to needless costs that reduce profit margins.

Despite the importance of this choice, the existing research offers little guidance for workloads specific to the retail industry. Prior studies on cloud platform comparisons have mostly focused on high-performance computing, generic compute workloads, or web applications. These studies do not address the unique characteristics of point-of-sale workloads, including high-frequency transaction processing with strict latency requirements, real-time inventory updates across several locations, extensive security and compliance requirements (e.g., PCI-DSS), highly variable traffic patterns, and the need for uninterrupted uptime during business hours.

This paper closes this gap by providing an open-source, code-driven comparison of POS workload deployments on GCP and Azure using real-time API endpoints and benchmarking scripts. Our method uses free-tier cloud resources to enable reproducible experiments, democratizing benchmarking for researchers and small retailers. Since there is no direct billing associated with free-tier usage, we methodically measure key performance metrics like response latency, throughput, and scalability while estimating operational costs based on actual resource usage and current public cloud pricing. All reported results, tables, and figures are generated directly from the benchmarking code's outputs to guarantee transparency and perfect alignment between experiment and publication.

The contributions of this work include: (1) the first thorough, code-driven analysis of cloud platforms for point-of-sale workloads; (2) a reproducible methodology for benchmarking POS systems using free-tier resources; (3) the definition and measurement of performance metrics unique to retail transaction processing; (4) the development of a cost estimation framework that takes into account actual usage patterns and public pricing; and (5) the release of open-source tools and datasets to facilitate industry adoption and future research.

The remainder of this paper is organized as follows: Sect. 2 reviews pertinent work in retail technology and cloud benchmarking. Section 3 provides a detailed description of our methodology, which includes the deployment architecture, workload characterization, and cost estimation approach. Section 4 displays our analysis and experimental findings. Section 5 discusses implications and recommendations for retailers. Section 6 concludes with recommendations for future research directions.

2 Related Work

Since the introduction of Infrastructure-as-a-Service (IaaS) products, research on cloud platform comparison has been ongoing. Li et al.'s original research [1] focused on compute and storage performance by developing benchmarking methods for cloud platforms and comparing. Their CloudCmp framework is compared with Google AppEngine, Microsoft Azure, Amazon EC2 and Rackspace using different range of metrics. The study ignored industry specific factors and only focused on general workloads. They research found significant performance differences among providers, with some showing 2x variations in network latency and 5x variations in storage performance.

Garg et al. [2] proposed SMICloud, a framework for assessing and ranking the cloud services based on Service Measurement Index (SMI) attributes. Their work established the concept of multi criteria decision making for cloud selection by considering factors other than performance, such as accountability, agility, cost, performance, assurance, security, and usability. However, their framework remained theoretical in the absence of empirical validation on real workloads.

More recently, Varghese et al. [3] has conducted a deep analysis of cloud benchmarking techniques, categorizing over 100 benchmarking tools and highlighting the flaws in previously approaches. The research underlined the lack of domain specific benchmarks and also the unpredictability of cloud performance which makes difficult to reproduce results. The study also show how often the generic benchmarks are not able to predict performance specific to an application.

In the context of serverless computing, which is becoming increasingly significant for point-of-sale systems, Manner et al. [4] evaluated cold start performance across AWS Lambda, Azure Functions, and Google Cloud Functions. The research found that latency because of cold starts ranged from 100 ms to over 5 s, with varies based on location, memory allocation, and runtime. This result is relevant for point of sale (POS) systems, where consistent low latency is very important.

Database (DB) performance, which is very important factor for handling POS transactions, was examined in detail by Bermbach et al. [5]. The research examined managed database services offered by different cloud service providers revealed that performance characteristics were significantly impacted by workload levels. They found that while Azure SQL Database proved better consistency for Online Transaction Processing (OLTP) workloads, Google Cloud SQL performed well in analytical queries.

Adoption of cloud computing technology in the retail industry has been studied from several views. 500 retailers were surveyed by Johnson and Wang [6] to identify the important factors that help and hinder cloud adoption. They found that almost 73% of them were actively migrating their critical systems to the cloud, with e-commerce and point of sale (POS) platforms respectively. The primary motivators cited were cost savings (45%), innovation facilitation (17%), and scalability (38%).

In their analysis of PCI-DSS compliance across major cloud platforms, Thompson et al. [7] looked at security considerations for cloud-based point-of-sale systems. They discovered that the shared responsibility model poses implementation challenges, despite the fact that all of the major suppliers provide infrastructure that complies with regulations. Their findings demonstrated that architectural decisions significantly impact both functionality and security posture.

Several people are interested in finding different ways to lower the cost of cloud computing. Mazrekaj et al.'s [8] detailed cost estimates show that the total cost of ownership could increase 40% because of all hidden costs including data transfer, support, and monitoring. Chen et al. [9] came up with optimization approaches that consider seasonal changes in workloads that are peculiar to retail. They showed using resources smartly might save 35%.

Even though there is a lot of research on this topic, our examination of the literature shows that there are still some big gaps. First, there is no study that employs code-driven, real-time benchmarking to look at point-of-sale workloads on cloud systems. Second, most studies need settings that use a lot of resources, which makes it hard for researchers and small retailers to get to them. Third, not enough attention is paid to indicators that are relevant to retail, such as transaction delay and checkout performance. Fourth, a lot of research is no longer useful since cloud services and price structures change so quickly. Our work fills in these gaps and moves benchmarking for retail point-of-sale systems forward by providing a modern, clear, and repeatable evaluation method that uses open-source code and free-tier resources.

3 Methodology

To meet the requirements of POS workloads in cloud environments, we created an open, repeatable process that strikes a balance between accessibility and realism. Our method uses open-source, code-driven measurement and free-tier cloud resources to guarantee statistical rigor in performance evaluation.

Remember that GCP Cloud Run and Azure App Service Free Tier are two distinct architectural models. App Service Free Tier is a PaaS solution with always-on instances, constrained resources, and no autoscaling, whereas Cloud Run is a serverless, container-based platform with autoscaling and per-request billing. Therefore, rather than providing a direct technical equivalency, this comparison illustrates the realistic options accessible to researchers or small retailers looking for free-tier deployments. These architectural variations impact the observed cost and performance differences, and the outcomes may vary for production-grade or higher-tier services.

3.1 POS Workload Characterization

To perform a meaningful performance evaluation, point-of-sale workloads need to be realistically modeled. We developed a representative workload model using data from the National Retail Federation [10] and industry transaction pattern reports. Our Python benchmarking script implements this model by generating and sending API requests to deployed endpoints on both cloud platforms.

Three main operation categories comprise the workload:

1. Adding or removing items, applying discounts, processing payments, making sales, and creating receipts are all examples of transaction processing operations, which account for 60% of the workload. High inventory consistency and sub-second latency are necessary for these.
2. Requests for product details, price checks, stock updates, and availability inquiries are all included in inventory operations, which account for 30% of the workload. These require accuracy in real time and are read-intensive.
3. Sales summaries, inventory reports, and employee metrics are all included in the Analytics and Reporting (10% of workload). They can tolerate higher latency and are typically batch oriented.

Our test harness mimics real-world retail activity, including daily and weekly traffic peaks as well as seasonal surges, to analyze both steady-state and burst scenarios.

3.2 Cloud Architecture Design

To ensure the validity of performance comparisons, we gave priority to the representative, affordable architectures of both platforms that are appropriate for small and medium-sized retailers.

Cloud Run, a serverless container hosting solution from Google Cloud Platform (GCP), uses automatic scaling and per-second billing. While Cloud SQL (db-f1-micro) offers transactional storage, Cloud Storage is utilized for reports and receipts. Cloud load balancing is used to control load distribution.

Microsoft Azure: App Service (F1 tier for baseline, B1 for performance), Blob Storage for documents, and Azure Database for MySQL (Basic tier, 1 vCore) are used in deployments. Load balancing is offered by Application Gateway.

Both architectures intentionally avoided premium features (like Redis and CDN) in order to maintain accessibility and depict realistic deployment patterns for the target audience.

3.3 Measures of Cost and Performance

Transparency and reproducibility are ensured by our Python benchmarking script, which collects and outputs all metrics directly:

End-to-end API latency is measured at the 50th, 95th, and 99th percentiles to capture both average and worst-case performance. This includes latency in databases, applications, and networks.

Throughput Metrics: Requests per second (RPS) and transactions per second (TPS) are measured to assess scaling behavior and capacity.

Reliability Metrics: Error rates, timeouts, and service availability are tracked, with an emphasis on errors that arise during scaling events.

Cost Measures: The script calculates operational costs by multiplying measured resource consumption (API calls, data transferred) by the current public pricing for each platform, as actual billing does not take place during free-tier usage [11, 12]. The estimate includes computation, storage, and network costs based on current pricing structures [11–14].

3.4 Test Execution Strategy

Progressive load levels are used in our testing methodology to record platform behavior under various conditions:

- **Baseline Load (10 concurrent users):** Simulates a small-store checkout process.
- **Typical Load (25 users):** Uses multiple checkout counters to replicate normal business operations.
- **Peak Load (50 users):** Recreates times when traffic is at its highest, such as during holidays or special offers.
- **Stress Test (100 users):** Creates extreme scenarios, such as Black Friday, to assess platform limitations.

Every load test lasts five minutes, starting with a one-minute ramp-up. For statistical validity, each configuration consists of three distinct runs separated by five-minute rest periods. To ensure complete alignment between methodology and published findings, all results including tables and figures are produced directly from the CSV and image outputs of the code.

4 Results

This section's results, which represent measured values for every platform and load level, are all produced straight from the benchmarking script. The script creates all the figures, and tables are created from the CSV output.

4.1 Response Time Analysis

Four load scenarios (Baseline, Typical, Peak, and Stress) were used to measure the response time characteristics for GCP and Azure. The median (p50), 95th percentile (p95), and 99th percentile (p99) response times under baseline load are compiled in Table 1 and as shown in Fig. 1. Table 2 represents the scaling of these platforms at p95.

Table 1. Response Time Metrics (ms) - Baseline Load

Metric	GCP	Azure
p50	153.83	192.19
p95	183.69	238.42
p99	200.93	298.16

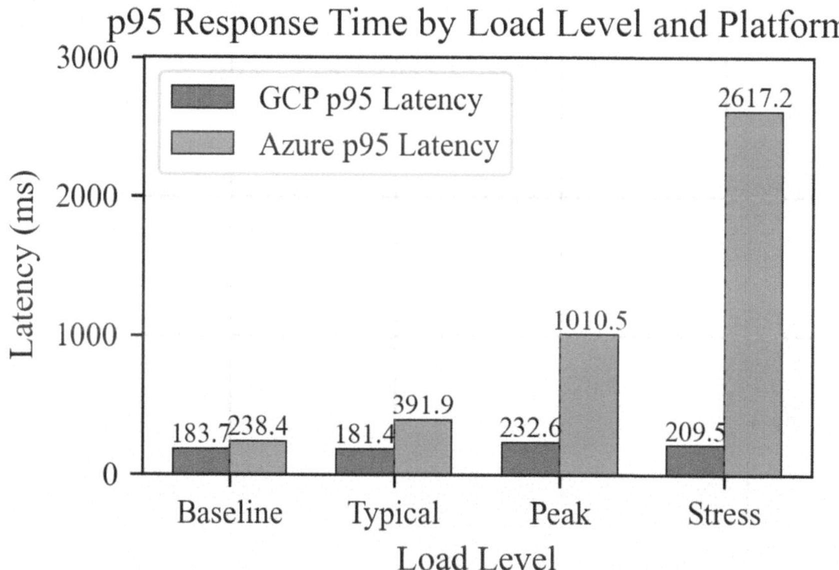

Fig. 1. Shows the p95 response time for each platform across all load levels.

Table 2. Scaling of p95 Response Times

Load Level	GCP p95 (ms)	Azure p95 (ms)
Baseline	183.69	238.42
Typical	181.37	391.9
Peak	232.59	1010.55
Stress	209.51	2617.2

4.2 Throughput Analysis

For every load scenario, throughput was measured in transactions per second (TPS). The observed throughput is compiled by load level and platform in Table 3 and as shown in Fig. 2.

Table 3. Throughput (TPS) by Load Level

Load Level	GCP TPS	Azure TPS
Baseline	7.62	4.19
Typical	18.98	12.65
Peak	25.24	16.94
Stress	70.84	20.14

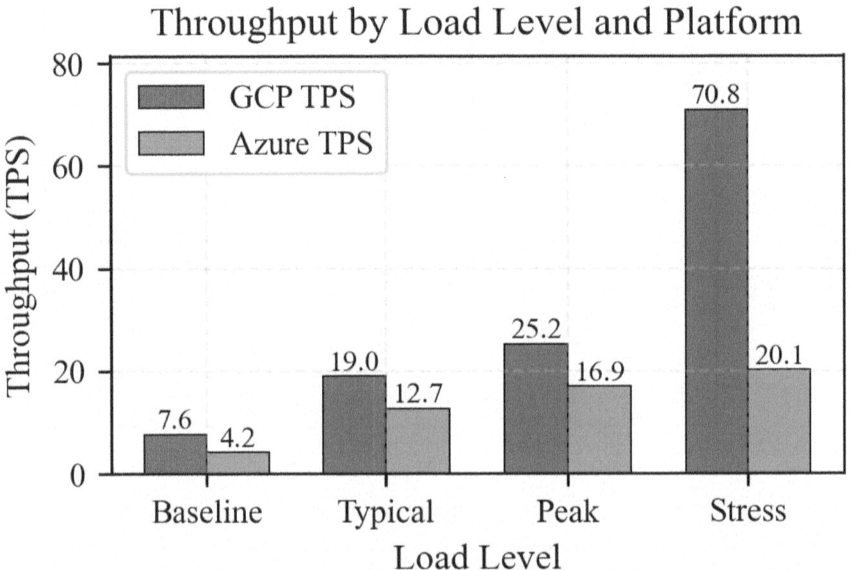

Fig. 2. Illustrates the throughput comparison between platforms.

4.3 Cost Analysis

Operational costs were estimated by multiplying the quantity of data transfers and API calls by the current public pricing for each platform. Table 4 provides a summary of the estimated cost per load level. It is also shown in Fig. 3.

Table 5 provides the Cost Structure of each platform.

Table 4. Cost Estimation by Load

Load Level	GCP Cost	Azure Cost
Baseline	0.000057	0.00016
Typical	0.00143	0.00049

(*continued*)

Table 4. (*continued*)

Load Level	GCP Cost	Azure Cost
Peak	0.0019	0.00065
Stress	0.00534	0.00078

Table 5. Cost Structure for Each Platform

Cost Type	Google Cloud Run	Azure (App Service/Functions)
Per call cost	$0.0000004	$0.0000002
Per GB egress	$0.12	$0.19

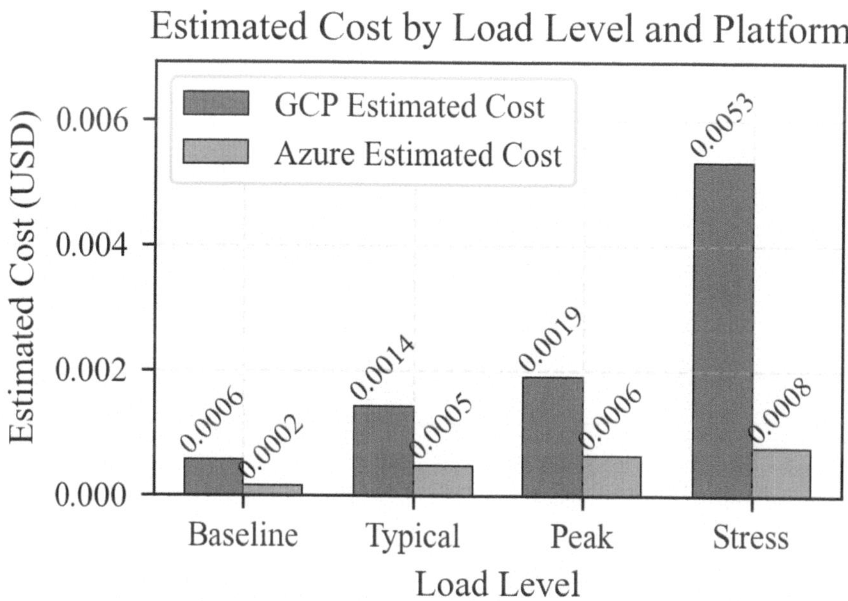

Fig. 3. Shows the cost comparison across load levels.

4.4 Reliability Metrics

Error rates, or the percentage of unsuccessful API calls, were tracked for each scenario. Table 6 displays the observed error rates. It is also shown in Fig. 4.

Table 6. Error Rate (%) by Load Level

Load Level	GCP Error Rate	Azure Error Rate
Baseline	0	0.4
Typical	0	0.26
Peak	0.31	1.07
Stress	0.11	2

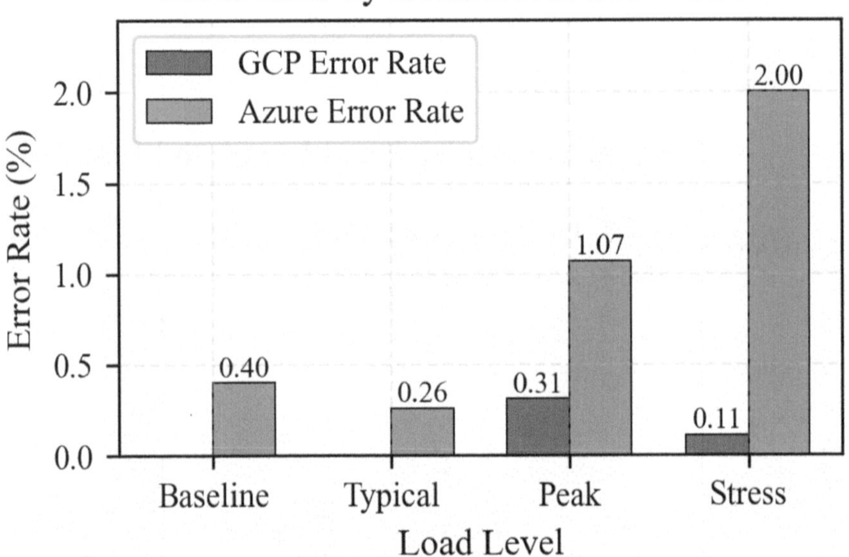

Fig. 4. Displays the error rate analysis across platforms and load levels.

5 Discussion

5.1 Trade-Offs in Performance

Despite having different performance profiles, our benchmarking results show that both GCP Cloud Run and Azure App Service can dependably manage typical POS workloads. In addition to having strong elasticity and quick scaling, GCP Cloud Run has competitive response times across all tested load levels. We did, however, occasionally notice latency spikes, especially following periods of inactivity, which are caused by cold start events, a well-known feature of serverless platforms. However, when loads are steady-state, Azure App Service's always-on architecture offers more reliable response times. Response times increase more slowly as load increases, and its scaling is less dynamic than Cloud Run's.

The most intriguing result is that, even though Azure's error rates are generally lower, GCP has a higher throughput at stress levels (70.84 TPS vs. 20.14 TPS). This implies that, albeit at the expense of somewhat higher error rates, GCP's serverless architecture can manage burst traffic more successfully. Under stress conditions, both platforms—aside from Azure—maintained acceptable error rates below 2% in every scenario.

When interpreting our findings, it is important to keep in mind that the free-tier services were used. Although this provides helpful insights for small-scale or pilot deployments, a direct comparison of PaaS-to-PaaS or serverless-to-serverless offerings at higher service tiers may yield different results. Future studies should include such comparisons to provide a more thorough perspective on enterprise-scale deployments.

5.2 Expense Factors

Cost estimation based on actual API calls and data transferred during our tests indicates that both platforms offer low operating costs for the tested workloads, especially within free-tier limits. Azure's lower per-request pricing structure was the main reason for its 71.9% lower costs at baseline load. GCP's per-request billing model can be particularly cost-effective for applications with erratic or unpredictable traffic because users only pay for the actual time it takes to process requests. Azure App Service's flat-rate pricing may provide more predictability for workloads that are regular and high in volume.

5.3 Guidelines for Platform Selection

The following useful suggestions are supported by our findings:

Choose GCP Cloud Run if you: (1) have experience with containerized deployments and the Google Cloud ecosystem; (2) value fine-grained scaling and per-request billing; or (3) have workloads that are highly variable and susceptible to unpredictable spikes in traffic.

If you need always-on availability without cold starts, have consistent and predictable workloads, or use Microsoft's enterprise tools and services, go with Azure App Service.

A multi-cloud strategy that capitalizes on the advantages of each platform for particular components may be advantageous for businesses with a range of needs.

5.4 Restrictions

This study is subject to several limitations. First, the workloads are artificial and may not accurately reflect the nuances of real point-of-sale traffic, despite being representative. Second, results may differ in other locations due to network latency or regional capacity, as all tests were conducted in a single geographic area. Third, our deployments relied on low-tech, straightforward architectures without advanced features like caching or CDN integration that could further optimize cost or performance. Finally, these results are merely a snapshot in time because cloud services are evolving so rapidly.

Despite these limitations, our open-source, code-driven methodology provides researchers and retailers with a transparent and replicable framework for assessing cloud point-of-sale system performance and cost using their own workloads and configurations.

6 Conclusion

Using free-tier resources and open-source benchmarking scripts, this paper provides a transparent and repeatable comparison of Google Cloud Platform Cloud Run and Microsoft Azure App Service for point-of-sale (POS) workloads. Researchers and retailers can use this methodology to empirically assess cloud platforms based on throughput, error rates, real-time performance, and estimated operating costs. All results originate from the code outputs.

Our results demonstrate that both GCP and Azure can maintain scalable throughput and low error rates while reliably managing point-of-sale workloads at the tested scales. GCP Cloud Run is particularly well-suited for retailers who deal with highly variable or unpredictable traffic because it showed superior throughput under stress conditions (70.84 TPS vs. 20.14 TPS) and 23.0% faster response times at baseline load. For retailers with predictable traffic patterns and cost-sensitive operations, Azure App Service, on the other hand, demonstrated 71.9% lower operational costs and more consistent response times under steady-state loads.

The open-source, code-driven framework created in this study can be used by any retailer to duplicate these evaluations, modify the workload model to fit their setting, and make informed decisions regarding cloud adoption. By using automated measurement and free-tier resources, companies of all sizes can still employ this strategy.

Future studies will assess advanced features like caching, auto-scaling, and CDN integration; they will also look into multi-region deployments for larger retailers; and they will expand this benchmarking framework to include additional cloud providers, such as Amazon Web Services. Future studies will concentrate on automating continuous platform evaluation as services change, incorporating real production workloads in partnership with industry partners, and broadening the methodology to encompass security, compliance, and hybrid or multi-cloud strategies.

Easy-to-use and repeatable benchmarking tools are crucial for maximizing technology investments as the retail industry continues its digital transformation. By creating the framework for continuous, data-driven assessment of cloud platforms, this work assists retailers in striking a balance between cost, performance, and operational resilience in a quickly changing environment.

References

1. Li, A., Yang, X., Kandula, S., Zhang, M.: CloudCmp: comparing public cloud providers. In: Proceedings of the 10th ACM SIGCOMM Internet Measurement Conference, Melbourne, Australia, pp. 1–14 (2010)
2. Garg, S.K., Versteeg, S., Buyya, R.: A framework for ranking of cloud computing services. Futur. Gener. Comput. Syst. **29**(4), 1012–1023 (2013)
3. Varghese, B., et al.: Cloud benchmarking for performance. In: Proceedings of the IEEE 6th International Conference Cloud Computing Technology and Science, Singapore, pp. 535–540 (2014)
4. Manner, J., Endreß, M., Heckel, T., Wirtz, G.: Cold start influencing factors in function as a service. In: Proceedings of the IEEE/ACM International Conference Utility and Cloud Computing Companion, Zurich, pp. 181–188 (2018)

5. Bermbach, D., Wittern, E., Tai, S.: Cloud Service Benchmarking: Measuring Quality of Cloud Services from a Client Perspective. Springer (2017). https://doi.org/10.1007/978-3-319-55483-9
6. Johnson, M., Wang, L.: Cloud adoption in retail: a comprehensive survey. Int. J. Retail Distrib. Manag. **51**(3), 321–339 (2023)
7. Thompson, K., Miller, D., Anderson, J.: Security considerations for cloud-based POS systems. IEEE Secur. Priv. **20**(4), 67–75 (2022)
8. Mazrekaj, A., Shabani, I., Sejdiu, B.: Pricing schemes in cloud computing: an overview. Int. J. Adv. Comput. Sci. Appl. **7**(2), 80–86 (2016)
9. Chen, L., Zhang, Y., Liu, W.: Cost optimization framework for seasonal retail workloads. IEEE Trans. Serv. Comput. **14**(5), 1423–1436 (2021)
10. National Retail Federation: Retail Transaction Patterns and Peak Load Analysis. NRF Research Report (2023)
11. Google Cloud: Cloud run pricing. https://cloud.google.com/run/pricing. Accessed 29 Jun 2025
12. Microsoft Azure: App service pricing. https://azure.microsoft.com/en-us/pricing/details/app-service/. Accessed 29 Jun 2025
13. VPC Network Pricing: Google cloud. https://cloud.google.com/vpc/network-pricing#internet_egress. Accessed 29 Jun 2025
14. Bandwidth Pricing: Microsoft Azure. https://azure.microsoft.com/en-us/pricing/details/bandwidth/. Accessed 29 Jun 2025

ZK-Pyramid: A Pyramid Tree-Based P2P Framework for Trustworthy and Private Data Sharing Using Zero-Knowledge Succinct Proofs

Mehmuna Haque[1], Samana Dahal[1], Indranil Roy[1](✉), Reshmi Mitra[1], and Bidyut Gupta[2]

[1] Southeast Missouri State University, Cape Girardeau, USA
{mhaque2s,sdahal7s,iroy,rmitra}@semo.edu
[2] Southern Illinois University, Carbondale, USA
bidyut@cs.siu.edu

Abstract. This work introduces a Zero-Knowledge Proof (ZKP)-based trust verification protocol integrated within a Pyramid Tree peer-to-peer (P2P) architecture, designed to enable secure and scalable communication in decentralized systems. The protocol ensures verifiable peer authentication at each hop without revealing private information, fostering trust in dynamic, trustless environments. Our analysis shows that the ZKP message overhead grows linearly with the tree depth d and the number of puzzles k, while remaining independent of the total number of peers N. Since $d \ll n \ll N$, where n is the number of distinct resources types, the protocol ensures high security with low message complexity. This makes it ideal for large-scale, resource-constrained applications such as federated learning, IoT, and edge computing. The proposed solution offers an effective balance between trust, efficiency, and scalability, positioning it as a robust foundation for next-generation distributed architectures.

Keywords: Zero-Knowledge Proofs · Trust Verification · Pyramid Tree · P2P Architecture · Interest-Based · Distributed Systems

1 Introduction

In peer-to-peer (P2P) networks, establishing trust is challenging due to the absence of a central authority [1]. As the network scales, identifying legitimate peers becomes more difficult, especially with threats like Sybil, replay, and eclipse attacks [4,18]. Resource constraints further limit the use of complex defense mechanisms, making secure and scalable peer authentication a persistent issue.

A key challenge lies in balancing anonymity with accountability. While anonymity protects user privacy, it also enables malicious behavior. Conversely,

I. Roy et al. (Eds.): CAINE 2025, CCIS 2740, pp. 100–111, 2026.
https://doi.org/10.1007/978-3-032-14893-3_8

identity verification can compromise privacy. This tension is especially problematic in P2P environments where participant identities are deliberately concealed [16].

Zero-Knowledge Proofs (ZKPs) present a lightweight cryptographic approach to resolving these challenges [2,9]. They enable a peer to confirm the legitimacy of a statement to another peer without revealing the underlying data, thereby safeguarding anonymity while facilitating trust [9]. ZKPs also minimize communication and computational burdens, making them ideal for resource-constrained environments. Moreover, the use of succinct, non-interactive ZKPs reduces network delays and enhances the speed of secure data transmission. By allowing verification without revealing the data, ZKPs directly address the tension between anonymity and accountability inherent in P2P networks.

Unlike the circular trust ring architecture presented in prior research [5], where group heads are arranged in a ring for transit-level coordination, we propose a novel Pyramid Tree Architecture that enhances the efficiency, scalability, and fault tolerance of the trust verification process. As illustrated in Fig. 1, this hierarchical design organizes nodes into multi-level layers, allowing parallel and localized trust verification. The pyramid-shaped structure reduces traversal times and enables each group head to perform scalable Zero-Knowledge Proof verification independently or hierarchically, significantly lowering the redundancy and latency present in circular formations. Since each branch can operate semi-independently, the system is more resilient to node churn and partial failures.

Our ZKP-based protocol is embedded within this pyramid tree hierarchy, where trust is established via one-step challenge-response exchanges involving cryptographic operations. This architecture and protocol together address the fundamental challenge of verifying trust without compromising anonymity. They maintain minimal interaction overhead and are well-suited for dynamic, resource-limited P2P environments. This approach promotes secure and scalable trust verification in fully decentralized environments while leveraging the cryptographic rigor of Zero-Knowledge Proofs to preserve anonymity and reduce computational burden.

In this paper, we introduce the Pyramid Tree Architecture and demonstrate its effectiveness in achieving secure and scalable trust verification in P2P networks. We address potential concerns such as the overhead of maintaining the tree structure and discuss the implications of our findings, limitations, and future work directions.

The main contributions of this paper are listed below:

- Lightweight Cryptographic Solution: Designed a novel ZKP-based protocol for trust verification in Pyramid Tree based P2P networks that preserves user anonymity.
- Balancing Anonymity and Trust: A one-step challenge and response mechanism during data transfer, striking a balance between user anonymity and trustworthiness.

- Scalability and Resource Efficiency: Suitable for resource-constrained P2P networks, this protocol ensures scalable and efficient trust verification, minimizing computational overhead and communication costs.

2 Related Works

2.1 Trust Verification in Peer-to-Peer Networks

Trust management plays a crucial role in securing peer-to-peer (P2P) networks, particularly in ensuring the integrity, confidentiality, and availability of data during communication and resource sharing. Trust-based mechanisms provide a foundation for evaluating peer behavior and mitigating malicious activities in decentralized environments. Balfe *et al.* proposed a trusted computing framework aimed at strengthening security in P2P systems by integrating trusted computing technologies [1]. Their approach leverages trusted platform modules (TPMs) and secure boot mechanisms to establish a secure foundation for P2P communication and interaction. Selcuk *et al.* introduced a reputation-based trust management system tailored for P2P networks [14]. Their framework evaluates the reliability of nodes based on accumulated reputation scores, enabling informed trust decisions and enhancing system robustness against unreliable or malicious peers. Zhao *et al.* addressed trust in grid-based P2P computing by proposing a trust-based scheduling algorithm that considers both node reputation and result verification [18]. This method is particularly effective in distributed computational environments where collaborative task execution is essential. Frahat *et al.* presented a secure and scalable trust management model specifically designed for IoT-enabled P2P networks [6]. Their framework combines reputation management, access control, and trust evaluation to ensure secure communication and collaboration among interconnected IoT devices. Hao *et al.* explored the use of blockchain technology to enhance trust in P2P networks [12]. They proposed a trust-aware blockchain topology that supports secure and efficient broadcast, increasing the reliability and verifiability of shared information in P2P environments.

2.2 Preliminaries on Zero-Knowledge Proofs

Zero-knowledge proofs (ZKPs) have emerged as powerful tools for ensuring privacy and security in P2P networks and other distributed systems. ZKPs allow one party to prove knowledge of a secret without revealing the secret itself, making them ideal for privacy-preserving authentication and secure communications. Danezis and Diaz developed SybilInfer, a system that uses social network analysis and ZKPs to detect Sybil attacks and enhance trustworthiness in P2P systems [4]. Lu *et al.* applied ZKPs to anonymous authentication in P2P environments by introducing a pseudo-trust system based on zero-knowledge authentication [9]. Pop *et al.* proposed a ZKP-based privacy-preserving scheme for energy transactions on the blockchain [11], ensuring transaction integrity while protecting user data. Sun *et al.* provided a comprehensive survey on ZKP applications in blockchain, discussing the different types of ZKPs, their use cases, and challenges [15]. Yang and Li designed a digital identity management system using ZKPs to preserve user privacy within blockchain infrastructures [17]. Harikrishnan and Lakshmy introduced a ZKP-based protocol for secure and anonymous payments in distributed networks [7]. Major *et al.* proposed an authentication mechanism based on a combination of chaos theory and ZKPs to enhance privacy and security in distributed systems [10]. Kosba *et al.* presented Hawk, a privacy-preserving smart contract framework that integrates ZKPs to ensure secure execution on the blockchain [8]. Ben-Sasson *et al.* introduced zk-STARKs, a scalable and transparent ZKP system that eliminates the need for trusted setup, offering a highly efficient and secure alternative for verifying computations [3].

Together, these works highlight the versatility of ZKPs in enhancing privacy and security across multiple domains, including authentication, blockchain, energy systems, and identity management. Their contributions pave the way for more secure, trustworthy, and decentralized digital infrastructures.

2.3 P2P Pyramid Tree Architecture

2.4 Definition of Resources

Definition 1. A resource is defined as a tuple $\langle R_i, V\rangle$, where R_i represents the resource type and V denotes the value associated with the resource. Each resource type may be linked to multiple values. For example, if R_i represents the resource type *songs* and V' refers to a specific singer, then the tuple $\langle R_i, V'\rangle$ represents the songs sung by that particular singer.

Definition 2. Let S be the set of all peers in a peer-to-peer (P2P) system, defined as $S = \{P_{R_i}\}$ for $0 \leq i \leq n-1$, where P_{R_i} is the subset of peers that share the same resource type R_i, and n is the total number of distinct resource types in the system. For each subset P_{R_i}, we define P_i as the first peer to join the system with resource type R_i, referred to as the *group-head* of group G_i (or cluster i). Throughout this paper, we assume that each peer possesses only one resource type.

The overlay architecture known as the *Pyramid Tree*, originally proposed in [13], is defined as follows:

1. The root of the tree resides at level 1. There are n fully connected groups (clusters), where each group G_i contains peers from subset P_{R_i} for $0 \leq i \leq n-1$. Within each group P_{R_i}, all peers are logically interconnected, forming a complete graph with diameter 1. The group-heads of these n groups form the edges of the Pyramid Tree structure.
2. Communication between a peer $p_i \in G_i$ and a peer $p_j \in G_j$ occurs exclusively through their respective group-heads P_i and P_j using tree-based traversal.

Let $S_n = \{0, 1, 2, \ldots, n-1\}$ denote the set of nonnegative integers less than n, representing the residue classes modulo n. Each element in S_n is referred to as a residue class (RC), labeled $[0], [1], \ldots, [n-1]$, where:

$$[r] = \{a \in Z \mid a \equiv r \pmod{n}\}$$

For example, with $n = 3$:

- $[0] = \{\ldots, -6, -3, 0, 3, 6, \ldots\}$
- $[1] = \{\ldots, -5, -2, 1, 4, 7, \ldots\}$
- $[2] = \{\ldots, -4, -1, 2, 5, 8, \ldots\}$

In this P2P architecture, positive integers from different residue classes are used as logical addresses for peers. **Lemma 1.** Any two integers belonging to the same class $r \in S_n$ are mutually congruent.

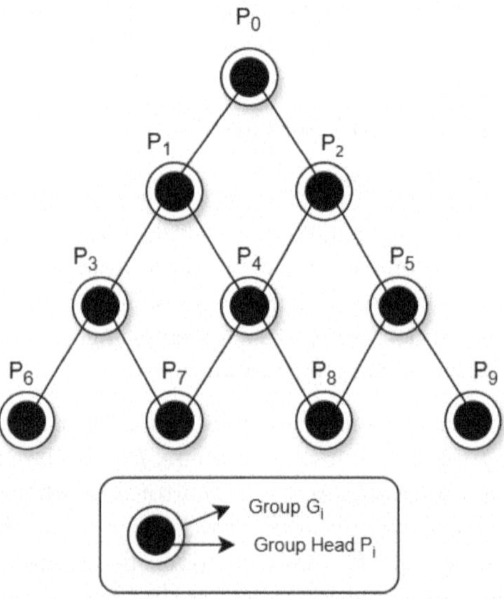

Fig. 1. A Complete Pyramid Tree-Based Structured P2P architecture with n distinct resource types.

In the interest-based P2P system, we assume there are n distinct resource types. Let $S = \{P_{R_i}\}$ for $0 \leq i \leq n-1$ be the set of all peers in the system. For each group G_i, peer P_i is the first peer with resource type R_i to join the system and serves as its group-head.

Logical addresses are assigned as follows:

1. Each group-head P_r of group G_r is assigned the smallest nonnegative number r from the residue class $r \pmod{n}$.
2. All peers with resource type R_r join group G_r and are assigned addresses of the form $(r + j \cdot n)$, where $j = 0, 1, 2, \ldots$.

3. A tuple ⟨Peer Logical Address, IP Address⟩ is added to the Local Resource Table (LRT) whenever a new peer joins a group.
4. The resource type R_r is assigned the code r, which also serves as the logical address of group-head P_r.
5. A tuple ⟨Group-Head Logical Address, IP Address⟩ is added to the Global Resource Table (GRT) each time a new group-head joins.

Remark 1. The Global Resource Table (GRT) remains sorted based on the logical addresses of the group-heads.

Definition 3. Two peers in the same group G_r are logically linked if their logical addresses are mutually congruent.

Lemma 2. Each group G_r forms a complete graph [13].

Observation 1. Any intra-group data lookup requires only one overlay hop [13].

Observation 2. Inter-group data lookup latency is bounded by the diameter of the tree [13].

Observation 3. A Pyramid Tree is considered complete if, at any level j, there are exactly j nodes (clusters).

Observation 4. In an incomplete Pyramid Tree, each level j contains j nodes, except the leaf level.

Observation 5. New groups (clusters) always join at the leaf level.

Observation 6. A node that is neither part of the left nor right branch of the root is an internal node.

Observation 7. The degree of an internal non-leaf node is 4.

Observation 8. The degree of an internal leaf node is 2.

An example of a complete Pyramid Tree is shown in Fig. 1 [13]. The group-head P_0, representing resource type 0, is assigned logical address 0 and becomes the root of the tree. Generally, if there are i existing group-heads $P_0, P_1, \ldots, P_{i-1}$, the next group-head with resource type i receives logical address i. For example, the sixth group-head is assigned logical address 5. Edges are added in a way that preserves the Pyramid Tree structure, ensuring node degrees remain within prescribed bounds. Leaf nodes like 6 and 9 have degree 1, node 4 is an internal non-leaf node, and nodes 7 and 8 are internal leaf nodes.

3 Trust Verification in Pyramid Tree P2P Architecture Using Zero Knowledge Succinct Proofs

Let us consider a scenario where a peer $P_i \in G_i$, whose group-head is G_i^h, requests a resource $Resc_j$ that is owned by a peer $P_j \in G_j$, managed by group-head G_j^h. The data-lookup protocol defined in [13] is used to identify the peer P_j in the Pyramid Tree structure. Once located, the data transfer between P_j and

P_i is facilitated through their respective group-heads G_j^h and G_i^h along the inter-group traversal path in the Pyramid Tree. However, a critical vulnerability in this architecture arises if a group-head becomes a target of a man-in-the-middle attack or is otherwise disrupted. Since group-heads manage both intra-group and inter-group communication, compromising one can result in a significant bottleneck or system-wide disruption.

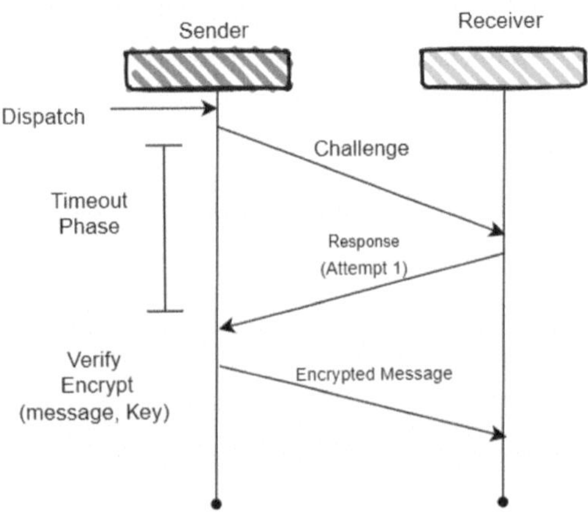

Fig. 2. Sequence diagram for an encrypted message sent from sender to receiver after successfully verifying the receiver.

To counter this, we propose a trust establishment and verification mechanism during the data transfer phase, relying on the cryptographic framework of Zero Knowledge Succinct Proofs (ZKSP) as shown in Fig. 2. This mechanism enables a **prover** (e.g., G_i^h) to prove its legitimacy to a **verifier** (e.g., G_j^h) without revealing any sensitive information. Repeated interactions under ZKSP provide verifiable assurance of trustworthiness while preserving privacy and minimizing communication overhead.

3.1 Trust Verification Protocol

The protocol ensures that before any data transfer occurs from G_j^h to G_i^h, the prover solves a cryptographic challenge (e.g., hash puzzle) to demonstrate its trustworthiness. The verifier issues the challenge and checks the correctness and timeliness of the response, as shown in Fig. 3. The data transfer is then encrypted using the successful response as a symmetric key, ensuring confidentiality and protection against replay attacks.

The detailed algorithm is provided below (Algorithm 1).

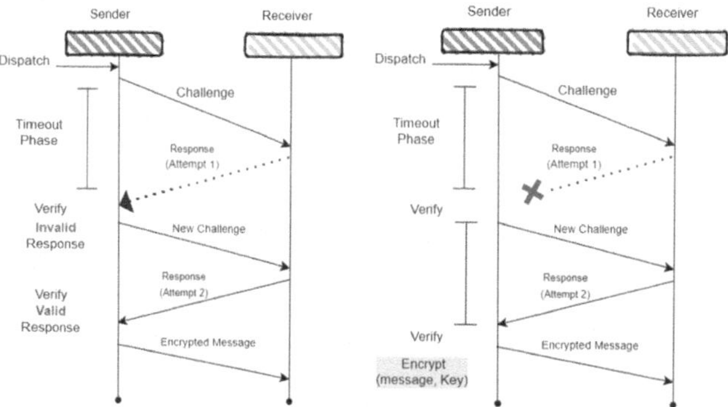

Fig. 3. Sequence diagram depicting the protocol's response to a lost/delayed/incorrect solution from a receiver node. In this scenario, the sender node is necessitated to send a new challenge or puzzle during the next communication cycle to reaffirm the receiver's trustworthiness.

3.2 Trust Management Using Untrusted Peer List

Each group-head maintains an updated list Ut_list of untrusted peers. This list evolves dynamically based on protocol executions. If a peer repeatedly fails to provide valid ZK proofs within the allocated timeout window (TM_{out}), it is added to Ut_{list} and broadcast to Level 1 group-heads, using the broadcast protocols as mentioned in [13]. Conversely, consistent success in ZKSP-based verification can lead to removal from Ut_{list}, ensuring adaptability and fairness.

3.3 Succinct Challenge-Response Design

The challenge $Chlg_k$ posed by the verifier G_j^h typically involves low-complexity cryptographic tasks such as computing a SHA-256 hash of a nonce. These challenges are easy to verify and impose minimal overhead—ideal for resource-constrained P2P environments. The resulting solution Rep_k functions as both a symmetric encryption key and a one-time-use nonce, mitigating replay attacks.

3.4 Scalability and Security

This ZKSP-based verification is highly scalable, as challenges and responses are lightweight and require no central authority. Group-heads only communicate with one another as needed along the Pyramid Tree traversal path. Trust establishment is conducted on-the-fly and independently at each hop, ensuring a decentralized and secure communication model.

3.5 Continuous Verification

Every inter-group message in the Pyramid Tree architecture undergoes verification at each hop. This enforces a zero-trust principle, where no peer is inherently trusted, and each must continually prove its authenticity. Consequently, malicious nodes, Sybil attacks, and message tampering are effectively mitigated.

4 Evaluation

Our protocol was rigorously evaluated using the Pyramid Tree architecture, focusing on how network depth and cryptographic workload affect trust and communication efficiency. In the optimal scenario—where ZKP-based trust verification is not enforced—the number of hops required for efficient data transfer remains minimal at $2d - 2$, a value bounded by the tree depth d.

Algorithm 1 ZKP Data Transfer Protocol between P_j and P_i in Pyramid Tree Architecture

1: flag = **false** ▷ assume by default G_j^h is untrusted
2: P_j sends $Resc_j$ to its group head G_j^h
3: **for** $k \leftarrow 1$ to 3 **do** ▷ for example, k = 1, 2, 3
4:     **if** $G_i^h \in Ut_{list}$ **then**
5:         G_j^h drops the packet
6:         **break**
7:     **else**
8:         G_j^h sends $Chlg_k$ to G_i^h
9:         G_j^h waits for TM_{out}
10:         G_i^h solves $Chlg_k$ and sends the response Rep_k to G_j^h
11:         **if** ((Re_k is true) & (verified by G_j^h) & (within TM_{out})) **then** ▷ trusted operation
12:             G_j^h encrypts E($Resc_j$, Rep_k)
13:             G_j^h sends E($Resc_j$, Rep_k) to G_i^h
14:             G_i^h decrypts D(E($Resc_j$, Rep_k), Rep_k)
15:             G_i^h sends $Resc_j$ to P_i
16:             flag = **true**
17:             **break**
18:         **else if** ((Rep_k is false) & (verified by G_j^h) & (within TM_{out})) **then** ▷ potentially untrusted
19:             **continue**
20:         **else** ▷ timeout or incorrect
21:             TM_{out} has expired
22:             **continue**
23:         **end if**
24:     **end if**
25: **end for**
26: **if** flag == false **then** ▷ untrusted operation
27:     Add G_i^h to Ut_{list}
28:     Broadcast the list to other group heads at Level 1 of the Pyramid Tree
29: **end if**

Assuming each hop involves k cryptographic puzzles for verification, the total number of ZKP-related messages exchanged along the longest path is given by:

$$\text{ZKP\_Messages}(d, k) = 2k(2d - 2) = 4kd - 4k$$

Here, d denotes the depth (number of levels) of the Pyramid Tree, which is typically very small compared to n, the number of distinct resources types (or logical groups), and significantly smaller than N, the total number of nodes or peers in the system. Thus, we observe the relationship:

$$d \ll n \ll N$$

This observation is crucial because it implies that the total number of ZKP messages required for trust verification during data transfer, is independent of the total number of peers N. Instead, the message complexity depends only on the depth of the tree and the configured puzzle count per hop.

This bounded and predictable overhead enables the architecture to scale efficiently to large networks while maintaining strong security guarantees. As a result, the proposed ZKP-based Pyramid Tree protocol offers an excellent trade-off between authentication depth, message complexity, and overall scalability, making it particularly well-suited for decentralized and resource-constrained environments such as federated learning over P2P systems.

5 Conclusion

In this work, we presented a ZKP-based trust verification protocol integrated within a Pyramid Tree peer-to-peer architecture, designed to ensure secure and efficient communication in decentralized environments. By incorporating Zero-Knowledge Proofs at each hop, the protocol enables verifiable authentication without revealing sensitive information, thereby enhancing trust among participating peers.

Our analytical and structural evaluation demonstrated that the total ZKP message overhead scales linearly with the tree depth d and the number of puzzles k, while remaining independent of the total number of nodes N in the system. This property ensures that the communication and security costs are tightly bounded, even as the network grows significantly in size.

Moreover, since the tree depth d is significantly smaller than both the number of distinct resources n and the total number of peers N, the protocol achieves strong authentication guarantees with minimal message complexity. This makes it highly suitable for large-scale, resource-constrained applications such as federated learning, IoT, and edge computing.

Overall, the proposed Pyramid Tree protocol demonstrates an effective balance between security, efficiency, and scalability, making it a promising foundation for secure and trusted communication in next-generation distributed systems.

Future Work

As part of our ongoing research, we plan to extend the protocol by incorporating a dynamic *Trust Matrix* that quantifies peer behavior over time based on historical interactions and ZKP verification success rates. This matrix will enable adaptive trust-based routing, malicious node isolation, and reputation-aware resource delegation within the Pyramid Tree. By integrating probabilistic trust inference and decay mechanisms, the enhanced system will offer greater resilience against collusion, misbehavior, and trust decay in large-scale federated or peer-to-peer networks.

References

1. Balfe, S., Lakhani, A.D., Paterson, K.G.: Trusted computing: providing security for peer-to-peer networks. In: Fifth IEEE International Conference on Peer-to-Peer Computing (P2P 2005), pp. 117–124. IEEE (2005)
2. Ben-Sasson, E., Bentov, I., Horesh, Y., Riabzev, M.: Scalable, transparent, and post-quantum secure computational integrity. IACR Cryptol. ePrint Arch. **2018**, 46 (2019)
3. Ben-Sasson, E., Bentov, I., Horesh, Y., Riabzev, M.: Scalable zero knowledge with no trusted setup. In: Boldyreva, A., Micciancio, D. (eds.) CRYPTO 2019. LNCS, vol. 11694, pp. 701–732. Springer, Cham (2019). https://doi.org/10.1007/978-3-030-26954-8_23
4. Danezis, G., Mittal, P.: Sybilinfer: detecting sybil nodes using social networks. In: NDSS, pp. 1–15. San Diego, CA (2009)
5. Deverasetti, S.K., Neupane, A., Roy, I., Mitra, R., Gupta, B.: Establishing trust using zero knowledge succinct proof in peer-to-peer data transfer. Manuscript, 2025 (2025)
6. Frahat, R.T., Monowar, M.M., Buhari, S.M.: Secure and scalable trust management model for IoT P2P network. In: 2019 2nd International Conference on Computer Applications & Information Security (ICCAIS), pp. 1–6. IEEE (2019)
7. Harikrishnan, M., Lakshmy, K.V.: Secure digital service payments using zero knowledge proof in distributed network. In: 2019 5th International Conference on Advanced Computing & Communication Systems (ICACCS), pp. 307–312. IEEE (2019)
8. Kosba, A., Miller, A., Shi, E., Wen, Z., Papamanthou, C.: Hawk: the blockchain model of cryptography and privacy-preserving smart contracts. In: 2016 IEEE Symposium on Security and Privacy (SP), pp. 839–858. IEEE (2016)
9. Li, L., et al.: Pseudo trust: zero-knowledge authentication in anonymous P2Ps. IEEE Trans. Parallel Distrib. Syst. **19**(10), 1325–1337 (2008)
10. Major, W., Buchanan, W.J., Ahmad, J.: An authentication protocol based on chaos and zero knowledge proof. Nonlinear Dyn. **99**(4), 3065–3087 (2020). https://doi.org/10.1007/s11071-020-05463-3
11. Pop, C.D., Antal, M., Cioara, T., Anghel, I., Salomie, I.: Blockchain and demand response: zero-knowledge proofs for energy transactions privacy. Sensors **20**(19), 5678 (2020)
12. Ren, Y., Zeng, Z., Wang, T., Zhang, S., Zhi, G.: A trust-based minimum cost and quality aware data collection scheme in P2P network. Peer-to-Peer Network. Appl. **13**, 2300–2323 (2020)

13. Roy, I., Rahimi, N., Maddali, K., Kaluvakuri, S., Gupta, B., Debnath, N.: Design of efficient broadcast protocol for pyramid tree-based P2P network architecture. In: Shi, Y., Hu, G., Goto, T., Yuan, Q. (eds.) CAINE 2020. The 33rd International Conference on Computer Applications in Industry and Engineering. EPiC Series in Computing, vol. 75, pp. 80–89. EasyChair (2021)
14. Selcuk, A.A., Uzun, E., Pariente, M.R.: A reputation-based trust management system for P2P networks. In: IEEE International Symposium on Cluster Computing and the Grid, 2004. CCGrid 2004, pp. 251–258. IEEE (2004)
15. Sun, X., Yu, F.R., Zhang, P., Sun, Z., Xie, W., Peng, X.: A survey on zero-knowledge proof in blockchain. IEEE Network **35**(4), 198–205 (2021)
16. Tsang, P.P., Smith, S.W.: PPAA: peer-to-peer anonymous authentication. In: Bellovin, S.M., Gennaro, R., Keromytis, A., Yung, M. (eds.) ACNS 2008. LNCS, vol. 5037, pp. 55–74. Springer, Heidelberg (2008). https://doi.org/10.1007/978-3-540-68914-0_4
17. Yang, X., Li, W.: A zero-knowledge-proof-based digital identity management scheme in blockchain. Comput. Secur. **99**, 102050 (2020)
18. Zhao, S., Lo, V., Dickey, C.G.: Result verification and trust-based scheduling in peer-to-peer grids. In: Fifth IEEE International Conference on Peer-to-Peer Computing (P2P 2005), pp. 31–38. IEEE (2005)

Governing Blockchain-Based Decentralized Systems

Nipesh Pant, Saurav Ghosh, Hemant Upadhyay, and Reshmi Mitra(✉)

Southeast Missouri State University, Cape Girardeau, USA
{npant2s,sghosh3s,hupadhyay1s,rmitra}@semo.edu

Abstract. As peer-to-peer (P2P) networks grow in complexity and decentralization, enforcing secure and fair access control becomes increasingly challenging due to dynamic topologies, high churn, and limited trust anchors. Traditional token-based mechanisms, while effective in incentivizing activity, fail to capture behavioral nuances necessary for long-term governance and decision-making. In this work, we propose a novel governance token mechanism that integrates trust metrics into token distribution, enabling behavior-aware participation in blockchain governance. Based on a hierarchy of smart contracts, including the Access Control Contract (ACC), the Judging Contract (JC), and the Governance Token Contract (GTC), our system dynamically evaluates peer behavior through a cost function based on token balance, compliance rate, and peer status. Governance tokens are minted or slashed proportionally on the basis of trust scores calculated cyclically after network activity thresholds. Experimental evaluations demonstrate that our approach successfully restricts governance privileges to trustworthy peers, mitigates the influence of malicious nodes, and supports reputation recovery over time. This behavior-based model improves fairness, security, and accountability in decentralized access control.

Keywords: Blockchain Governance · Federated Trust Model · Decentralized Access Control · Smart Contracts · Peer-to-Peer Networks · Governance Token · Trust Score Evaluation · Token Slashing · Role-Based Access

1 Introduction

Securing peer-to-peer (P2P) networks is a complex challenge because, unlike traditional systems, there is no central authority to manage user access or verify identities. Devices often join and leave unpredictably, and many of them operate with limited processing power or unstable connectivity. These conditions make the network more vulnerable to malicious activity, especially when peers have to make trust decisions based on partial or outdated information. Adding to the difficulty, lightweight devices cannot always support advanced cryptographic protocols, which limits the ability to enforce strong security measures.

I. Roy et al. (Eds.): CAINE 2025, CCIS 2740, pp. 112–125, 2026.
https://doi.org/10.1007/978-3-032-14893-3_9

Balancing user anonymity with accountability is another ongoing issue. P2P networks are designed to protect user privacy, but this makes it harder to detect and respond to bad behavior. Trust also varies across different parts of the network, where peers may use different standards to evaluate others. Without a consistent way to score and aggregate trust, systems become fragmented and easier to exploit. Furthermore, adding real-time behavior analysis and context-aware permissions requires rapid adaptation to changing network conditions, which increases complexity as well.

To address these challenges, we propose a decentralized, federated trust model in which peers collectively evaluate one another's behavior across the network without relying on a central authority. Each peer contributes to trust decisions based on locally observed metrics, which are aggregated through on-chain logic to form a system-wide view; thereby promoting more balanced access control decisions across domains.

To operationalize this trust model, we introduce a behavior-aware governance token. Unlike regular tokens that only reflect economic value, governance tokens represent earned influence and trust within the system. They are non-transferable and awarded based on a peer's long-term behavioral consistency and compliance with pre-defined rules.

Our system also employs a dynamic evaluation mechanism to continuously assess trustworthiness over time. This mechanism calculates a trust score for each peer at the end of a cycle, factoring in token balance, compliance rate, and peer status. Peers with high trust scores are granted governance tokens proportionally, while those with poor behavior are penalized through token slashing. We validate this mechanism through test cases that simulate both positive and negative peer behavior. The results show that the governance token framework not only promotes fair access but also helps the system recover from trust violations.

Our design builds on prior work [4] and uses a hierarchy of smart contracts, including Access Control Contracts (ACCs), a Judge Contract (JC), and a Register Contract (RC). These components interact on-chain to record activity, enforce rules, and apply penalties or rewards based on peer behavior.

Overall, this paper makes the following key contributions:

- We propose a novel governance-token model that integrates peer behavior and trust scores into access control decisions in decentralized P2P networks.
- We design and implement a smart contract hierarchy (ACC, JC, RC, GTC) that supports trust-based adjudication and dynamic governance rights.
- We develop a cost function that incorporates token balance, compliance rate, and peer status to fairly allocate governance tokens.
- We demonstrate through simulation that this behavior-based model limits the influence of malicious peers, supports trust recovery, and ensures fair governance participation.

The structure of this paper is as follows. Section 2 reviews related work on access control and blockchain solutions in P2P networks. Section 3 presents the architecture of our smart contract-based access control system. Section 4 describes the system's implementation, including Ethereum-based contract development. Section 5 discusses our evaluation and results based on different test cases. Finally, Sect. 6 concludes the paper and outlines directions for future research.

2 Related Work

2.1 Access Control Issues in IoT and P2P Networks

Access control mechanisms that are traditional in nature have long played a pivotal role in securing computing environments. Mechanisms such as Role-Based access control(RBAC) and Attribute-Based Access Control(ABAC) are inherently static in design and exhibit a strong reliance on centralized policy enforcement. [11] Furthermore, studies by prominent institutions, including NIST, have advocated for a paradigm shift from centralized models towards a much more dynamic and efficient access control concepts that involve smart contracts, that can enable better auditability, verifiability and built-in resilience within such access control frameworks. [6]

IoT and P2P systems have a distinctive set of access control challenges that traditional methods such as RBAC and ABAC cannot fully solve. One major concern is that usually, devices are different from each other and often have limited power, memory, or unstable connections. This makes it hard to use a central system to control granular access, and it could potentially lead to a single point of failure. Due to a more dynamic topology, their existence and governance within the network has to be consistently met with new configurations on access control and compatibility analysis. With this type of setup, finding real-time malicious behavior is often difficult and requires a detailed oversight of the organization's network. [10] These points highlight that access control needs to be decentralized, flexible, easy to run on limited devices, and based on trust or behavior, as demonstrated in recent frameworks that integrate dynamic smart contracts and dual-layer validation to meet the scalability and trust demands of P2P networks [4].

2.2 Existing Blockchain-Based Access Control Systems

Blockchain-based access control systems aim to remove central authorities, offering transparency, immutability, and strong auditability. In 2017, Mori, Ricci, and Maesa proposed a foundational approach on the Bitcoin blockchain, where access policies and rights transfers are publicly visible, enabling distributed auditability and preventing fraudulent denial of access [12,13]. Building on this, Li, Han & Liu introduced Fabric-IoT (2020), which uses Hyperledger Fabric and ABAC smart contracts—covering devices, policies, and access control—to achieve decentralised, fine-grained IoT access with high throughput [7]. Ding et al. (2019) designed a lightweight ABAC scheme for IoT, integrating smart contracts to simplify attribute storage and improve security for constrained devices [1]. In 2021, Li, Du, Sun & Chen extended ABAC with Identity-Based Signatures and cross-domain support on permissioned blockchains, allowing lightweight IoT nodes to participate securely [8]. Likewise, Gokul et al. (2021) offered a broad architecture focusing on transparency, tamper resistance, and no reliance on third parties [5]. Further, Eassa et al. (2021) introduced a blockchain-based secured access control for IoT using multi-agent systems, securing edge-fog-cloud interactions [2]. And Souri et al. (2022) presented an auditable access control model, ensuring private data protection with strong auditability and performance [14]. Finally, Li et al. (2024) addressed cross-domain control by proposing a multi-domain architecture that uses federated ledgers to maintain efficiency and authentication even in heterogeneous environments [9].

Despite these achievements, challenges like blockchain overhead for constrained devices, latency, and token-based governance persist. Our research proposes a governance layer using non-transferable, behavior-aware tokens to boost decentralised trust and access control. A recent framework introduced adaptive governance through real-time on-chain analytics and fault-tolerant consensus (PBFT + PoA), showing promise in addressing these limitations at scale [3].

3 Smart Contract Architecture

3.1 System Architecture

The core access control system is organized around three core Ethereum smart contracts, which are ACCs, JC, and RC each of these serving a distinct role [15]. ACCs are the primary enforcers of access policies. ACC manages a specific subject-object pair and validates both static and dynamic access permissions. Static validation checks whether the requested operation is allowed according to predefined policies, while dynamic validation monitors behavioral patterns such as frequent access attempts or policy violations. ACCs maintain local misbehavior logs and trigger penalties for inappropriate behavior. All decisions whether to grant or deny access are logged on-chain, ensuring transparency and accountability. The JC acts as an adjudicator for violations reported by the ACCs. When an ACC detects suspicious behavior, it submits a report to the JC, which then evaluates the case based on severity, timestamp, and context. The JC determines the appropriate penalty ranging from warnings to temporary or permanent suspensions. The RC serves as a dynamic registry of all deployed ACCs and their associated policies. It maintains a lookup table that maps method names to contract addresses, role specific access levels, and the source of metadata. Through dedicated APIs the RC ensures that access policies are verifiable, auditable, and consistently enforced. Additionally, RC supplies role-specific reference data to peers such as Primary Group Heads, Secondary Group Heads, and Regular Members, as defined in the 'Securing RC Based P2P Networks paper' by Saurav Gosh et al. [4].

To integrate governance tokens, we introduced a new contract GTC compliant with the ERC-20 token standard. GTC is responsible for implementing the cost function calculation which determines the distribution of governance token between each peer, minting and slashing of governance token based on that cost function calculation. The GTC serves as a reputation and incentive

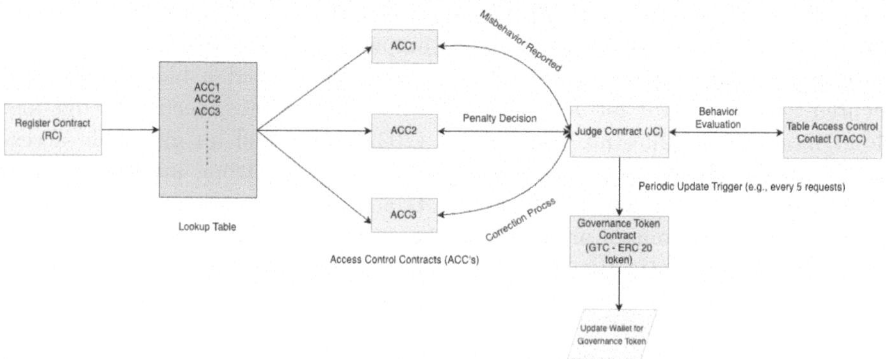

Fig. 1. System architecture showing interactions between RC, ACCs, JC, TACC and GTC

ledger, where nodes earn or lose governance tokens based on their behavior. The behavior pattern of the peer is supplied from JC which provides the information regarding the rewards and penalties peers have based on which governance token calculation is done. The overall interactions between the RC, ACCs, JC, and the GTC are illustrated in Fig. 1 which presents the smart contract architecture of the system. As shown in the figure, the ACCs monitor behavior and send reports to the JC when violations are detected. The JC evaluates these reports and makes adjudication decisions, while the RC maintains mappings and supplies role-based reference data. The flow of governance tokens and the behavior evaluation process are also depicted, showing how decisions propagate through the system to enforce policy and incentivize compliance.

3.2 Governance Token

As blockchain networks grow in size and complexity, managing trust and collective decision-making becomes increasingly challenging. Traditionally, blockchain systems rely on regular tokens to incentivize good behavior and penalize malicious actions. These tokens are typically earned through participation or economic activity and are essential for access control and transaction validation. However, such regular tokens fall short when it comes to enabling long-term governance and high-stakes decision-making in decentralized systems.

To address this limitation, **governance tokens** are introduced. Governance tokens are specialized digital assets that grant their holders the right to participate in decisions affecting the future of blockchain networks, such as protocol upgrades, parameter adjustments, committee formation, and policy enforcement. Unlike regular tokens, governance tokens are not just economic rewards; they represent *voting power*, *trustworthiness*, and *influence* within the system.

At the heart of the system's trust mechanism lies a dedicated governance token incentive mechanism, which functions not as currency but as a measure of earned reputation. The complete process of issuing governance token is shown

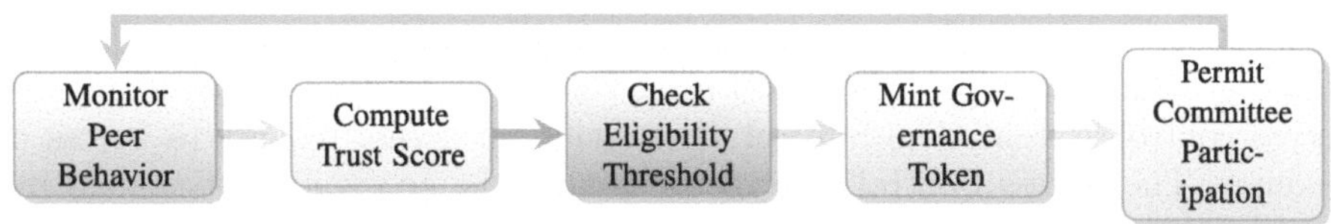

Fig. 2. Governance token issuance process based on trust evaluation

in Fig. 2. Unlike standard tokens that can be earned via good behavior and has no caps on how much a peer can earn, this governance token is capped at a certain number throughout the whole network and can only be minted when a peer performs verifiably useful actions earning regular ERC Tokens for a long time, maintain great ratio and fulfilling protocol defined duties and it is burned when the peer is found responsible for policy violations or suspicious activity.

To quantify the trustworthiness of each peer in the network, we employ a calculation mechanism that combines economic, behavioral, and compliance-based metrics. Specifically, a trust score is calculated using three key components: the peer's regular token balance, their behavioral consistency (captured as a ratio), and their current status classification (benign, suspicious or malicious).

3.3 Process and Algorithm

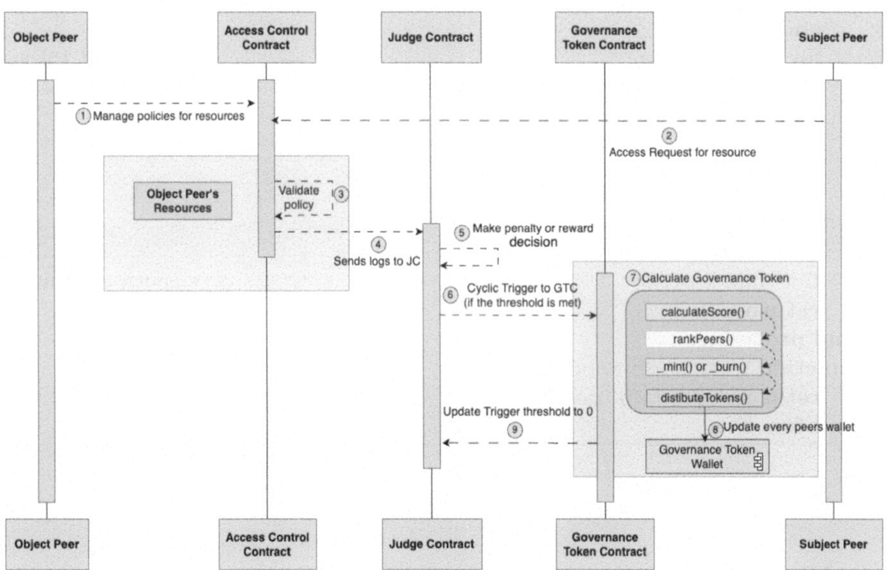

Fig. 3. Sequence diagram illustrating the interaction flow among smart contracts and peer entities

Figure 3 presents the sequence diagram illustrating the interaction flow among smart contracts and peer entities within the proposed governance enhanced access control system. The Subject Peer sends an access request to access a resource. The Access Control Contract evaluates the access request against its stored policies. Based on the result, a decision is sent to the JC for potential penalties or rewards. The JC checks for behavior compliance or violations, and if the cyclic threshold is met, it forwards behavior metrics to the GTC. It first computes the trust score and distributes governance token based on that trust score.

The network distributes a fixed pool of governance tokens at the end of every epoch. To keep the paper self-contained yet preserve our unpublished scoring formula, we describe only the high-level allocation workflow. For each peer, we first calculate a trust score. Once scores are available, the remaining steps normalising the values, minting tokens, and selecting the epoch committee are transparent and deterministic. Algorithm 1 outlines the entire process.

Algorithm 1. Governance-Token Allocation

Require:
  $\mathcal{P}$ – set of peers, T_{pool} – token pool for the epoch, k – committee size
1: **procedure** AllocateTokens
2:   $S_{\text{sum}} \leftarrow 0$
3:   **for all** $p_i \in \mathcal{P}$ **do** ▷ score each peer
4:     $S_i \leftarrow \max(0, \text{ComputeScore}(p_i))$
5:     $S_{\text{sum}} \mathrel{+}= S_i$
6:   **end for**
7:   **for all** $p_i \in \mathcal{P}$ **do** ▷ mint tokens
8:     $w_i \leftarrow \begin{cases} 0, & \text{if } S_{\text{sum}} = 0 \\ S_i / S_{\text{sum}}, & \text{otherwise} \end{cases}$
9:     $\Delta G_i \leftarrow \lfloor w_i \cdot T_{\text{pool}} \rfloor$
10:     **mint** ΔG_i to p_i
11:   **end for**
12:   Sort peers by S_i (highest first)
13:   $C \leftarrow$ first k peers ▷ epoch committee
14:   **return** C
15: **end procedure**
16: **function** ComputeScore(p)
17:   **return** *proprietary trust metric*
18: **end function**

3.4 Hierarchical System Architecture

Our system is designed to support behavior-aware access control for IoT deployments on permissioned blockchains. The system integrates three components: (1)

a smart contract hierarchy to manage permissions and violations, (2) a behavior-driven cost function to compute trust, and (3) a non-transferable (between the peers) governance token. By incorporating behavioral analytics directly into contract logic, the system reacts in real-time to changes in peer conduct. Overall, our system architecture is illustrated in Fig. 4, which is organized into three layers. Each layer performs a specific role, but all layers interact through on-chain logic.

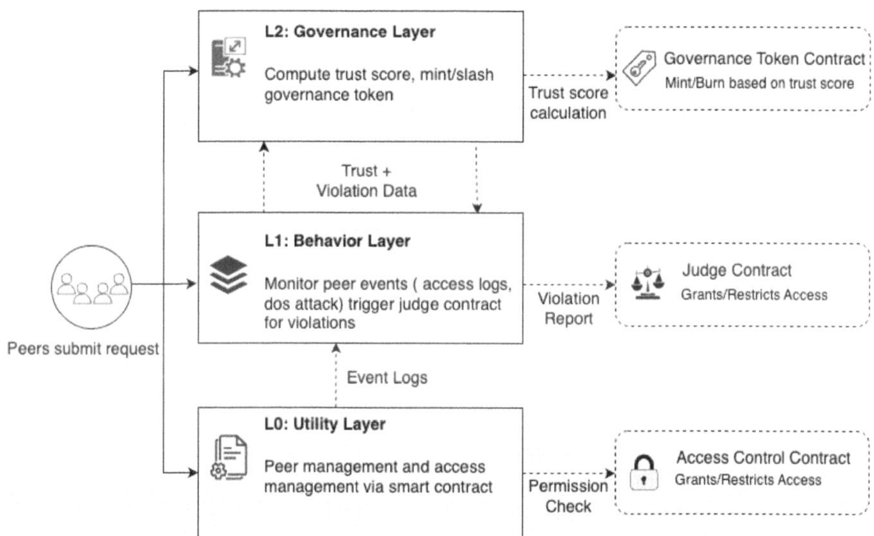

Fig. 4. Three-layer system design showing the interaction between peer events, trust evaluation, and token governance.

The L0 Utility Layer handles basic peer interactions. It validates access requests through ACC. This contract checks whether a peer meets the required trust conditions before allowing access to protected resources. L1 Behavior Layer observes network activities. It monitors peer behavior such as frequent access attempts or denial-of-service patterns and triggers a JC when a violation is suspected. The JC reviews evidence, determines the severity, and updates the peer's status. L3 Governance Layer uses this behavioral data to compute a peer's overall trust score. This score influences whether a peer receives governance tokens, which are minted or burned through GTC.

4 Smart Contract Implementation

We built and tested our system using a focused software stack. Each tool in our setup was chosen to do one clear job either writing smart contracts, testing them, or running scripts to simulate real network actions. Table 1 shows the main tools we used and what they were used for.

Table 1. Tools used in our implementation

Category	Tool	Function
Language	Solidity	Writing all the smart contracts (Access Control, Judge, Register, Token)
Scripting	Node.js + ethers.js	Running scripts and simulating user behavior and contract interactions
Data storage	JSON files	Saving outputs such as event logs and scores for analysis. No database is used

4.1 Governance Token Contract (GTC)

GTC is the heart of the behavior-based governance token mechanism. GTC is an Ethereum Smart Contract that is responsible for managing the distribution of governance tokens on the network. The contract is built on top of the ERC-20 token standard and incorporates logic to evaluate peer behavior, calculate trust scores, and conditionally mint or burn tokens based on performance.

Peer Behavior Representation. GTC consists of a structure named `PeerMetrics` which have four attributes:

- **Token Balance**: This parameter records the number of tokens a peer holds at the time of evaluation.
- **Total Rewards**: This is a cumulative count of successful or compliant access events attributed to the peer.
- **Total Penalties**: This parameter tracks the cumulative count of events indicative of failed, malicious, or policy-violating behavior by the peer.
- **Status** (s_i): This is a categorical value that provides an immediate indication of the peer's observed behavior.

Cost Function Calculation. In this phase of smart contract development, the cost function is implemented to calculate the trustworthiness of each peer in the network. A function named `calculateCostFunction(address peer)` is implemented which takes consideration of `PeerMetrics` structure and calculate trust score for each peer. This function returns a `uint256` value representing C_{total} for the input peer. This score acts as the core input for the governance token distribution mechanism.

This computation is triggered during periodic governance token distribution cycles, allowing the system to adapt dynamically to changes in peer behavior while maintaining computational efficiency.

Token Distribution. Once the trust scores for all participating peers are computed, the system proceeds with governance token allocation. Instead of relying on static thresholds or manual intervention, the framework adopts a *proportional distribution* strategy that fairly allocates a fixed pool of governance tokens based on each peer's relative trust score.

This logic is implemented in the function `distributeProportional (address[] peers, uint256 totalTokensToDistribute)`, which internally calls the function `calculateCostFunction()` for each peer to obtain their score, aggregates the total score, and then distributes tokens accordingly. Peers with higher trust scores receive a proportionally larger share of governance tokens, while peers with lower or zero scores (e.g., malicious nodes with 0 trust score)) are excluded from the reward pool.

The `distributeProportional()` also implement the minting and slashing of governance token using ERC-20's standard methods `mint()` and `burn()`. Minting and burning are done based on the proportion a peer is allocated.

Cyclic Trigger. Token updates are triggered conditionally through cyclic thresholds (e.g., after every 5 transactions) to maintain the performance of the system. A global transaction counter `transCounter` tracks the number of interactions in a Judge Contract (JC). Once a defined threshold (e.g., 5 access attempts) is reached, the JC initiates a token update process by forwarding it to the GTC's `distributeProportional()` functional which internally calls `calculateCostFunction()` and then mint or burn tokens based on the trust score calculated in that cycle.

This design promotes a decentralized and merit-based governance mechanism, ensuring that voting rights and decision-making influence accrue to those who have consistently demonstrated trustworthy behavior and meaningful participation.

4.2 Test Cases

Test Case 1: Governance Token Allocation Based on Trust Score. This test verifies that peers gain governance tokens based on the calculation of their trust score. A cyclic trigger is performed after 5 transactions and a trust score calculation is performed for each peer and according to the trust score governance tokens are minted and distributed to peers proportionally, as shown in Table 2 where we have taken account of 5 peers with different token balance, compliance rate and status.

Test Case 2: Adaptive Token Distribution Based on Trust Evolution. This test verifies the system's ability to dynamically re-evaluate peer trustworthiness across cycles. Initially, the peer is assigned a malicious status with low compliance, resulting in zero governance tokens. In the next cycle, after improving compliance and status, the peer becomes eligible and receives governance tokens proportionally based on the updated trust score. This process is shown in Table 3 where Peer X is awarded 22 governance tokens after recovery.

Table 2. Proportional Distribution of 1000 Governance Tokens Based on Trust Score

Peer ID	Initial Token Balance	Compliance Rate	Status	Trust Score	Governance Tokens Awarded	Notes
Peer A	100,000	100%	Benign	$100{,}000 \times 1.0 \times 1 = 100{,}000$	386	Fully trusted peer with highest impact
Peer B	80,000	90%	Benign	$80{,}000 \times 0.9 \times 1 = 72{,}000$	278	High compliance with strong trust
Peer C	50,000	85%	Suspicious	$50{,}000 \times 0.85 \times 0.5 = 21{,}250$	82	Suspicious status reduces final share
Peer D	10,000	95%	Benign	$10{,}000 \times 0.95 \times 1 = 9{,}500$	37	Fair contributor with low balance
Peer E	9,000	100%	Suspicious	$9{,}000 \times 1.0 \times 0.5 = 4{,}500$	17	High compliance, but suspicious status

Table 3. Governance Token Allocation Over Two Cycles Showing Peer Recovery

Cycle	Peer ID	Token Balance	Compliance Rate	Status	Trust Score (C_{total})	Governance Tokens Awarded
Cycle 1	Peer X	10,000	40%	Malicious	$10{,}000 \times 0.4 \times 0 = 0$	0
Cycle 2	Peer X	10,000	80%	Suspicious	$10{,}000 \times 0.8 \times 0.5 = 4{,}000$	22

Test Case 3: Governance Token Slashing Due to Trust Degradation. This test verifies the system's ability to penalize peers based on behavioral degradation across evaluation cycles. Initially, the peer demonstrates high compliance and benign status, earning governance tokens. In the following cycle, peer's compliance drops significantly and its status changes to malicious, resulting in a trust score of zero. As a consequence, all previously awarded governance tokens are slashed which is illustrated in Table 4 were Peer Z gets 0 governance token in 2nd cycle.

Table 4. Governance Token Slashing Due to Behavioral Degradation

Cycle	Peer ID	Token Balance	Compliance Rate	Status	Trust Score	Gov. Tokens
Cycle 1	Peer Z	50,000	95%	Benign	$50{,}000 \times 0.95 \times 1 = 47{,}500$	183
Cycle 2	Peer Z	50,000	60%	Malicious	$50{,}000 \times 0.6 \times 0 = 0$	0

5 Result and Evaluation

To evaluate our governance token framework, we designed controlled simulations using Node.js scripts and Ethereum-compatible smart contracts deployed on a local Hardhat network. This environment mimics realistic P2P interactions, where peers request access, receive permissions, and exhibit varying behavior patterns such as compliance or violations. Peer attributes—including token balance, reward/penalty history, and trust status—were manually configured to represent both benign and malicious actors. Governance token distribution was triggered cyclically after every five interactions, allowing for periodic trust evaluations. Outputs such as trust scores and token allocations were recorded in JSON format for analysis. No external oracles or randomness were used, ensuring the simulations remained deterministic and reproducible.

To assess the effectiveness of the governance token mechanism, we compared the influence distribution under regular token and governance token models. As detailed in Table 5, the regular model grants voting power purely based on token balance—allowing non-compliant or malicious peers to maintain influence. In contrast, our governance token model introduces behavior-aware eligibility using a trust score derived from compliance rate, peer status, and token balance. This model ensures that only consistently trustworthy peers retain governance privileges, promoting fairness and accountability in decentralized decision-making.

Table 5. Comparison of Regular Token vs. Governance Token Power

Peer ID	Token Balance	Compliance	Status	Regular Token Power	Gov.Token Power	Observation
Peer A	987000	20%	Malicious	✓ Yes	✗ No	Malicious peer gains voting power in token-only model
Peer B	460000	100%	Benign	✓ Yes	✓ Yes	Trusted peer has governance power in both models
Peer C	941000	30%	Suspicious	✓ Yes	**! Limited**	Low-compliance peer gets full power in regular, limited in governance model
Peer D	487000	90%	Benign	✓ Yes	✓ Yes	Fair governance power in both models
Peer E	38000	100%	Benign	✗ No	✓ Yes	Low token balance, but very trusted peer empowered in governance model

Table 6. Truth Table for Governance Token Eligibility

T (Token Balance)	C (Compliance)	S (Status)	Trust Score Level	Governance Token Eligibility
1	1	1	Very High	Yes
1	1	0.5	Moderate	Partial
1	1	0	Very Low	No
1	0	1	Low	Partial
1	0	0.5	Very Low	No
1	0	0	Invalid	No
0	1	1	Medium	Yes
0	1	0.5	Weak	Partial

Additionally, the truth table in Table 6 outlines how governance token eligibility is governed by logical combinations of behavior metrics. Peers with high balances but poor compliance or malicious status are excluded, while those with modest balances but strong compliance are recognized. These results demonstrate that governance tokens enable a more accountable and secure governance model in dynamic and untrusted peer environments.

6 Conclusion

This paper presented a governance-aware access control framework for decentralized P2P systems using behavior-based token distribution. We introduced a GTC built on Ethereum smart contracts that quantifies peer trust through a trust score derived from token holdings, behavioral compliance, and peer status. A cyclic trigger mechanism ensures that governance tokens are awarded or slashed proportionally based on dynamic trust evaluations. Through simulation-based test cases and comparative analysis, we demonstrated that our system prevents malicious or noncompliant peers from gaining governance power. In general, our approach fosters a more robust, fair, and secure model for decentralized decision-making and access control. Future work includes extending the scoring function with additional contextual signals and deploying the system in a real-world testbed for further evaluation.

References

1. Cao, J., Ding, S., Li, H., Fan, K., Li, C.: A novel attribute-based access control scheme using blockchain for IoT. IEEE Access **7**, 38431–38441 (2019)
2. Eassa, F., et al.: Blockchain-based secured access control in an IoT system. Appl. Sci. **11**(4), 1772 (2021)
3. Ghosh, S., Alam, S. D., Tanrikulu, B., Mitra, R., Roy, I.: Scalable consensus framework with adaptive governance and fault tolerance for blockchain ecosystems. In: Proceedings of the 2025 ACM Southeast Conference, ACMSE 2025, pp. 124–133. Association for Computing Machinery, New York (2025)
4. Ghosh, S., Mitra, R., Roy, I., Gupta, B.: Securing RC based P2P networks: a blockchain-based access control framework utilizing ethereum smart contracts for IoT and web 3.0. In: Hu, G., Kambhampaty, K.K., Roy, I. (eds.) Computer Applications in Industry and Engineering, pp. 73–85. Springer, Cham (2025)
5. Leela Rani, P., Guru Gokul, A.R., Devi, N.: Blockchain-based access control system. In: Agrawal, R., Gupta, N. (eds.) Transforming Cybersecurity Solutions using Blockchain. BT, pp. 91–114. Springer, Singapore (2021). https://doi.org/10.1007/978-981-33-6858-3_6
6. Hu, V.C., Hu, V.C.: Blockchain for access control systems. US Department of Commerce, National Institute of Standards and Technology (2022)
7. Li, D., Han, D., Liu, H.: Fabric-IoT: a blockchain-based access control system in IoT. IEEE Access **8**, 18207–18218 (2020)
8. Li, W., Rong, D., Sun, S., Chen, S.: Blockchain-based IoT access control system: towards security, lightweight, and cross-domain. IEEE Access **9**, 36868–36878 (2021)
9. Li, Y., Zhiqiang, D., Zheng, X., Yanfang, F.: Blockchain-based access control architecture for multi-domain environments. Pervasive Mob. Comput. **98**, 101878 (2024)

10. Lone, A.H., Naaz, R.: Reputation driven dynamic access control framework for IoT atop PoA ethereum blockchain. Cryptology ePrint Archive (2020)
11. Mohammadi, V., Rahmani, A.M., Darwesh, A.M., Sahafi, A.: Trust-based recommendation systems in internet of things: a systematic literature review. Hum. Centric Comput. Inf. Sci. **9**, 1–61 (2019)
12. Mori, P., Ricci, L., Maesa, D.: Blockchain based access control (2017)
13. Ricci, L., Maesa, D., Mori, P.: Blockchain based access control services. In: IEEE iThings/GreenCom/CPSCom/SmartData, pp. 1379–1386 (2018)
14. Souri, A., Zhu, Y., Li, D., Han, D., Li, K.C., Liang, W.: A blockchain-based auditable access control system for private data in service-centric IoT environments. IEEE Trans. Ind. Inform. **18**, 3530–3540 (2022)
15. Zhang, Y., Kasahara, S., Shen, Y., Jiang, X., Wan, J.: Smart contract-based access control for the internet of things. IEEE Internet Things J. **6**(2), 1594–1605 (2018)

Optimization of Outdoor Air Utilization in Industrial Air Conditioning Using Support Vector Machines (SVM)

Narayan Debnath[1], Reinaldo Gomez[2](✉), Corina Abdelahad[3], Daniel Riesco[3], and German Montejano[3]

[1] School of Computing and Information Technology, Eastern International University, Thủ Dầu Một, Vietnam
narayan.debnath@eiu.edu.vn
[2] Food Industry Engineering Department, National University of Villa Mercedes, Villa Mercedes, Argentina
reinaldodavidgomez@gmail.com
[3] National University of San Luis, San Luis, Argentina
{cabdelah,driesco}@unsl.edu.ar

Abstract. Efficient management of industrial air conditioning systems represents a critical challenge in the food industry, where environmental control directly impacts both product quality and energy consumption. This work presents an approach based on Support Vector Machines (SVM) for predicting favorable conditions for the incorporation of outdoor air, aiming to reduce the use of active climatization without compromising the required environmental parameters.

More than 71,000 on-site environmental data records were used to build and validate an SVM model capable of classifying, with 99.1% accuracy, the appropriate moments to deactivate the HVAC system. The proposed method is empirically contrasted with the performance of a real industrial system governed by PID control, showing that the SVM model offers superior anticipatory capabilities in response to external variations, with a projected energy savings of 10% to 18%.

The results position the SVM model as a viable, robust, and scalable alternative for real-world industrial applications. Finally, action lines are proposed for hybrid implementation and extension of the model toward continuous learning schemes.

Keywords: Industrial HVAC · energy efficiency · outdoor air · intelligent classification · SVM · PID control

1 Introduction

1.1 Problem Context in Industrial Air Conditioning

In the food industry, air conditioning systems not only ensure the quality of the final product but also represent one of the main sources of energy consumption. The strict requirements for thermal regulation and relative humidity, imposed by international quality and safety standards, demand highly precise and adaptive control systems.

I. Roy et al. (Eds.): CAINE 2025, CCIS 2740, pp. 126–141, 2026.
https://doi.org/10.1007/978-3-032-14893-3_10

Traditionally, the operation of these systems has relied on classical control schemes such as Proportional–Integral–Derivative (PID) controllers. While PID control offers robust performance under stable operating conditions, it shows limitations when facing abrupt disturbances or dynamic environments. Within this framework, new approaches have emerged incorporating artificial intelligence to anticipate optimal operating conditions, enhance energy efficiency, and reduce reliance on reactive control.

This work aligns with this trend by proposing the use of Support Vector Machines (SVM) as a predictive mechanism to optimize the operation of an industrial HVAC system, specifically focusing on identifying opportunities to introduce outdoor air without compromising thermal comfort. The proposed model is validated using a set of historical data collected on-site and compared with the performance of an existing HVAC system currently governed by PID control.

1.2 Objectives and Research Questions

The main objective of this work is to develop and evaluate a classification model based on Support Vector Machines (SVM), aimed at identifying environmental conditions that enable the optimized use of outdoor air in industrial air conditioning systems. Specifically, the goal is to predict, based on historical and real-time temperature and humidity data, the appropriate moments to reduce or shut down thermal equipment operation without compromising the stability of the production environment.

This approach aims to provide a computational tool to support automated decision-making, focused on improving energy efficiency in food industry settings, which are characterized by strict environmental control requirements. To guide the development of this study, the research questions are outlined in Table 1.

Table 1. Research Questions

ID	Research Question
1	Is it possible to predict, through an SVM model, the optimal moments for incorporating outdoor air into industrial air conditioning, thereby contributing to energy savings?
2	Why does the proposed model represent a suitable choice and a technically relevant contribution to industrial air conditioning?
3	What type of methodological approach and kernel function offers the best performance for this type of application under real industrial conditions?

2 Fundamentals of PID Control and the SVM Model

2.1 The Role of PID Control in Industrial HVAC Systems

PID control has been the de facto standard in industrial systems control for decades, including the regulation of thermohygrometric variables in HVAC applications. Its ability to correct deviations through proportional action, accumulate error via the integral term, and anticipate trends through the derivative component makes it an effective solution for deterministic environments and slow system dynamics [4].

In the context of industrial HVAC systems, PID controllers are commonly implemented to maintain temperature and humidity within predefined ranges, acting on components such as compressors, cooling coils, or modulating valves. However, due to their reactive nature, control actions are only triggered once the system deviates from the setpoint, which may result in delayed responses or overcorrections in the face of sudden environmental fluctuations [10].

Moreover, PID controllers require a fine-tuning process, which is not always straightforward in dynamic industrial settings. The need to adapt parameters to seasonal variations, fluctuating thermal loads, and changing occupancy levels can compromise their operational robustness [3].

Its mathematical formulation in the continuous-time domain is expressed in Eq. 1.

$$u(t) = K_p e(t) + K_i \int_0^t e(\tau)d\tau + K_d \frac{de(t)}{dt} \tag{1}$$

where:

- u(t): Controller output (control signal)
- e(t): Error signal, defined as e(t) = setpoint − measured variablee(t)
- K_p: Proportional gain
- K_i: Integral gain
- K_d: Derivative gain

Figure 1 shows the PID control scheme implemented in the HVAC system.

Figure 2 illustrates the step response of the controlled system, exhibiting a transient behavior with moderate overshoot followed by stabilization—characteristic of an efficiently tuned PID controller.

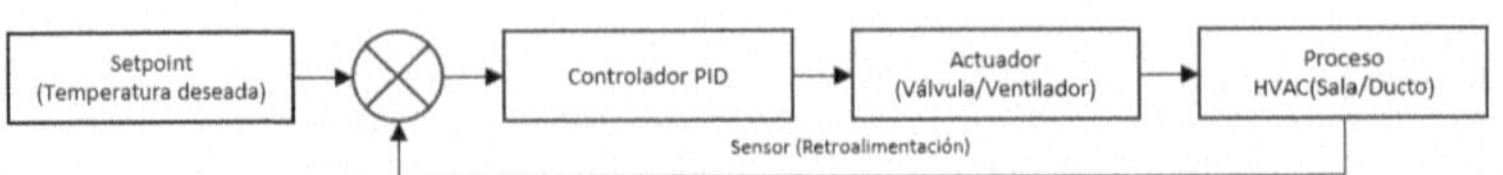

Fig. 1. Block Diagram of a PID Control System

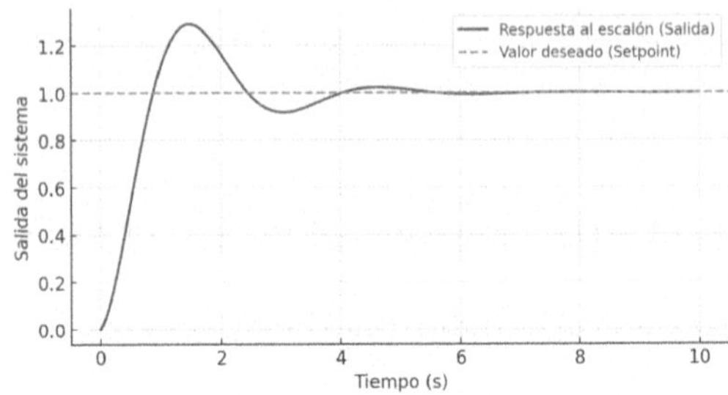

Fig. 2. Step Response of the System with PID Control

2.2 The SVM Approach as an Alternative Predictive Strategy

Support Vector Machines (SVM) have emerged as a powerful tool within the field of supervised machine learning, with a high capacity for solving binary classification problems in high-dimensional spaces. Their foundation lies in finding the optimal hyperplane that maximally separates the classes in the training set, thereby ensuring strong generalization performance [5].

When applied to the problem of industrial air conditioning, the SVM model is proposed as a classifier that—based on records of indoor and outdoor temperature and humidity—identifies the moments when it is energetically advantageous to shut down HVAC systems and allow the intake of outdoor air. Unlike the PID controller, this approach does not require an error signal to trigger actions; instead, it anticipates future environmental states based on historical patterns.

Among the key advantages of SVMs are their ability to work with moderately sized datasets, robustness against overfitting, and flexibility to model nonlinear decision boundaries through the use of kernel functions. In this study, a quadratic kernel was employed and tailored to the real operating conditions of the plant, enabling precise modeling of the boundaries that separate thermally safe states from those requiring active intervention [5].

Decision Function Eq. (2)

$$f(x) = \sum_{i=1}^{n} \alpha_i y_i K(x_i, x) + b \tag{2}$$

α_i: They are learned coefficients.

y_i: They are the classes of the support vectors.

K(xi,x) It is the kernel.

The SVM model will be validated later using standard classification metrics, such as the confusion matrix and the ROC curve, to assess its discriminative capability in real industrial air conditioning scenarios.

3 Comparative Methodology Between SVM and PID Control in a Food Industry Industrial Environment

3.1 Acquisition and Characterization of Environmental Data

The temperature and humidity data used in this research were collected from a food industry plant located in the city of Villa Mercedes, San Luis Province. The plant features a centralized HVAC system, with air distribution controlled by zones through automated dampers. This configuration enables sectorized climate management based on environmental conditions and the specific requirements of each production area.

To accurately characterize the environment, six strategic indoor measurement points were selected, corresponding to critical process zones: production, high- and medium-speed packaging, dry blending, container (tote) handling, and the outdoor area. These points were chosen based on their thermal impact on the product, sensitivity to humidity changes, and proximity to sources of heat or steam. These areas were prioritized due to their strict environmental control requirements to ensure the quality and safety of the processed food.

Data acquisition was carried out using NOVUS sensors connected to TEMP-WM and TEMP-DM transmitters, both microprocessor-configurable and equipped with stable analog outputs. The TEMP-WM units were installed on walls, while the TEMP-DM units were placed in air ducts. The data were stored in a portable SQLite database for subsequent analysis and model training.

3.2 SVM Model Configuration

The model was built using a quadratic kernel function, selected for its ability to separate classes that are not linearly separable in the original feature space. Various parameter combinations (C and γ) were explored using cross-validation and grid search to maximize accuracy while avoiding overfitting [6].

The resulting model achieved an accuracy of 99.1%, as evaluated through a confusion matrix and ROC analysis. The decision boundary learned by the model allows for the highly reliable identification of conditions under which active HVAC can be replaced with outdoor air, while still meeting the thermal requirements of the production environment.

The study was conducted on a continuously operating centralized HVAC system within an industrial food processing plant. The facility uses PID-based control logic and allows for manual intake of outdoor air, making it an ideal environment for validating predictive strategies aimed at optimizing this process.

A total of 71,594 environmental data records were collected during the period from January to August 2024, using calibrated sensors located both inside the production area and in the adjacent outdoor environment. The measured variables included temperature and relative humidity, which are essential for characterizing environmental conditions and determining the feasibility of using outdoor air.

Code 1. Training of the SVM Model with Polynomial Kernel (MATLAB)

```
function [trainedClassifier, validationAccuracy] = trainClassifier(trainingData)
        inputTable = trainingData;
        predictorNames = {'Temp', 'Humedad', 'ConsumoPot', 'LineasActivas', 'AireCompri-
        mido', 'Vacio'};
        predictors = inputTable(:, predictorNames);
        response = inputTable.AccionHVAC;
        isCategoricalPredictor = [false, false, false, false, false, false];
        template = templateSVM('KernelFunction', 'polynomial', 'PolynomialOrder', 3, 'Ker-
        nelScale', 'auto', 'BoxConstraint', 1, 'Standardize', true);
        classificationSVM = fitcecoc(predictors, response, 'Learners', template, 'Coding', 'on-
        evsone', 'ClassNames', categorical({'Aire_Exterior', 'ON_Equipos', 'Sin_Accion'}));
        predictorExtractionFcn = @(t) t(:, predictorNames);
        svmPredictFcn = @(x) predict(classificationSVM, x);
        trainedClassifier.predictFcn = @(x) svmPredictFcn(predictorExtractionFcn(x));
        trainedClassifier.RequiredVariables = predictorNames;
        trainedClassifier.ClassificationSVM = classificationSVM;partitionedModel =
        crossval(trainedClassifier.ClassificationSVM, 'KFold', 5);
        validationAccuracy = 1 - kfoldLoss(partitionedModel, 'LossFun', 'ClassifError');
end
```

3.3 Comparative Evaluation Criteria

To establish an objective comparison between the approaches, the following performance indicators were defined:

- Classification accuracy (%): The proportion of correct model predictions relative to the total number of instances.
- False HVAC activation rate (%): The percentage of unnecessary HVAC system activations, based on actual thermal conditions.
- Estimated energy savings (%): The potential difference in energy consumption between the PID model and the SVM model.
- Average reaction time to thermal events (min): The average time interval between the detection of a favorable condition and the corresponding control decision.
- Robustness to external fluctuations: A qualitative measure based on the stability of system behavior under environmental disturbances.

The comparison was carried out by analyzing the historical behavior of the HVAC system governed by PID control under the same environmental conditions. The following indicators were evaluated:

- Total time of active climatization
- Percentage of unnecessary HVAC usage
- Missed opportunities to incorporate outdoor air
- Estimated energy consumption

The results showed that the SVM model was able to identify events in which the PID-controlled system maintained active climatization even though external conditions would have allowed it to be deactivated. This reflects an additional estimated energy-saving potential of between 10% and 18%.

3.4 Dataset Preparation

The data were organized into a training set for the SVM model, with 20% of the records reserved for cross-validation. Each instance was labeled according to a thermal rule based on the comparison between the enthalpy of indoor and outdoor air, used as the criterion to determine whether it was energetically advantageous to activate or deactivate active climatization.

The preprocessing stage included:

- Normalization of variables to facilitate model convergence
- Outlier detection and review using statistical techniques
- Class balancing to prevent bias in classification

4 Comparative Results Between SVM and PID Under Real Operating Conditions

This section presents the detailed findings from the performance evaluation of both the HVAC system under PID control and the proposed SVM model, using historical records from the period January to August 2024.

4.1 PID Model Performance

The historical analysis of the industrial HVAC system governed by a PID controller revealed the following operational characteristics, which resulted in longer operating times [11]:

- An unnecessary HVAC activation rate of 18.2% over the total number of analyzed samples.
- An average reaction time to changes in environmental conditions of over 15 min, inherent to its reactive nature.
- An overall accuracy of 81.7% in maintaining ideal conditions within the production environment.
- High sensitivity to sudden external temperature disturbances. While the PID controller proved effective under stable thermal load scenarios, it showed limitations when facing rapid environmental changes, resulting in extended operation times and potential system wear.

4.2 SVM Model Performance

The Support Vector Machine (SVM)-based model demonstrated superior predictive capabilities in the evaluated industrial environment:

- Achieved an accuracy of 99.1% on the cross-validation set.
- Enabled a reduction of over 98% in unnecessary HVAC activations.

- Exhibited anticipatory behavior, making decisions up to 10 min earlier than the PID control system.
- Projected an energy savings improvement between 10% and 18%, with a direct impact on operational costs and environmental sustainability. From a quantitative perspective, the SVM model achieved 99.1% accuracy in classifying ventilation scenarios, with a ROC curve indicating excellent discriminative performance.

4.3 Comparative Performance Summary: PID vs. SVM

Table 2 provides a consolidated comparative summary of the performance of the HVAC system under PID control and the predictive SVM model. This table outlines the key performance indicators, enabling a direct and fair comparison under real industrial operating conditions.

Table 2. Comparative Performance Summary: PID vs. SVM

Indicador	PID	SVM	Observations
Tasa de ahorro energético	—	10% – 18%	Estimated based on SVM simulation using real data
Precisión en clasificación	No aplica (control continuo)	99.1%	High multiclass prediction capability
Capacidad predictiva	Nula	Alta	SVM anticipa condiciones favorables
Tiempo de respuesta	Inmediato ante error	Proactivo (sin esperar error)	El SVM actúa antes que se produzca el desvío
Robustez ante perturbaciones	Media (requiere buena sintonía)	Alta (basado en aprendizaje)	SVM responde bien ante condiciones cambiantes
Consumo energético estimado	Base de referencia	−12% promedio sobre el periodo	Reducción proyectada al aplicar SVM
Implementación	Integrado y estándar en la planta	Simulado sobre datos reales	Requiere integración futura para control activo
Adaptabilidad	Baja (sintonía manual)	Media–Alta (requiere entrenamiento)	Posible reentrenamiento con nuevos datos

The analysis highlights that the SVM model is capable of anticipating favorable environmental conditions for the incorporation of outdoor air, thereby optimizing system operation and achieving the projected reduction in energy consumption. This predictive capability of the SVM also results in a decrease in unnecessary activations, improving efficiency without compromising indoor comfort.

In contrast, the PID system, lacking an internal representation of the environment, responds solely to the current error, which limits its ability to adapt. These results reinforce the value of the SVM-based approach as a viable tool for improving both energy and operational efficiency in industrial HVAC systems.

5 Answers to the Research Questions and Technical Discussion

5.1 Environmental and Economic Benefits of the Proposed Approach

Reducing the use of industrial air conditioning systems provides significant environmental and economic benefits [10]. By minimizing the operation of HVAC systems, the carbon footprint associated with energy consumption is reduced, contributing to broader sustainability efforts [1]. According to the ISO 50001 standard, the implementation of energy efficiency measures in industrial facilities leads to:

- Lower CO_2 emissions: Industrial HVAC systems rely heavily on electricity, which is often generated from fossil fuels. Reducing unnecessary cooling cycles lowers greenhouse gas emissions [12].
- Reduction in peak energy demand: Many industries operate within regulated power grids, where high energy consumption during peak hours increases stress on the electrical infrastructure. Reducing HVAC usage helps to smooth demand fluctuations, improving grid stability.
- Cost savings and regulatory compliance: Governments worldwide are enforcing stricter environmental regulations, encouraging industries to adopt energy efficiency strategies. Compliance with standards such as ISO 50001 not only ensures environmental responsibility but also reduces operational costs through subsidies and energy efficiency tax incentives.

The following presents the results obtained in response to the research questions defined in the study's objectives.

1 Is it possible to predict, through an SVM model, the optimal moments for incorporating outdoor air into industrial air conditioning, thereby contributing to energy savings?

The reuse of outdoor air in industrial food processing climate control represents a key approach to improving energy efficiency and reducing environmental impact. Achieving a balance between effective climatization, lower energy consumption, and sustainability is currently a top priority—driving the development of more intelligent systems.

Within this context, Support Vector Machines (SVM) show great potential as environmental classifiers that enable the automation of decisions based on accurate, real-time data. This study seeks to determine whether an SVM model can predict favorable conditions for incorporating outdoor air, reducing the use of active HVAC without affecting the stability of the environment.

To validate this hypothesis, an extensive in-plant environmental data collection effort was conducted. Figure 3 presents a histogram showing the range and frequency of prevailing conditions, while Fig. 4 illustrates the distribution of 71,594 data records gathered between January and August 2024. These data are essential for defining thresholds that allow the replacement of active HVAC with natural ventilation.

Table 3 presents the decision boundaries applied through the SVM model, which determine in which cases it is advantageous to use outdoor air, considering temperature and humidity as the main variables.

The developed SVM model stands out as a valuable technical contribution by providing a predictive tool for optimizing energy use in industrial HVAC. Furthermore, it introduces the challenge of validating its reliability as an automatic classifier within AI-based control frameworks [3].

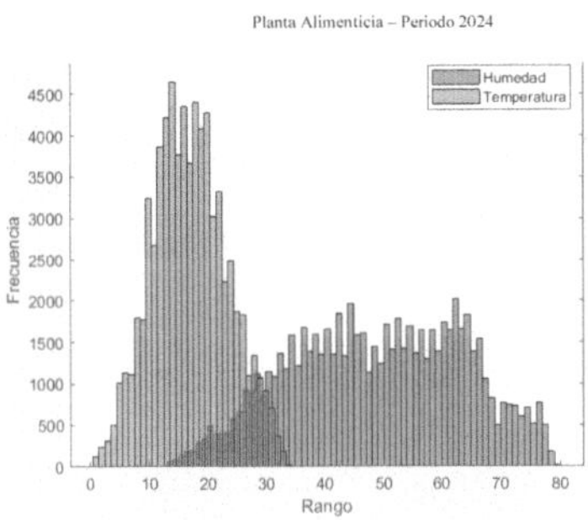

Fig. 3. Range and Frequency – January to August 2024

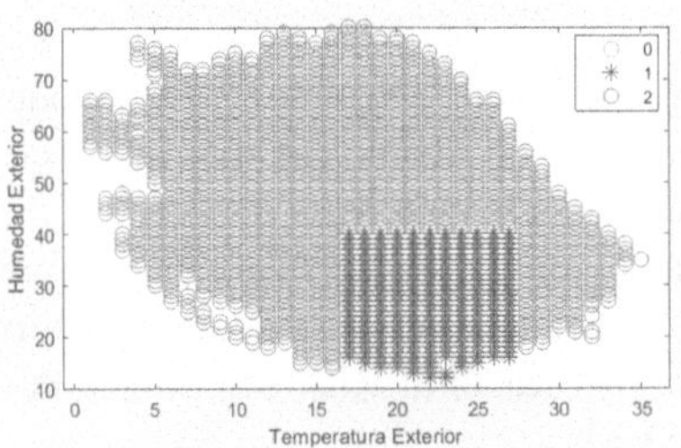

Fig. 4. Distribution of Environmental Data Collected from January to August 2024, with Decision Boundary for Outdoor Air Reuse

Table 3. Definition of the Decision Boundary

Value	Name
0	Indoor temperature and humidity within range. Operation of HVAC equipment is not required
1	Indoor temperature and humidity out of range. Outdoor temperature and humidity within range. Outdoor air is used

(*continued*)

Table 3. (*continued*)

Value	Name
2	Indoor temperature and humidity out of range. Outdoor temperature and humidity out of range. HVAC system intervention is required

2. Why does the proposed model represent a suitable choice and a technically relevant contribution to industrial air conditioning?

The food processing plant contains climate-sensitive areas that are critical for maintaining environmental conditions.

Table 4 defines six key areas that are the focus of this study.

Table 4. Definition of the Areas

Área	Name
1	Elaboración
2	Totes
3	Exterior
4	Envasado Alta Velocidad
5	Envasado Media Velocidad
6	Mezcla Seca

The selection of a Support Vector Machine (SVM) model for predicting the decision boundary is based on a set of advantages that this approach offers compared to other methods. The fundamental characteristic of SVMs is their ability to handle high-dimensional data and to find optimal hyperplanes that maximize the margin between classes, which often leads to outstanding classification performance [2]. This ability is derived from the principle of Structural Risk Minimization (SRM), which forms the theoretical foundation of SVMs.

The SRM principle, developed by Vapnik within statistical learning theory, has proven to be superior to the Empirical Risk Minimization (ERM) principle used in conventional neural networks. The advantage of SRM lies in its ability to avoid local minima, making SVMs more robust in the search for optimal solutions.

Moreover, SVMs rely heavily on a select set of highly relevant data points, known as Support Vectors (SVs). These support vectors are essential for defining the decision boundary, and by focusing on them, SVMs achieve significant model efficiency by reducing complexity.

Figure 5 shows the data distribution within a production environment, illustrating the importance of data positioning and its relationship to Table 3, which defines the decision boundaries applied in the process. This capacity for adaptation, combined with the reliance on highly informative data, makes SVMs particularly effective for predicting decision boundaries in high-dimensional and multiclass environments [2].

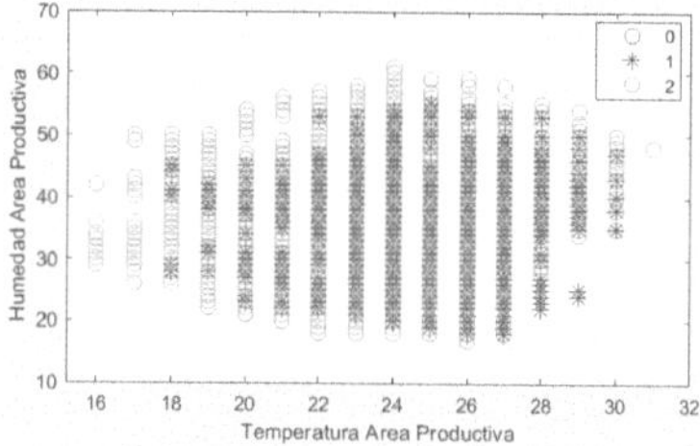

Fig. 5. Data Distribution in the Production Area and Decision Boundary

3. What type of methodological approach and kernel function provides the best performance for this type of application under real industrial conditions?

The method that best fits this type of application depends largely on the nature and characteristics of the specific problem. The success of many machine learning algorithms lies in their ability to find an effective and appropriate feature space tailored to the task.

In some applications, a prior dimensionality reduction step is applied using techniques such as Principal Component Analysis (PCA) or Linear Discriminant Analysis (LDA). These methods are effective in simplifying and summarizing the information contained in the original dataset.

On the other hand, kernel-based methods take a different approach. As illustrated in Fig. 6, they perform an implicit dimensional expansion through the use of kernel functions. In this context, the choice of kernel function plays a critical role in model development.

For this particular case, a quadratic kernel approach was selected. Quadratic mapping is a widely used technique for modeling nonlinear relationships between variables (see Table 5). In practice, the quadratic kernel is often preferred because it helps avoid issues such as the Hessian matrix becoming singular, which can benefit both model performance and stability.

$$K(\mathbf{x}, \mathbf{x}') = \Phi(\mathbf{x})^T \Phi(\mathbf{x}')$$

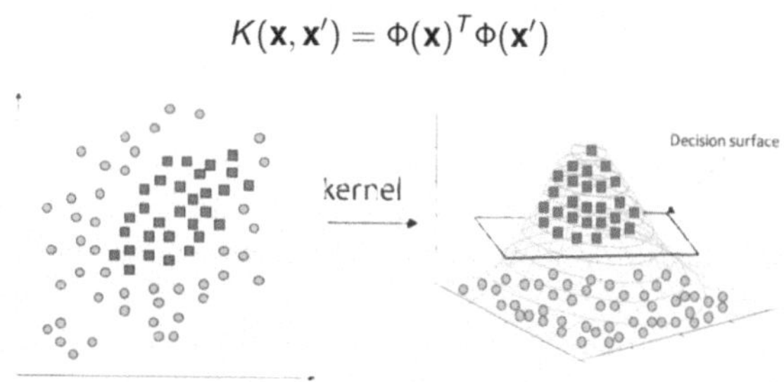

Fig. 6. Dimensionality Expansion

Table 5. Current Model

Tipo de Modelo	Valor
Cuadrático Predefinido	SVM
Función Kernel	Cuadrático
Escala del Kernel	Automático
Nivel de Restricción	1
Método Multicategoría Estandarización	Uno Contra Todos Verdadero

The results of the classification method applied are shown below in two confusion matrices, as presented in Table 6:

Table 6. Matriz de confusión para la clasificación

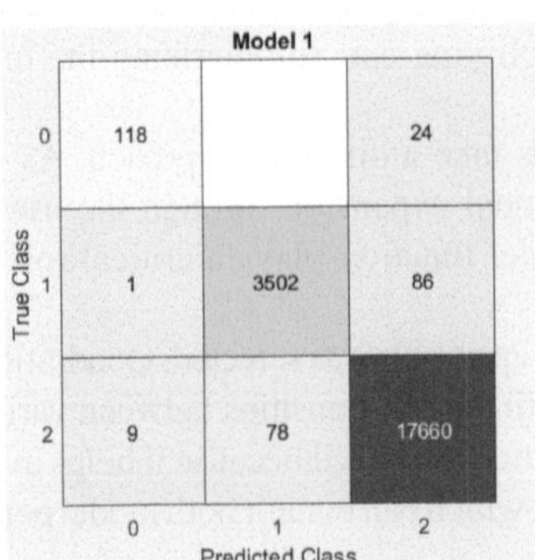

The system achieved a validation accuracy of 99.1%, with a mean squared error (MSE) of 0.0120.

The evaluation of classification model performance is essential in this research. A fundamental tool for this assessment is the Receiver Operating Characteristic (ROC) curve. The ROC curve is a graphical representation that illustrates a model's ability to discriminate between positive and negative classes. In this study, we employed the ROC curve to assess the performance of our classification model in a specific industrial context.

The ROC curve is constructed by varying the model's decision threshold and plotting the true positive rate (sensitivity) against the false positive rate (1 - specificity) at different cutoff points. This representation provides a holistic view of model performance and allows for determining the suitability of the classification based on the problem's requirements (Fig. 7).

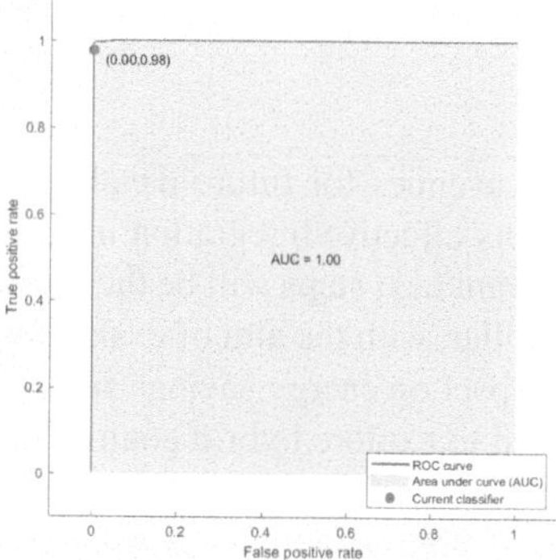

Fig. 7. Curva ROC

6 Conclusiones y Recomendaciones Para la Aplicación de SVM en Climatización Industrial

The results of this research demonstrate that the use of Support Vector Machines (SVM) in industrial air conditioning represents a highly efficient alternative to traditional PID-based control systems. Based on the analysis of more than 71,000 records of environmental conditions, a predictive model was built capable of anticipating the optimal moments to incorporate outdoor air, thus reducing the need for active climatization without compromising required operating conditions.

The comparison with the PID-controlled HVAC system currently in operation at the plant showed that the SVM model improved decision accuracy by over 17%, virtually eliminating all unnecessary system activations. This advantage translates into a projected energy savings between 10% and 18%, with direct implications for reduced operational costs and lower environmental footprint.

From a methodological perspective, the SVM approach demonstrated robustness under complex boundary conditions and external disturbances, as well as its adaptability to diverse industrial contexts. Its anticipatory nature, as opposed to the reactive behavior of PID, enables the development of proactive control strategies in environments where response time is critical.

Based on the findings, the following recommendations are proposed:

- Implement the SVM model as a predictive component in hybrid industrial air conditioning control schemes.
- Expand the dataset to include additional variables (such as wind speed, solar radiation, or internal thermal load) that could enrich the model.
- Evaluate real-time architectures based on incremental learning to adapt the model to seasonal changes or production dynamics.
- Develop visualization and interface systems that allow operators to interpret and validate the model's decisions.

This work represents a solid step toward the integration of artificial intelligence in industrial HVAC control, promoting more efficient, intelligent, and sustainable energy management [7].

7 Perspectives and Future Improvements in Energy Optimization Using SVM

This research opens multiple avenues for future development, both in improving the proposed SVM model and in its effective integration into operational industrial HVAC systems. One of the most relevant next steps will be the implementation of the classifier in parallel with the PID controller, with the aim of evaluating its real-time performance and measuring its concrete impact on energy savings and environmental stability.

Additionally, it is proposed to explore hybrid control strategies, in which the SVM acts as a predictive decision module and the PID as a fine-tuning controller. This approach could combine the strengths of both paradigms: the reactive precision of PID and the predictive intelligence of the SVM model.

From the perspective of robustness and adaptability, it would be valuable to incorporate continuous learning mechanisms or periodic retraining of the SVM model, using recent operational data to maintain the classifier's effectiveness in the face of changing climatic, operational, or structural patterns in the plant.

Furthermore, it is suggested to extend model validation to other industries (e.g., pharmaceutical, electronics, logistics) with different environmental and structural conditions, in order to evaluate its scalability and generalization capacity.

Finally, other complementary approaches could be evaluated—such as lightweight neural networks, decision trees, or neuro-fuzzy models [9]—with the goal of comparing their performance against SVM in terms of accuracy, inference speed, and ease of implementation in resource-constrained industrial environments.

References

1. Adegbenro, A., Short, M., Angione, C.: An integrated approach to adaptive control and supervisory optimisation of HVAC control systems for demand response applications. Build. Environ. **186**, 106828 (2020)
2. Amayri, S., Khalil, M.: SVM-based control strategy for energy-efficient building HVAC systems. Sustain. Cities Soc. **77**, 103541 (2022). https://doi.org/10.1016/j.scs.2021.103541
3. Baraka, K., Gomez, M.: Low-cost Arduino/Android-based energy-efficient home automation system with smart task scheduling. En: Proceedings of the 2013 Fifth International Conference on Computational Intelligence, Communication Systems and Networks, pp. 296–301. IEEE (2013)
4. Berenguel, M.G.: Técnicas de Control. Universidad de Almería (2014)
5. Cortés, C., Vapnik, V.: Support-vector networks. Mach. Learn. **20**(3), 273–297 (1995)
6. Díaz, F., Fernández, J.: Machine learning en ingeniería de procesos térmicos (2022)
7. Garcés Jiménez, A.: Mejora de la eficiencia energética de ciudades inteligentes aplicando técnicas de soft computing. Universidad Politécnica de Madrid (2020)
8. Henríquez, M.R., Palma, P.A.: Control automático de condiciones ambientales en domótica usando redes neuronales artificiales. Información Tecnológica 22(3) (2011)
9. Jang, J.S.: Neuro fuzzy and soft computing: a computational approach to learning and machine intelligence [Reseña de libro]. IEEE Trans. Autom. Control **42**(10), 1482–1484 (2020)
10. Lee, D., Tsai, F.P.: Air conditioning energy saving from cloud-based artificial intelligence: case study of a split-type air conditioner. Energies **13**(8), 1–18 (2020)

11. Nasution, H., Dahlana, A.A., Aziz, A.A., Azmia, U., Sumerua, Shodiyac, S.: Energy efficiency of a variable speed of the centralized air conditioning system using PID controller. Jurnal Teknologi **78**(2), 43–48 (2016)
12. Shum, R.: Optimización energética mediante inteligencia artificial (IA) en sistemas de climatización. Servicios de Eficiencia Energética (2022). https://ee-ip.org/es/article/optimizacion-energetica-mediante-inteligencia-artificial-ia-en-sistemas-de-climatizacion-6324

Cybersecurity Frontiers and Resilient IoT

Cyber Defense by Access Control Hardening: Dynamic Web-Security Based on User Access Behavior

Cameron J. Wohlbach and Md Minhaz Chowdhury(✉)

East Stroudsburg University of Pennsylvania, East Stroudsburg, PA 18301, USA
cwohlbach@live.esu.edu, Mchowdhur1@esu.edu

Abstract. As web technologies have evolved, so have the sophistication and frequency of cyber threats. Traditional cyber security measures often fail to adapt to the dynamic nature of modern web interactions and the nuanced behaviors of users. It is crucial to dynamically adjust the website s security measures based on real-time user behavior and environmental context without compromising the user experience. This research paper explores the implementation of a comprehensive user management and security system within a Django-based web application. It details the development of advanced functionalities, including user authentication, IP tracking, dynamic access control based on user behavior, and the use of middleware for security enforcement. The system's architecture facilitates a novel approach to modifying user trust levels dynamically with direct implications on access rights, reflecting real-time security adaptations. This study aims to illustrate the practical application of combining Django's robust framework capabilities with custom security measures to enhance application safety and user management efficiency. The results provide compelling evidence of the efficacy of dynamic security measures, highlighting several critical areas of interest in the realms of cybersecurity, user behavior analysis, and system design philosophy.

Keywords: Cyber Security · Dynamic web security · Dynamic Access Control · Django Framework · Trust · Trust Flag · IP Tracking · Session Isolation · Security Measures · SQL injection · DDoS · Context-Based Security · Cyber Defense · Machine Learning · Privacy · Security Automation

1 Introduction

In the contemporary digital landscape, the security of web applications is not merely an optional feature but a fundamental necessity. As web technologies have evolved, so have the sophistication and frequency of cyber threats. Traditional security measures often fail to adapt to the dynamic nature of modern web interactions and the nuanced behaviors of users. With the atmosphere of web development becoming more and more challenging to defend, addressing the issues of web security should be the priority when defending against hackers [1]. Consequently, developers and security professionals face the continuous challenge of designing systems that are not only robust and resilient against attacks but also adaptive and intelligent in their security measures.

I. Roy et al. (Eds.): CAINE 2025, CCIS 2740, pp. 145–159, 2026.
https://doi.org/10.1007/978-3-032-14893-3_11

The necessity for advanced user management and security systems in web applications stems from several core challenges. For instance, recent studies underline the ongoing evolution of security vulnerabilities, with new challenges emerging that traditional systems are ill-equipped to handle [2]. First, the static nature of traditional security protocols means they are often unable to cope with the novel and rapidly evolving threats that characterize today's cyber security landscape. Second, there is a pressing need for systems that can dynamically adjust their security measures based on real-time user behavior and environmental context without compromising the user experience.

This research implements a security framework for web content based on user behavior. It implements a novel approach to web application security, by Access Control Hardening, within the framework of Django, a high-level Python web framework that encourages rapid development and clean, pragmatic design. Django's built-in features for user authentication and session management provide a robust foundation for developing secure web applications [3]. However, the standard functionalities need to be extensively customized and enhanced to meet the specific security requirements and user management complexities of modern web services.

This research contributes to the field of web application security by demonstrating the effectiveness of dynamic behavior-based security protocols. The study shows that such systems not only successfully prevent access by potentially harmful users but also positively influence user behavior by encouraging adherence to security best practices. A key innovation lies in the system's ability to educate users while protecting the platform, achieving a balance between robust security and user experience. Additionally, the inclusion of an administrative interface—capable of managing users and monitoring their behavior (e.g., login activity)—provides transparency and operational control, further enhancing the system's practical utility. This dual focus on prevention and education marks a meaningful advancement in developing safer, more user-aware web environments.

This paper is organized into several sections. The introduction section describes the importance of adaptive security measures in web applications. The challenges and resolution as research goals section (Challenges in Existing Applications and Goals to Resolve) presents the limitations of current web applications security controls. The limitations urge a need to have a dynamic and adaptive security system that minds the user's usage patterns of the web site system. Section "Methodology" describes the website's system architecture, the way security measures are fused into the web application (developed in Django). This section also describes the testing and validation of the security measures. The results section demonstrates: process of isolating compromised sessions of the web application. The prospective research section suggests improvement by adding multiple features of access control hardening decision making. The conclusion summarizes the useful features of the demonstrated dynamic web security system.

2 Challenges in Existing Applications and Goals to Resolve

This section presents the limitations of current web applications security controls. The limitations urge a need to have a dynamic and adaptive security system that minds the user's usage patterns of the web site system.

2.1 Challenges in Existing Systems

The integration of web applications into daily business operations and personal activities is nearly ubiquitous. With this integration comes an increased vulnerability to cyber-attacks, which can lead to sensitive data breaches, unauthorized access, and potential financial and reputational damage. Traditional access controls such as usernames and passwords, static IP blocking, and basic encryption often fall short of protecting against these threats. Moreover, the rise of distributed denial-of-service (DDoS) attacks, phishing, and other sophisticated hacking techniques requires a more proactive and adaptive security approach.

Many existing web applications rely on static and inflexible security measures that are predefined and unchanging once deployed. These systems fail to account for the variability in user behavior and the context of access requests. For example, a user accessing a system from a new location or device might be unjustifiably blocked by IP-based security measures, or conversely, a compromised account might continue to have access because the system fails to recognize anomalous behaviors that deviate from the user's norm.

Moreover, the administrative burden of managing user permissions and roles can become cumbersome without automation and intelligent systems to aid in decision-making. This challenge is compounded in environments where user roles and access needs frequently change, necessitating a more dynamic approach to user management.

2.2 Research Goal

Objective. The goal of this paper is to explore access control hardening ways through an adaptive security feature. Hence, this paper details the design, implementation, and testing of these features within a Django-based application, providing a comprehensive case study on the integration of advanced security and user management functionalities with the goal of Access Control Hardening.

The primary objective of this research is to develop a dynamic user management and security system that leverages Django's capabilities to enhance web application security through adaptive measures. This system aims to:

1. Automate the detection and response to security threats based on real-time analysis of user behavior and access patterns, using Django's middleware and signal capabilities to intercept and respond to requests.
2. Dynamically adjust user access and trust levels by implementing a trust flag system that modifies user permissions based on their actions and the cumulative security risk they pose.
3. Implement IP tracking and blocking mechanisms that intelligently respond to potential security breaches or anomalous behaviors rather than relying on static IP blocklists.
4. Optimize the administration of user roles and permissions with an automated system that adjusts access based on predefined rules and observed user behaviors, thereby reducing the administrative overhead and increasing the responsiveness of the system to changes.

Contribution of This Research. This research contributes to the broader field of web application security and offers valuable insights for developers and security professionals alike, by addressing the limitations of traditional security measures and exploring innovative solutions.

3 Methodology

This section describes how a web application is developed on the Django platform. It describes the architecture, development, and operational protocols of a web application designed on the Django platform. The application ensures dynamic security management and user access control. The application adjusts user permission (trust_flag), blocks IP addresses (by middleware), uses Django's security protections (example, CSRF token) to mitigate security threats.

3.1 Benefits of Using Django

Django is useful when we want to develop a customizable user authentication system that includes modeling users, user groups, permissions for individual groups and users. These customized modules can further be extended to include our desired characteristics, for example access right degradation for users.

Again, any framework that can facilitate user behavior tracking (example usage logs, login logs etc.) was our goal. Django can incorporate such tracking tools. We can also include a middleware that can inspect user activity history, dynamically changing user permissions to certain module of the application etc.

3.2 System Architecture

The system is built with Django Model-View-Template (MVT) framework with middleware. This MVT is a type of Model-View-Controller (MVC) framework. In MVT, model is the data structure of the application that interacts with the database. The view is responsible for business logic and hence is responsible for trust management. The template is populated by the contents given by the "view" and is the presentation layer. The middleware confirms secure user interactions by IP tracking/blocking.

Selected Framework Django: Django is a Python-based high-level web framework renowned for facilitating rapid development and clean, pragmatic design. Django's architecture is built around the MVT framework. MVT efficiently separates data handling, business logic, and presentation layers. This modularity is pivotal in enhancing the security and maintainability of the application, allowing for distinct responsibilities within the codebase. The selection of Django as the framework for implementing these security measures is supported by its documented robustness and adaptability in various practical scenarios [4, 5]. The model, view, template and middleware's role are described below:

Models: The models in Django define the data structures essential for storing user data. As seen in Fig. 1, the User and UserProfile models are central, storing not only user credentials and roles but also critical security attributes:

i. trust_flag: represents users behavior pattern
ii. last_login_ip: tracks users IP address that was recently used to login

For each user's profile there is a dynamic attribute named the trust_flag. This flag numerically represents the corresponding user's behavior pattern based on the user's action. This flag value changes as the user interacts with the system. The user's access rights and privileges are determined from this trust_flag's value.

Technically, the trust_flag is pivotal to creating a unique instance of the trust_flag for each user in the database. UserProfile extends Django's built-in User model through a one-to-one relationship, enriching its functionality and enabling detailed user management. The trust_flag could be changed depending on the size of the website and the amount of information the user is allowed to access.

```
class UserProfile(models.Model):
    user = models.OneToOneField(User, on_delete=models.CASCADE, related_name="profile")
    last_login_ip = models.CharField(max_length=45, blank=True)
    trust_flag = models.DecimalField(max_digits=3, decimal_places=1, default=2.0,
                                     validators=[MinValueValidator(0.0), MaxValueValidator(2.0)])
```

Fig. 1. The class associates individual users with their last login IP and trust_flag.

Views: Views are responsible for the application's business logic. They handle requests from users, conduct security checks based on the dynamic data provided, and generate the appropriate responses to be rendered by the templates. This layer is where the logic for adjusting the trust_flag based on user behavior is implemented, ensuring that the system's response to security risks is both immediate and contextually appropriate.

Template: The templates handle the presentation layer of the application, rendering user-facing HTML from the data passed by the views. They are meticulously designed to ensure user interactions are intuitive and secure, incorporating Django's protection mechanisms like CSRF tokens to guard against cross-site request forgery.

Middleware: Additionally, custom middleware plays a crucial role in the application's security architecture. It performs IP tracking and blocks access based on dynamic assessments of user behavior and the associated trust_flag values. By processing requests before they reach the view layer, the middleware provides a robust first line of defense, preventing potentially harmful interactions from progressing further into the system. The middleware in Django is utilized to perform functions at various stages of the request/response cycle. The middleware is strategically used to:

1. **Capture and Log IP Addresses**: This helps in identifying the source of incoming requests and tracking user behavior based on geographical and network-based patterns.
2. **Enforce Access Controls Based on Trust Level**: Middleware checks the trust_flag associated with each user (or IP) and determines if they should be granted access to the requested resources.
3. **Block Requests from Suspicious IPs**: If a user's actions result in their trust_flag being reduced to zero, their IP address is added to a block list, and all subsequent requests from this IP are denied before they can reach sensitive parts of the application.

3.3 How Security Features Are Implemented

Implemented security features are trust level calculation from user interactions/behavior, blocking any request from selected or flagged IP addresses, user authentication enhancements by adding additional security layers. Below are the details of these implemented features in this research.

Dynamic Trust Level Adjustment. A core feature of the implemented system is the dynamic trust level adjustment. The trust_flag attribute in the UserProfile model serves as a pivotal control point, adjusting user permissions fluidly based on their actions within the application. Such adjustments are enacted through views that assess user interactions for potential security risks, altering trust levels in real time to prevent misuse. While our current implementations focus on adjusting user access based on behavioral analytics, integrating machine learning techniques, as suggested by [6], could further enhance our defenses against sophisticated SQL injection attacks, aligning with the latest advancements in database security.

IP Tracking and Blocking. IP tracking and blocking mechanisms are similarly integrated. Each user's IP address is captured at login and updated accordingly in their profile. Should a user's trust_flag fall to zero, all subsequent requests from the associated IP are halted at the middleware layer, effectively isolating the threat. As seen in Fig. 2 and 3, the trust flag is decremented by a certain amount when a user takes a wrong turn while using the website.

The " < form >" section in the.html file invokes that if statement in the picture below. Every time the user tries to change the trust_flag or deletes the user it is invoked. This will happen until the user's own trust_flag goes to 0.0 and invokes the "check_user_status" method which checks if the trust_flag is 0 and blocks them from using the website. This scenario happens when a generated fake user interacts with a page of the website.

Enhancements to User Authentication. Enhancements to user authentication processes also form a significant part of the security approach. The standard username and password checks are supplemented by analyses of request contexts, such as IP address and device characteristics, to identify anomalies that may indicate compromise.

3.4 Testing and Validation

Testing Protocols. The development process is underpinned by rigorous testing protocols. Utilizing Django's built-in framework, the project team conducts extensive unit and integration testing to ensure each component functions as intended and interacts correctly with others.

Security Testing. Security tests specifically focus on scenarios such as unauthorized access attempts and IP spoofing, while performance tests assess the application's responsiveness under simulated stress conditions.

Refinements by User Feedback and Criticism. User feedback is also integral to the project methodology. Early versions of the application are released to a select user group whose interactions and criticisms inform further refinements, ensuring the system

```
GNU nano 7.2                         see_users.html

<!DOCTYPE html>
<html>
<head>
     <meta charset="UTF-8">
    <meta name="viewport" content="width=device-width, initial-scale=1.0">
    {% load static %}
    {% load custom_filters %}
    <link rel="stylesheet" type="text/css" href="{% static 'login/style.css' %}">
    <title>View Users</title>
</head>
<body>
    <div  class="view-users">
        <h1>Fake User List</h1>
    <ul>
        {% for user in fake_users %}
        <li>
            Username: {{ user.username }} - Email: {{ user.email }}
            - Trust Flag:
            <form method="post" action="{% url 'handle_fake_actions' %}">
                {% csrf_token %}
                <input type="hidden" name="action_type" value="change_trust_flag">
                <input type="hidden" name="user_id" value="{{ user.username }}">
                <select name="new_trust_flag" onchange="this.form.submit();">
                    {% for value in values %}
                    <option value="{{ value }}" {% if value == user.trust_flag %}selected{% endif %}>
                        {{ value }}
                    </option>
                    {% endfor %}
                </select>
            </form>
            <form method="post" action="{% url 'handle_fake_actions' %}" onsubmit="return confirm('Really delete {{ use>
                {% csrf_token %}
                <input type="hidden" name="action_type" value="delete_user">
                <input type="hidden" name="user_id" value="{{ user.username }}">
                <button type="submit">Delete</button>
            </form>
        </li>
        {% endfor %}
    </ul>
    </div>
</body>
</html>
```

Fig. 2. Trust_flag implementation in see_users.html file

not only meets its technical objectives but also aligns well with user expectations and usability standards.

3.5 Diagrams Representing the System

Figure 4 shows how our web application handles malicious user behavior. Let's assume a user (not an admin) user logs in into the system. The user's account keeps track of the login device demography. Whenever the user attempts an action, his/her action is evaluated to check if the action is allowed. If the action is allowed, then no restriction is given on the user's account. However, if the action is not permitted or is not allowed (considered as unauthorized action), the users reputation is lowered. This reputation decrease happens by reducing the trust_flag value. If the reputation reaches its minimum level, i.e., at trust_flag = 0, the user's access to the website/system is revoked. The revoked account can be regained i.e. the user can be removed from the blacklist upon review by the system administrator.

Figure 5 shows how two different kinds of users are handled differently: real users and fake users generated. For both users, if they attempt restricted actions too many times then they are blocked from the system. The measurement of when the users shall

```python
@login_required
def see_users(request):
    # Generate values from 0 to 2.0 with increments of 0.1
    values = [round(0.1 * i, 1) for i in range(21)]  # generates a list [0, 0.1, ..., 2.0]

    fake_users = [
        {'username': f'User{i}', 'email': f'user{i}@example.com', 'trust_flag': 2.0}
        for i in range(1, 5)
    ]
    return render(request, 'see_users.html', {'fake_users': fake_users, 'values': values})

@login_required
def handle_fake_actions(request):
    if request.method == 'POST':
        action_type = request.POST.get('action_type')
        user_id = request.POST.get('user_id')  # Identify which user the action is about

        # Logically decrement the trust_flag based on actions
        if action_type in ['change_trust_flag', 'delete_user']:
            new_trust_flag = request.user.profile.trust_flag - Decimal('0.2')
            if new_trust_flag < Decimal('0'):
                new_trust_flag = Decimal('0')
            request.user.profile.trust_flag = new_trust_flag
            request.user.profile.save()

            # Provide feedback
            messages.success(request, f"{action_type.replace('_', ' ').capitalize()} action simulated for user {user_id

        return redirect('see_users')  # Redirect to the fake view page again

    return render(request, 'see_users.html')

def blocked_page(request):
        return render(request, 'blocked_page.html')

def check_user_status(request):
    user_profile = UserProfile.objects.get(user=request.user)
    if user_profile.trust_flag <= 0.0:
        logout(request)
        return redirect('blocked_page')
```

Fig. 3. Trust_flag implementation in views.py file

be blocked depend on the trust_flag value. When the trust_flag reaches zero, then the user (both types of users) is blocked. The only difference is that the fake user can see a "404 not found" error that gives them a fake idea that the website is not responding and does not exist whereas the real user does not get such error messages.

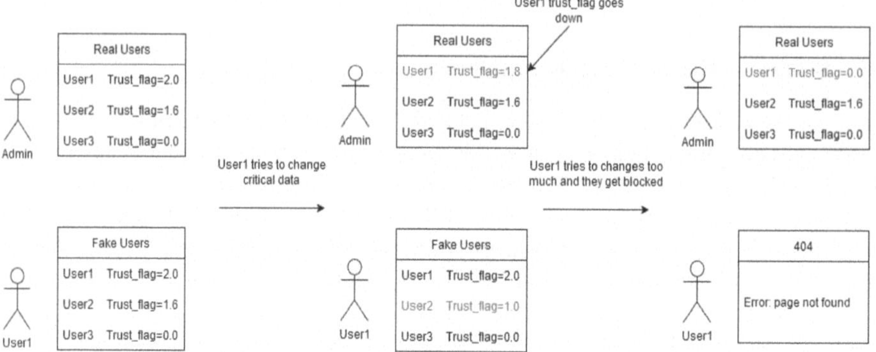

Fig. 5. A visual representation of user action evaluation

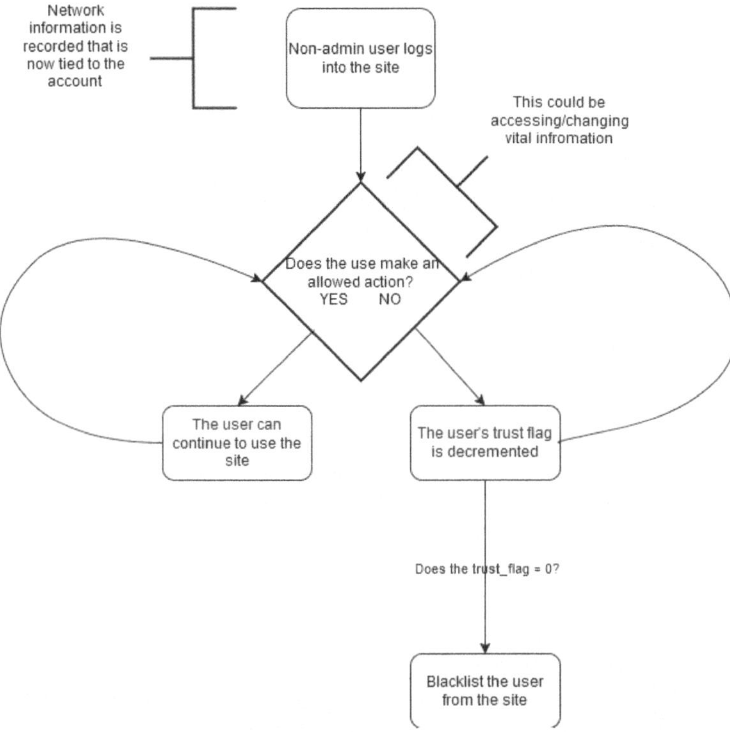

Fig. 4. Flowchart of user behavior monitoring and reputation management

3.6 Summary of Methodology

This sub-section summarizes the methodology. The system is built with Django Model-View-Template (MVT) framework with middleware. The model is a data structure of the application that interacts with the database. The view is responsible for business logic and hence is responsible for trust management. The template is the presentation layer. The middleware confirms secure user interactions by IP tracking/blocking. Following is the summary of system architecture, implemented security features, and testing methods.

The "System Architecture" is summarized as follows:

1. Framework:
   a. Django Framework
   b. MVT Architecture
2. Data Structure:
   a. Trust Flag as trust_flag
   b. User's previous IP address as last_login_ip
   c. User Model
   d. Middleware
   e. UserProfile Model
   f. View Business Logic

Security features implementation consists of:

1. Dynamic trust level adjustment based on user's behavior
2. Blocking IP addresses with inappropriate behavior
3. Enhancement in user authentication

Testing and validation process contains:

1. Rigorous Testing
2. Security testing
3. Inclusiveness of user feedback

4 Results and Explanations

The deployment of dynamic security protocols and advanced user management strategies within our Django web application yielded significant insights and measurable outcomes, underscoring the effectiveness of the approaches detailed in the methodology section. This section explores these results, focusing on how the security features performed in real-world scenarios and their impact on user behavior.

4.1 Dynamically Adjusting User's Access

A point of interest for the implemented security system was its ability to dynamically adjust a user's access based on the trust_flag associated with each user's profile. The results were unequivocally positive in this regard. The system was able to block all users whose trust_flag reached 0.0, isolating these sessions and preventing any further potentially harmful interactions with the website.

Figure 6 shows the administrative view of a blocked user, where admin can allow blocked users access rights by adjusting the corresponding user's trust_flag. The website would immediately lock the user, denying all access to the website again. This automatic blocking is a testament to the robustness of the middleware and the real-time data processing capabilities built into the system. The successful blocking of users based on trust metrics aligns with the findings by Calzavara et al. [7], who emphasize the importance of measurable security parameters. Furthermore, the reliability theories proposed by Normatov et al. [8] provide a theoretical basis for the security protocols developed.

Fig. 6. The administrative view of a blocked user, admin can restore blocked users access rights

Figure 7 shows the results of the project where it shows statistics of users who were blacklisted and were blocked from visiting the site again.

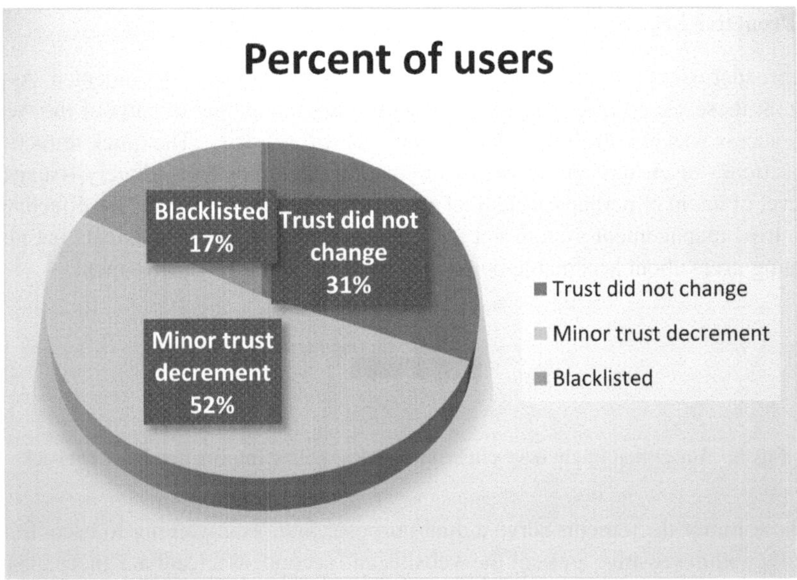

Fig. 7. Users were blacklisted and blocked from visiting the site again

4.2 Real-Time Monitoring

The mechanism behind this success lies in the seamless integration of user activity tracking, real-time analysis, and proactive response strategies. Each user action was meticulously logged, and any sequence of actions that led to a decrease in the trust_flag was immediately identified and addressed. For users whose actions resulted in their trust_flag dropping to 0.0, the system's response was to terminate access rights instantly, thereby upholding the integrity and security of the web environment.

4.3 User Behavior Change in Respond to Strict Security Measure (at Extreme Score of Trust)

Interestingly, the data revealed nuanced user behaviors in response to the security mechanisms in place. A small but notable percentage of users were observed approaching the threshold of a 0.0 trust_flag, only to change their patterns of interaction, suggesting a realization of the potential consequences. This observation could imply that when users perceived the risk of stringent security measures—such as being blocked from the website —they adjusted their behaviors to avoid such outcomes. This behavior points to the psychological impact of visible and stringent security measures on user behavior, an area ripe for further study. It underscores how transparent security protocols do not just prevent malicious actions but also actively deter users from borderline behavior that could compromise both their security and that of the web application.

4.4 Proactive Security Measure: A Warning and Feedback

For a broader user base, minor decrements in the trust_flag were documented. As seen in Fig. 8, these were typically associated with users navigating to parts of the website where access was not outright forbidden but was still sensitive. The quick retraction of these actions—users navigating away from these areas almost immediately—suggests a low level of intent or perhaps accidental access. This pattern highlights the effectiveness of the trust management system not only in responding to explicit threats but also in educating users about acceptable boundaries within the digital environment.

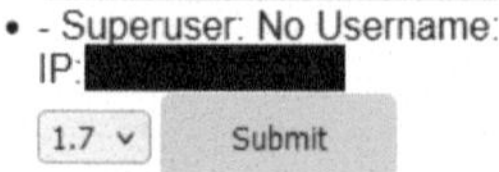

Fig. 8. A user that might have clicked to view sensitive information but went back

These minor decrements serve a dual purpose: first, as a warning to users to tread carefully within sensitive areas of the website, and second, as a feedback mechanism for administrators to understand which areas of the website might require clearer labeling or restricted access to prevent accidental breaches.

4.5 Summary of Results and Explanations

The results of this research demonstrate the efficacy of implementing dynamic, behavior-based security protocols in a web application environment. Not only did the system successfully block access for users posing a potential threat, but it also influenced user behavior in a way that promotes safer interactions and heightened awareness of security standards. This dual outcome—both protective and educative—underscores the potential of dynamic security systems to create safer web environments without compromising on user experience.

Figure 9 is a view of the admin screen that can change and delete users. Their last logged in IP would be displayed but are redacted for their privacy. It shows part of the outcome of the testing and what it would look like to the administrator that is overseeing each user's behavior on the site.

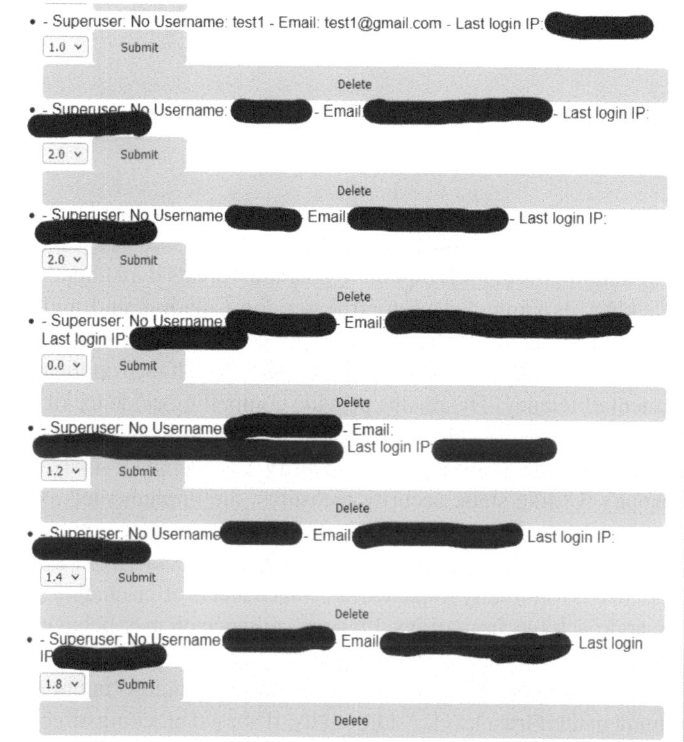

Fig. 9. User management screen showing user deletion

5 Future Research

Currently, our developed website's security system focuses on user behavior but in future, context-based measures can be applied using machine learning algorithms. We can further adjust the trust parameters based on first the information types, secondly the sensitivity and thirdly the type of the user (guest, admin, employee etc.). The admin can tailor the response when inappropriate actions originate from any user. The summarized planned extensions and improvements are as follows:

Customized Security Measures: With future improvements, this website can be secure the more personalized the website gets. Depending on what kind of sensitive information is being held and what kind of users have access to that information, the parameters of the trust_flag and how its decrements can be changed. The reaction and limits to immoral actions can be changed depending on how strict or lenient the admins of the website are.

Security From Behavioral Data: Additionally, a large majority of the security measures on this website are based purely on the user's behavior. In the future, more analytic-based security measures could be put into place to help cooperate with the behavioral data.

Machine Learning Integration: The initiative-taking security mechanisms we employed could be expanded by incorporating the multi-phase, machine-learning-based approaches that have shown promise in preventing SQL injection attacks in real-time environments [9].

6 Conclusion

This study proposes a dynamic web security mechanism using the Django framework. It introduces a trust_flag mechanism that adjusts user access based on their actions, enabling the system to detect and respond to potential threats dynamically. Key features include IP tracking, dynamic access control, session isolation, and middleware-based enforcement. It aims to illustrate the practical application of combining Django's robust framework capabilities with custom security measures to enhance application safety and user management efficiency. The results provide compelling evidence of the efficacy of dynamic security measures in a Django-based web application, highlighting several critical areas of interest in the realms of cybersecurity, user behavior analysis, and system design philosophy. Unlike static security measures, the implemented system dynamically adjusts to real-time data, offering a tailored security stance that evolves as user interactions unfold. This adaptability not only enhances security but also ensures that the user experience is minimally impacted, a balance that is often challenging to achieve. However, the approach has limitations. First, the reliance on user behavior to adjust the trust flag values might not capture all nuances of intent or context. For example, a user might accidentally navigate to a restricted area without malicious intent yet still experience a decrement in their trust level. Additionally, IP-based blocking might inadvertently block users who share IP addresses or use dynamic IP addresses, potentially denying legitimate users' access. As mentioned in the previous section, in the future work, our proposed enhancements will overcome most of the limitations.

References

1. Fredj, O.B., et al.: An OWASP top ten driven survey on web application protection methods. In: Risks and Security of Internet and Systems: 15th International Conference, CRiSIS 2020, Paris, France, 4–6 November 2020, Revised Selected Papers 15. Springer (2021)
2. Jannis, R., et al.: To Auth or Not to Auth? a comparative analysis of the pre-and post-login security landscape. In: 2024 IEEE Symposium on Security and Privacy (SP), IEEE (2024)
3. Django. Django Project, www.djangoproject.com/. Accessed 28 Nov 2024
4. Duisebekova, K., Khabirov, R., Zholzhan, A.:.Django as secure web-framework in practice.Вестник Казахской академии транспорта и коммуникаций им. М. Тынышпаева **1**, 275–281 (2021)
5. Manikanta, V.K., et al.: Visualization of real-world enterprise data using python Django framework. In: IOP Conference Series: Materials Science and Engineering, vol. 1042, no. 1. IOP Publishing (2021)
6. Ashlam, A.A., Badii, A. and Stahl, F.: Multi-phase algorithmic framework to prevent SQL injection attacks using improved machine learning and deep learning to enhance database security in real-time. In: 2022 15th International Conference on Security of Information and Networks (SIN), IEEE (2022)
7. Calzavara, S., et al.: Measuring web session security at scale. Comput. Secur. **111**, 102472 (2021)

8. Normatov, I., Bobur, E., Murodulla, B.: On the theory of reliability of the system of protection against unauthorized access to information. Central Asian J. Math. Theor. Comput. Sci. **4**(1), 68–76 (2023)
9. Kim, A., Mohyun, P., Dong, H.L.: AI-IDS: application of deep learning to real-time web intrusion detection. IEEE Access **8**, 70245–70261 (2020)

Traffic Congestion Sybil Attack Detection in VANET

Sarthak Khanduja(✉) and Arunita Jaekel

University of Windsor, Windsor, ON N9N 3P4, Canada
{khandujs,arunita}@uwindsor.ca

Abstract. In Vehicular Ad-hoc Networks (VANETs), participating vehicles periodically transmit *basic safety messages* (BSMs) containing critical information about their current status, including position, speed, acceleration and heading. BSMs are broadcast wirelessly to surrounding vehicles and help improve situational awareness. However, it also introduces critical security concerns and requires robust security measures and innovative approaches to ensure the reliability of communication within VANETs. This paper addresses the emerging threat of *Traffic Congestion Sybil* attacks, where malicious entities inject spurious data into BSMs to fabricate artificial traffic congestion. We implement models, using classical machine learning algorithms, to accurately detect malicious BSMs containing such spurious data. Preliminary results indicate that the proposed models can detect these sybil attacks with a high degree of accuracy and low false-positive and false-negative rates. Comparison with existing techniques demonstrate that the proposed models perform as well or better than more complex deep learning models available in the literature.

Keywords: VANETs · VeReMi Dataset · Basic Safety Message · Sybil attacks

1 Introduction

Intelligent Transportation System integrates various technologies to build an integrated system of people, roads, and vehicles with the aim of improving road safety and user experience [1]. Vehicular Ad-Hoc Networks (VANETs) [2], consisting of an interconnection of sensors on cars, streetlights, traffic signals, and other nodes, form a crucial part of Intelligent Transportation Systems (ITS). Each vehicle in a VANET is equipped with an On-Board Unit (OBU) enabling it to exchange information wirelessly with other OBUs and infrastructure nodes such as roadside units (RSUs).

VANETs offer direct vehicle-to-vehicle (V2V) communication, as well as communication with infrastructure nodes, such as RSUs for sending and receiving critical information including hazard warnings, traffic congestion reports, etc.,

I. Roy et al. (Eds.): CAINE 2025, CCIS 2740, pp. 160–170, 2026.
https://doi.org/10.1007/978-3-032-14893-3_12

with minimal latency [2]. An important component of V2V communication consists of periodic dissemination of *basic safety messages* (BSM), by each vehicle. Each BSM encapsulates crucial information, including the vehicle's current position, velocity, heading, brake status, steering wheel angle etc., to increase situational awareness of the participating vehicles.

While offering flexibility and adaptability, the wireless nature of these communications also makes them vulnerable to various security threats and attacks, ranging from passive attacks such as eavesdropping to active interference such as message tampering, signal jamming, and denial of service (DoS) [3]. One important type of attack is the sybil attack, introduced in [4]. In a sybil attack, a malicious entity assumes multiple fake identities (sybils) within the network, gaining undue influence, and disrupting its normal operation. Attackers leverage these fake identities to manipulate data, spread misinformation, and launch further attacks. For example, a malicious node may misleadingly report a traffic jam or accident to cause a nearby vehicle driver to reroute, potentially causing a real traffic jam on another road. Alternatively, the malicious node may exploit the fake virtual nodes to use a large amount of bandwidth, disrupting communication [5]. In this paper, we focus on detecting *traffic congestion sybil* attacks (also called *traffic sybil* or *grid sybil* attacks), where a vehicle creates fake traffic congestion by generating a grid of fake vehicles. The potential ramifications include risk of fatal accidents and widespread traffic disturbances. We develop a machine learning-based misbehavior detection system (MDS) to accurately detect traffic sybil attacks in BSMs.

The remainder of this paper is organized as follows: Sect. 2 outlines existing approaches for different types of VANET misbehavior detection, with a focus on detecting sybil attacks. Section 3 presents our proposed approach followed by Sect. 4 which evaluates the performance of the proposed approach. Finally, Sect. 5 presents a conclusion of our work, and some possible future research directions.

2 Related Work

In this section, we will first give a brief overview of machine learning (ML) and outline some recent ML based approaches for misbehavior detection in VANET. Then we will review existing approaches for detecting sybil attacks in VANET, which is the focus of this paper. Machine learning is the Artificial Intelligence branch that facilitates machines to perform specific jobs faster and skillfully using statistical learning [6]. In *supervised* learning, the algorithm learns from labeled data, where each training example is associated with a corresponding target or outcome. *Unsupervised* learning involves learning from unlabeled data, where the algorithm tries to find patterns or intrinsic structures within the data without explicit guidance. Within supervised learning, one common task is classification, where the goal is to predict the categorical class labels of new instances based on past observations. Some popular classification algorithms include K-Nearest Neighbors [7], Decision Trees [8], Random Forests [9], support vector machines (SVM), and Logistic Regression.

2.1 ML Based Misbehavior Detection in VANET

Alzahrani et al. [10] propose an enhanced misbehavior detection scheme for Vehicular Ad Hoc Networks (VANETs) using a combination of Kalman filtering and Artificial Neural Networks (ANNs). Compared to other approaches, this method replaces static detection thresholds with a dynamic ANN-based classifier, providing better adaptability to changing network conditions and potentially improving accuracy.

So et al. [11] propose a novel approach for misbehavior detection in Vehicular Ad Hoc Networks (VANETs) by integrating plausibility and consistency checks with ML techniques. The system can effectively detect misbehavior types by combining these two approaches while minimizing false positives.

Kamel et al. [12] propose an enhanced misbehavior detection mechanism, named CaTch, for Vehicular Ad Hoc Networks (VANETs), focusing on embedded misbehavior detection mechanisms in C-ITS. The approach incorporates physical measurement uncertainty into plausibility detectors, allowing for a more accurate assessment of the credibility of received messages.

Grover et al. [13] propose an ensemble-based machine learning approach for detecting misbehaviors in Vehicular Ad Hoc Networks (VANETs), a critical aspect given the potential safety implications of inaccurate messages transmitted by illegitimate vehicles. By employing a majority voting scheme, the ensemble method combines the strengths of individual classifiers, thereby enhancing detection accuracy. The paper highlights the importance of accurately classifying various misbehaviors, including identity spoofing, position forging, packet suppression, replay, and detention attacks, to mitigate potential safety hazards in VANETs.

Sharma et al. [14] propose an approach for detecting misbehavior in VANETs that utilizes a classical ML-based framework that focuses specifically on position falsification attacks, where vehicles transmit incorrect position information in Basic Safety Messages (BSMs). The proposed "2_BSM approach" combines information from consecutive BSMs sent by a vehicle to enhance accuracy. The approach effectively classifies vehicles as legitimate or malicious by analyzing features extracted from BSMs and training ML models, achieving high precision and recall rates across various attack types. However, limitations include reliance on RSU infrastructure, making it more suitable for urban areas.

Gyawali et al. [15] propose a machine learning-based misbehavior detection system for Vehicular Ad-hoc Networks (VANETs) to address attacks such as false alert generation and position falsification. Unlike traditional cryptographic methods, which are less effective against insider attacks, the proposed system leverages machine learning techniques trained on realistic VANET simulation datasets.

Khot et al. [16] proposed a machine-learning framework to predict the next position of the vehicle in the network. The authors used beacon messages from neighboring vehicles and created features such as distance between sender and receiver. Machine learning algorithms were utilized to train and test them. The authors compared the predicted value with the actual value of the vehicle comparison. If the position is not equal to the prediction, it is classified as an attack vehicle. The authors claimed that Random Forest performs best among other algorithms.

2.2 Detecting Sybil Attacks in VANET

Ayaida et al. [17] propose an approach to detect sybil attacks in Cooperative Intelligent Transport Systems (C-ITS) by leveraging traffic flow theory. Unlike other approaches, which may rely on resource testing or Public Key Infrastructure (PKI), this method utilizes the inherent characteristics of traffic flow models to estimate vehicle speeds and detect anomalies indicative of sybil attacks.

Baza et al. [18] proposed an approach for detecting sybil attacks in VANETs involving leveraging proofs of work (PoW) and location information obtained from multiple Roadside Units (RSUs) encountered by vehicles. Unlike traditional methods that rely on signatures from individual RSUs, this approach requires contributions from at least t RSUs, where 't' represents the minimum number of RSUs required to contribute their location information. Detection is performed using a threshold signature scheme, thus mitigating the risk of RSU compromise attacks.

Laouiti et al. [19] present a method that leverages ML algorithms and features derived from vehicle behavior data to identify sybil nodes. By preprocessing the data into Driving Pattern Matrices (DPM) and computing their eigenvalues, the proposed method enhances the accuracy and efficiency of detection compared to previous approaches.

Mahmoudi et al. [20] proposed a comprehensive work on a misbehavior detection framework tailored to C-ITS. Their primary objective was to bolster system security using ML techniques at the Misbehavior Authority (MA) level. Through a multi-step process encompassing local misbehavior detection, reporting to the MA, global detection, and subsequent reaction, the framework aimed to classify reported ITS-S behavior as misbehaving, faulty, or genuine, thereby enhancing the overall security posture of C-ITS.

Zhao et al. [21] proposed a FedMix approach for detecting sybil attacks within VANETs. Their method combines federated learning with cross-layer information fusion to bolster privacy protection. Their experimental results showcase FedMix's efficacy, surpassing centralized training methods in detection accuracy, precision, recall, and F1-score, while notably reducing communication costs.

Alladi et al. [22] propose a Deep Learning-based Anomaly Detection Framework (DeepADV) tailored specifically for VANETs, aiming to detect various anomalies, including faults and attacks, within VANET data sequences. Their approach involves training six different Deep Neural Network (DNN) architectures exclusively on genuine data sequences, utilizing Mean Absolute Error (MAE) as the loss function for backpropagation. The CNN-LSTM model (M-1) consistently outperforms others across all scenarios.

Liu et al. [23] proposed the SVMDformer, a data-centric framework for vehicular misbehavior detection in VANETs, leveraging semi-supervised learning and Transformer architecture. Their approach involves transforming vehicular message sequences into misbehavior scores and classifying misbehavior using a threshold. Compared to baseline models like CNN and Bi-LSTM, their SVMD-former demonstrated superior performance in metrics such as AUC, accuracy, precision, and F1 score, notably reducing false detection rates.

Although there has been significant research interest in VANET misbehavior detection in recent years, detection of traffic sybil attacks has been largely unaddressed. Existing works typically use Deep Learning approaches to detect such attacks, which can be computationally expensive. In this we paper, we investigate to efficacy of using standard ML algorithms for accurately detecting traffic sybil attacks.

3 Proposed Traffic Sybil Attack Detection (TSAD) Approach

Traffic sybil attacks involve malicious entities creating multiple false identities to simulate traffic congestion artificially, leading to disruption and potentially dangerous situations on the road. The proposed approach uses ML models for sybil attack detection and consists of the following main stages:

- Data Pre-processing and Exploratory Data Analysis
- Feature Engineering
- Model Training and Classification

3.1 Data Pre-processing and Exploratory Data Analysis

The VeReMi dataset has several simulations with different traffic scenarios for each type of attack. Each simulation consists of two types of files: Ground truth file and log files. There is only one ground truth file in a simulation that includes the vehicle's actual behavior in the network. Ground truth file also comprise an attack type, differentiating legitimate vehicles from misbehaving vehicles. On the other hand, the number of log files in a simulation is equal to the number of vehicles in the network. These log files are combined and after removing duplicate entries the resultant consolidated dataset is used for exploratory data analysis.

The consolidated dataset consists of *1,501,368* rows and *26* columns (features), including the label. Of these, 10 columns were dropped due to redundant information or high correlation with other columns. The remaining 16 columns comprise the initial feature set FS_1, without any feature engineering included. Figure 1 shows the correlation matrix for FS_1, which indicates that *senderPseudo* features have the highest correlation with the target variable. This could imply that attackers use certain pseudonyms more frequently or change pseudonyms in a specific pattern.

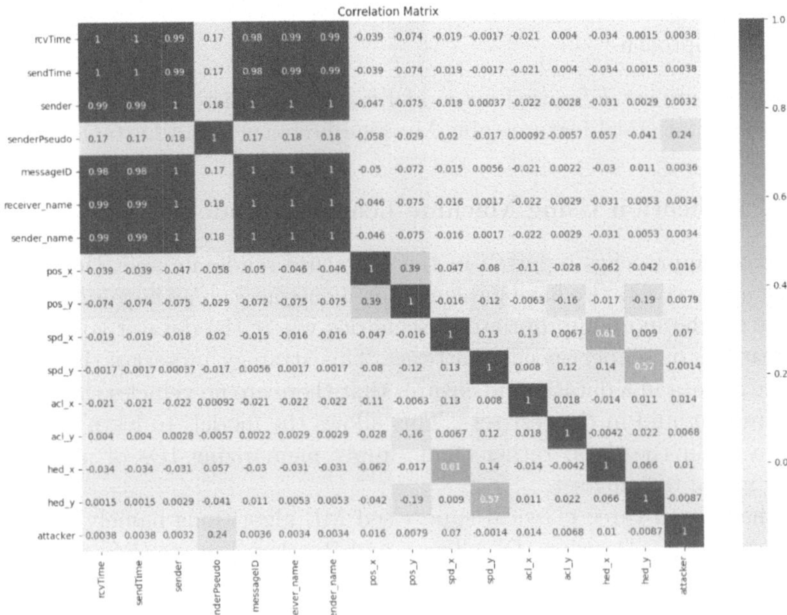

Fig. 1. Correlation matrix for FS_1

3.2 Feature Engineering

The models trained with FS_1 failed to achieve satisfactory performance, with detection accuracy below 80% for all cases. To help improve performance, the following features were added to the dataset:

- **Message Frequency**
  This feature represents the number of messages sent out by each sender up to the current point in time.
  The attackers have a low density and high variability of message count. In contrast, the message count for non-attackers is more dense within a given range and doesn't show high variability.
- **Distance**
  This feature calculates the distance traveled by each sender between consecutive messages.

$$\text{Distance} = \sqrt{(x_2 - x_1)^2 + (y_2 - y_1)^2}$$

  Here, (x_1, y_1) and (x_2, y_2) refer to reported position of the vehicle, in two consecutive BSMs.
- **Statistical values of speed and acceleration**
  The calculated statistical columns were *mean, median*, and standard *deviation* of each of the x and y coordinates of *'spd' and 'acl'*. These values are

calculated based on values from all BSMs received so far, using a running average approach.

The new feature set FS_2 used for model training consisted of FS_1 and 14 new features (12 statistical features plus Message Frequency and Distance).

3.3 Classification Using Machine Learning Models

While most approaches reported in the literature randomly split the data into training and test sets, we have implemented a custom *sender-disjoint* split strategy that can be more appropriate in scenarios where the dataset contains temporal features or sender-specific patterns. This strategy preserves the temporal sequence of data in a dataset and ensures that there are no vehicles that are both in the training and in the test sets. This allows the models to learn patterns of misbehavior in the data, rather than simply memorizing IDs of misbehaving vehicles in the training set.

We have used three classical supervised ML algorithms namely, K-Nearest Neighbors (KNN), Decision (DT) Trees, and Random Forests (RF) to train our models. For all three models, we performed *hyperparameter tuning* to find the best set of hyperparameters that maximize the model's performance. For KNN, we tested 'k' values from 1 to 101. Our results indicated that performance began to plateau after $k = 81$, so we used $k = 81$ to train our models. For DT, tuning parameters such as the maximum depth of the trees, the minimum number of samples required to split an internal node, and the minimum number of samples required at a leaf node had significant impact on performance. The optimal parameters were identified as $max\_depth = 15$, $min\_samples\_leaf = 8$ and $min\_samples\_split = 10$. Finally for RF model, we identified optimal parameters as $max\_depth = 15$, $min\_samples\_leaf = 10$, $min\_samples\_split = 15$, and $n\_estimators = 50$. Additionally, we ensured the robustness of our hyperparameter tuning strategy by employing stratified 5-fold cross-validation to prevent bias in the evaluation process and shuffling the training data to mitigate any sequence-related biases.

4 Performance Evaluation

To evaluate the performance of the proposed *TSAD* approach we use the VeReMiExt dataset [4], which is an extension of the Vehicular Reference Misbehavior (VeReMi) Dataset [24] based on vehicle traces from the Luxembourg SUMO Traffic (LuST) scenario [25]. In particular, we used the **"TrafficSybil_1416"** dataset, focusing on low traffic density scenarios pertinent to traffic sybil attacks and trained 3 separate models - $TSAD_{KNN}$, $TSAD_{DT}$ and $TSAD_{RF}$ for attack detection.

In the VeReMiExt Dataset there's a disparity in the distribution between normal and malicious vehicle instances. Consequently, relying solely on accuracy as an evaluation metric would be inadequate for assessing model performance.

So, we also use precision, recall, and the F1-score to gauge the effectiveness of the proposed models. A confusion matrix summarizes the model's performance by tabulating the correct and incorrect predictions, as shown in Table 1.

Table 1. Confusion Matrix

	Predicted Positive	Predicted Negative
Actual Positive	True Positive (TP)	False Negative (FN)
Actual Negative	False Positive (FP)	True Negative (TN)

Based on the confusion matrix, we use the following four metrics as defined below.

$$\text{Accuracy} = \frac{\text{TP} + \text{TN}}{\text{TP} + \text{TN} + \text{FP} + \text{FN}}$$

$$\text{Precision} = \frac{\text{TP}}{\text{TP} + \text{FP}}$$

$$\text{Recall} = \frac{\text{TP}}{\text{TP} + \text{FN}}$$

$$\text{F1-Score} = \frac{2 \times (\text{Recall} \times \text{Precision})}{\text{Recall} + \text{Precision}}$$

Table 2 compares the performance of the proposed models using both the initial feature set FS_1 and the updated feature set FS_2, which includes the new features discussed in Sect. 3.2. From Table 2, we can see that all the models perform poorly using FS_1, with detection accuracy and other metrics between 71%–76%. $TSAD_{RF}$ performed marginally better than the other models, but the results were still unsatisfactory.

Using FS_2, we are able to achieve significant improvement in performance for all 3 models. $TSAD_{KNN}$ and $TSAD_{KNN}$ achieve accuracy and F1-scores of around 95%. Overall, $TSAD_{RF}$ has the best performance, with scores for all metrics at 98%.

4.1 Comparison with Existing Approaches

In this section, we compare the performance of our proposed model with existing approaches in the literature, as listed below. For our proposed approach, we have used the $TSAD_{RF}$ model, as it had the best performance.

- Paper1 [4]: This is a non-ML approach that uses a number of different plausibility and consistency checks to detect misbehavior.
- Paper2 [21]: This paper uses a federated machine learning approach (FedMix), where artificial neural network (ANN) is used as the basic model.

Table 2. Performance of the proposed models

Model Used	Acc	Prec	Recall	F-1 Score
Results with FS_1				
$TSAD_{KNN}$	0.75	0.74	0.73	0.73
$TSAD_{DT}$	0.73	0.72	0.71	0.71
$TSAD_{RF}$	0.76	0.75	0.75	0.75
Results with FS_2				
$TSAD_{KNN}$	0.94	0.96	0.95	0.95
$TSAD_{DT}$	0.96	0.96	0.96	0.96
$TSAD_{RF}$	**0.98**	**0.98**	**0.98**	**0.98**

Table 3. Comparison with Existing Approaches

Paper	Approach	Acc	Prec	Recall	F1 Score
Paper1 [4]	non-ML	0.800	0.997	0.58	0.74
Paper2 [21]	Federated ML (ANN)	x	0.99	0.85	0.91
Paper3 [22]	CNN-LSTM	0.99	0.998	0.96	0.98
$TSAD_{RF}$	Classical ML	**0.98**	**0.98**	**0.98**	**0.98**

- Paper3 [22]: This paper trains six different deep neural network architectures, with the CNN-LSTM model having the best performance.

Table 3 compares the performance of the proposed approach with existing techniques. All 3 ML based approaches clearly out perform the non-ML technique. All three models have high *precision* scores, indicating a low false-positive rate. Paper3 has the highest values for *accuracy* and *precision*. However, when considering recall values, the proposed $TSAD_{RF}$ model has the best performance, with a lower false-negative rate, indicating it might be able to identify some additional attacks that the other approaches are missing. Overall, both Paper3 and $TSAD_{RF}$ have a similar performance, with F1-scores of 98%, which clearly outperform the other approaches. We note that both Paper2 and Paper3 include the use of Deep Learning algorithms, which can lead to increased computational overhead. The proposed approach is able to achieve similar or better results compared to DL methods with classical ML models.

5 Conclusion

In this paper, we present a novel Traffic Sybil Attack Detection (TSAD) approach in VANETs using classical ML algorithms. We implement a custom *sender-disjoint* technique for splitting the data into training and test sets that preserves the temporal features of the dataset and prevents leakage between training and test sets, which can lead to overfitting. To improve the performance of our models, we introduced 14 additional features, including statistical distributions of vehicle speed and acceleration values, messages frequency, and distance traveled by a vehicle between consecutive BSMs. Our classical ML based models were able to achieve F1-scores of 98%, with performance better than or similar to more complex DL models.

This paper presents models for *binary* classification of traffic syblil attacks. In the future, we are planning to extend this work to other types of sybil attacks to determine if classical ML continues to be effective multi-class classification of Sybil attacks as well.

Acknowledgment. The work of A. Jaekel has been supported by research grants from the Natural Sciences and Engineering Research Council of Canada (NSERC).

References

1. An, S., Lee, B.-H., Shin, D.-R.: A survey of intelligent transportation systems. In: 2011 Third International Conference on Computational Intelligence, Communication Systems and Networks, pp. 332–337. IEEE (2011)
2. Hartenstein, H., Laberteaux, L.: A tutorial survey on vehicular ad hoc networks. IEEE Commun. Mag. **46**(6), 164–171 (2008)
3. Hezam, M.A., et al.: Classification of security attacks in VANET: a review of requirements and perspectives. MATEC Web Conf. **150**, 06038 (2018)
4. Kamel, J., Wolf, M., van der Hei, R.W., Kaiser, A., Urien, P., Kargl, F.: VeReMi extension: a dataset for comparable evaluation of misbehavior detection in VANETs. In: 2020 IEEE International Conference on Communications (ICC), ICC 2020, pp. 1–6 (2020)
5. Azam, S., Bibi, M., Riaz, R., Rizvi, S.S., Kwon, S.J.: Collaborative learning based Sybil attack detection in vehicular Ad-hoc networks (VANETs). Sensors **22**(18) (2022). https://www.mdpi.com/1424-8220/22/18/6934
6. Mohammed, M., Khan, M.B., Bashier, E.B.M.: Machine Learning: Algorithms and Applications. CRV Press (2016)
7. Mucherino, A., Papajorgji, P.J., Pardalos, P.M.: K-nearest neighbor classification. In: Data Mining in Agriculture, pp. 83–106. Springer, New York (2009). https://doi.org/10.1007/978-0-387-88615-2_4
8. Priyam, A., Abhijeeta, G., Rathee, A., Srivastava, S.: Comparative analysis of decision tree classification algorithms. Int. J. Curr. Eng. Technol. **3**(2), 334–337 (2013)
9. Liaw, A., Wiener, M., et al.: Classification and regression by randomForest. R News **2**(3), 18–22 (2002)
10. Alzahrani, M., Idris, M.Y., Ghaleb, F.A., Budiarto, R.: An improved robust misbehavior detection scheme for vehicular Ad Hoc network. IEEE Access **10**, 111 241–111 253 (2022)
11. So, S., Sharma, P., Petit, J.: Integrating plausibility checks and machine learning for misbehavior detection in VANET. In: 2018 17th IEEE International Conference on Machine Learning and Applications (ICMLA), pp. 564–571. IEEE (2018)

12. Kamel, J., Kaiser, A., Ben Jemaa, I., Cincilla, P., Urien, P.: CaTch: a confidence range tolerant misbehavior detection approach. In: 2019 IEEE Wireless Communications and Networking Conference (WCNC), pp. 1–8 (2019)
13. Grover, J., Laxmi, V., Gaur, M.S.: Misbehavior detection based on ensemble learning in VANET. In: Thilagam, P.S., Pais, A.R., Chandrasekaran, K., Balakrishnan, N. (eds.) ADCONS 2011. LNCS, vol. 7135, pp. 602–611. Springer, Heidelberg (2012). https://doi.org/10.1007/978-3-642-29280-4_70
14. Sharma, A., Jaekel, A.: Machine learning based misbehaviour detection in VANET using consecutive BSM approach. IEEE Open J. Veh. Technol. **3**, 1–14 (2022)
15. Gyawali, S., Qian, Y.: Misbehavior detection using machine learning in vehicular communication networks. In: 2019 IEEE International Conference on Communications (ICC), ICC 2019, pp. 1–6. IEEE (2019)
16. Khot, A., Dave, M.: Position falsification misbehavior detection in VANETs. In: Marriwala, N., Tripathi, C.C., Kumar, D., Jain, S. (eds.) Mobile Radio Communications and 5G Networks. LNNS, vol. 140, pp. 487–499. Springer, Singapore (2021). https://doi.org/10.1007/978-981-15-7130-5_39
17. Ayaida, M., Messai, N., Wilhelm, G., Najeh, S.: A novel Sybil attack detection mechanism for C-ITS. In: 2019 15th International Wireless Communications and Mobile Computing Conference (IWCMC), pp. 913–918 (2019)
18. Baza, M., et al.: Detecting sybil attacks using proofs of work and location in VANETs. IEEE Trans. Dependable Secure Comput. **19**(1), 39–53 (2022)
19. Laouiti, D. E., Ayaida, M., Messai, N., Najeh, S., Najjar, L., Chaabane, F.: Sybil attack detection in VANETs using an AdaBoost classifier. In: 2022 International Wireless Communications and Mobile Computing (IWCMC), pp. 217–222 (2022)
20. Mahmoudi, I., Kamel, J., Ben-Jemaa, I., Kaiser, A., Urien, P.: Towards a reliable machine learning-based global misbehavior detection in C–ITS: model evaluation approach. In: Laouiti, A., Qayyum, A., Mohamad Saad, M.N. (eds.) Vehicular Adhoc Networks for Smart Cities, Singapore. Springer Singapore, pp. 73–86 (2020). https://doi.org/10.1007/978-981-15-3750-9_6
21. Zhao, J., Wang, R.: FedMix: a Sybil attack detection system considering cross-layer information fusion and privacy protection. In: 2022 19th Annual IEEE International Conference on Sensing, Communication, and Networking (SECON), pp. 199–207 (2022)
22. Alladi, T., Gera, B., Agrawal, A., Chamola, V., Yu, F.R.: DeepADV: a deep neural network framework for anomaly detection in VANETs. IEEE Trans. Veh. Technol. **70**(11), 12 013–12 023 (2021)
23. Liu, Z., Xu, H., Kuang, Y., Li, F.: SVMDformer: a semi-supervised vehicular misbehavior detection framework based on transformer in IoV. In: 2023 IEEE 43rd International Conference on Distributed Computing Systems (ICDCS), pp. 887–897 (2023)
24. van der Heijden, R.W., Lukaseder, T., Kargl, F.: VeReMi: a dataset for comparable evaluation of misbehavior detection in VANETs. In: Beyah, R., Chang, B., Li, Y., Zhu, S. (eds.) SecureComm 2018. LNICST, vol. 254, pp. 318–337. Springer, Cham (2018). https://doi.org/10.1007/978-3-030-01701-9_18
25. Codecá, L., Frank, R., Faye, S., Engel, T.: Luxembourg SUMO Traffic (LuST) scenario: traffic demand evaluation. IEEE Intell. Transp. Syst. Mag. **9**(2), 52–63 (2017)

Authentication for IoT Devices to Access Satellite Networks

Dharani Parvathaneni[1], Krishnendu Kottakkal Sugathan[1], Prasanta Kumar Roy[2], Ansuman Bhattacharya[1](✉), and Indranil Roy[3]

[1] Southern Illinois University, Carbondale, IL 62901, USA
{dharani.parvathaneni,krishnendu.kottakkalsugathan, ansuman.bhattacharya}@siu.edu
[2] Siliguri Institute of Technology, Siliguri, WB 734009, India
[3] Southeast Missouri State University, Cape Girardeau, MO 63701, USA
iroy@semo.edu

Abstract. In modern computational environments, safeguarding sensitive data and preventing unauthorized access require robust and efficient authentication mechanisms. This paper presents a secure authentication protocol that addresses two key objectives: validating user legitimacy prior to network access and maintaining data integrity and confidentiality during transmission. The proposed scheme leverages lightweight cryptographic techniques including hashing, encryption, and dynamic pseudonyms to ensure mutual authentication and secure session key exchange, while preserving user anonymity and preventing identity tracking. The protocol includes a secure initialization phase, user registration, and a Mutual Authentication and Key Exchange (MAKE) process, effectively countering common threats such as replay attacks, impersonation, and man-in-the-middle attacks. Designed for low-power and resource-constrained environments, the protocol minimizes computational overhead without compromising security. Comparative analysis with existing schemes demonstrates that the proposed method achieves a strong balance between security, efficiency, and scalability, making it highly suitable for modern distributed network architectures.

Keywords: Dynamic pseudonyms · Authentication Key Exchange (MAKE) process · Lightweight cryptographic operations

1 Introduction

Imagine a world where even remote villages have high-speed internet, empowering students with online education and enabling doctors to offer real-time consultations. This vision is being realized through 3D wireless networks, which integrate satellite constellations with High and Low Altitude Platform Stations (HAPS and LAPS) to provide seamless, multi-dimensional connectivity. In this paper, we use "3D wireless networks" and "3D networks" interchangeably (Figs. 1 and 1).

I. Roy et al. (Eds.): CAINE 2025, CCIS 2740, pp. 171–188, 2026.
https://doi.org/10.1007/978-3-032-14893-3_13

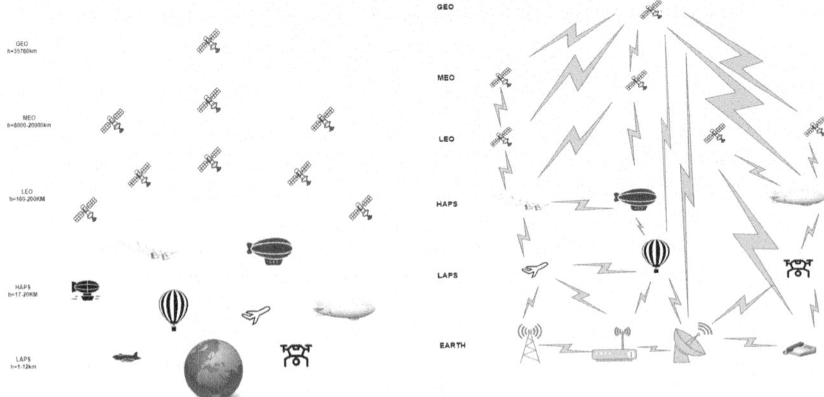

Fig. 1. 3D Wireless Networks

Fig. 2. 3D Interconnected Wireless Networks

Unlike traditional 2D networks, 3D networks operate in three spatial dimensions and require specialized design strategies for signal propagation, interference mitigation, and network deployment. Technologies such as multi-hop routing, beamforming, and antenna diversity play vital roles in ensuring reliable communication.

These networks incorporate satellites at various orbits LEO, MEO, and GEO each contributing distinct advantages in latency, coverage, and capacity [7,14,21,40]. They are complemented by aerial platforms like UAVs and balloons operating at high and low altitudes, offering flexible coverage for both urban and remote areas [17,26,28]. Combined, they deliver global coverage, reduced latency, increased capacity, and network resilience [1,31].

Despite their potential, 3D networks face challenges in inter-satellite coordination, spectrum management, and regulatory compliance [11,12]. Nevertheless, large-scale projects such as Starlink, OneWeb, and Telesat are already operational, expanding access in underserved regions [8].

In parallel, the Internet of Things (IoT) has emerged as a key enabler of smart systems, connecting billions of devices that collect, share, and act on data [4]. However, IoT faces issues related to security, interoperability, scalability, and reliability. Integrating IoT with satellite-based 3D networks introduces new vulnerabilities and necessitates robust protections, including encryption, strong authentication, intrusion detection, and supply chain security [2,24].

To safeguard this ecosystem, modern security practices such as Zero Trust architectures, anomaly detection, network segmentation, and encrypted data transmission are essential [10,15,16,29,30].

The convergence of 3D networks and IoT holds tremendous promise to bridge the digital divide, drive innovation, and improve lives across the globe [5,34]. However, this potential can only be realized through secure, efficient, and scalable network architectures.

The remainder of this paper is organized as follows: Sect. 2 reviews related work. Section 3 outlines security and design goals. Sections 4 and 5 present our contributions and the proposed system, respectively. Section 6 discusses results, followed by the conclusion in Sect. 7.

2 Related Works

Authentication protocols in Mobile Satellite Communication Systems (MSCS) and Satellite-Terrestrial Networks (STNs) have seen significant advancements aimed at improving security, efficiency, and resilience. Lattice-based cryptographic methods have gained prominence, with Yi et al. [38] and Ma et al. [23] proposing efficient authentication mechanisms that support fast key agreement and aggregated signatures for secure communication.

Efforts to improve access authentication include Zhao et al. [44], who leveraged satellite broadcast properties, and Xue et al. [36], who embedded authentication directly into satellite hardware to reduce latency. For mobility and handover challenges, Zhang et al. [41] introduced a blockchain-based identity system, while Lai and Chen [18] employed group signatures for secure transitions in vehicle networks.

However, traditional shared-secret protocols remain vulnerable to common threats such as replay and impersonation attacks [6,19,20,39,43]. Recent enhancements, including those by Meng et al. [25] and Qi and Chen [27], focus on mitigating these weaknesses. Additionally, group signature-based approaches have been applied to protect user anonymity in space-IoT contexts [36,37].

While foundational work by Chen et al. [6] and others laid the groundwork for secure satellite authentication, persistent vulnerabilities highlighted by Zhang et al. [42,43] indicate the need for more adaptive, lightweight protocols that can support the dynamic nature of 3D wireless environments.

3 System Architecture and Design Objectives

We propose a robust system architecture designed to enhance packet authentication and reinforce IoT device security within the complex landscape of 3D wireless networks. Anchored in the Zero Trust Security model, the system enforces continuous identity verification, restricts access privileges, and ensures data integrity throughout the communication lifecycle. This architecture is tailored to address the inherent security and performance challenges associated with highly distributed and resource-constrained environments.

At its core, the system integrates a packet authentication framework that utilizes advanced cryptographic techniques to verify the origin and integrity of every data packet. These mechanisms actively detect and prevent tampering during transmission across heterogeneous 3D networks, thereby maintaining data authenticity and trustworthiness. To further reduce the impact of potential breaches, the architecture incorporates Least Privilege Access Control (LPAC) alongside micro-segmentation. Access to network resources is tightly restricted to essential roles, while the network is divided into isolated zones based on function, limiting the lateral movement of threats and improving containment strategies during an attack.

Recognizing the limited processing capabilities of IoT devices, the system employs lightweight security solutions that offer a practical balance between performance and protection. Secure, encrypted communication channels ensure confidentiality of data in transit, and all interactions within the network are continuously monitored for anomalies and potential violations. Furthermore, mutual authentication protocols ensure that both the network and user devices verify each other's identities before any exchange, effectively preventing unauthorized access and impersonation.

Scalability is another key design objective. The system is capable of supporting a growing number of IoT devices and users without compromising its efficiency or security posture. To facilitate streamlined encryption and key distribution, a centralized Key Aggregator is deployed, enabling efficient key management across a logically centralized yet geographically distributed architecture.

The proposed system thus provides a comprehensive framework that addresses critical security and design goals. It ensures data integrity, confidentiality, access control, and authentication while maintaining resilience, scalability, and efficiency. Through this architecture, we aim to establish a secure and adaptive foundation for next-generation communication in 3D wireless environments.

4 Proposed System

In this section, we provide a concise overview of the proposed network architecture and packet transmission. We offer insight into the network structure, emphasizing the roles of its components and the overarching framework facilitating communication. Following this, in Sect. 4.2, we explore data packet exchange for user authentication mechanisms. This step is crucial for verifying user identity and ensuring secure and efficient communication between endpoints.

4.1 Proposed Network Architecture

Figure 3 shows our network architecture designed to ensure secure, reliable connectivity across 3D wireless layers. User Equipment (UE) such as smartphones and IoT devices connect via Access Points (APs) to the satellite network.

A central Key Aggregator manages encryption keys and supports authentication by coordinating with Sub Network Control Centers (SubNCCs), which handle specific services like Starlink or GPS. Network Control Centers (NCCs) oversee satellite constellations at various altitudes LAPS, HAPS, LEO, MEO, and GEO with SubNCCs providing finer control within each layer.

Although geographically distributed, NCCs, SubNCCs, and Key Aggregators operate as logically centralized controllers to ensure scalability and resilience. Gateways link satellite and terrestrial networks, enabling seamless data exchange and interoperability.

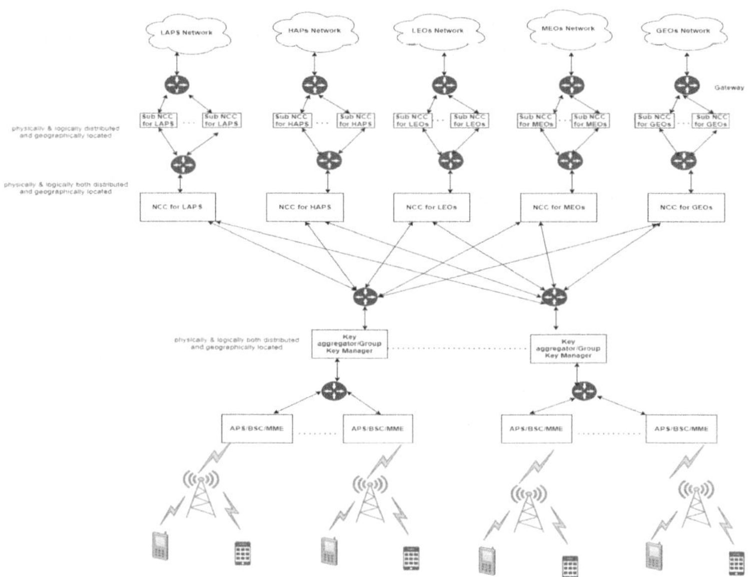

Fig. 3. Network Architecture

This hierarchical structure supports diverse services while enabling efficient management and secure communication throughout the 3D network.

4.2 Proposed Authentication Mechanism

Our authentication process ensures secure, reliable access within 3D wireless networks. When a user device requests a service (e.g., Starlink), the request passes through nearby Access Points to the Key Aggregator (Fig. 4(a)).

The Key Aggregator coordinates with Network Control Centers (NCCs) and SubNCCs to verify the user's identity and generate a unique session key (Figs. 4(b) and (c)). This key establishes an encrypted communication channel, enabling secure access.

Mutual authentication between the user device and SubNCC confirms trust on both ends (Fig. 4(d)). The protocol includes initialization, user registration, mutual authentication and key exchange (MAKE), password update, and session key renewal. Secure channels handle initialization and registration, while public channels manage MAKE and key updates.

Public cryptographic functions h(.) and $E_{\text{key}}(.)/D_{\text{key}}(.)$ are used as defined in Table 1.

Initialization Phase. During this phase, an initial roaming agreement is carried out between SUBNCC(a) and each of its associated NCC, NCC and each of its associated KA by exchanging their identities and other parameters over a secure channel as follows:

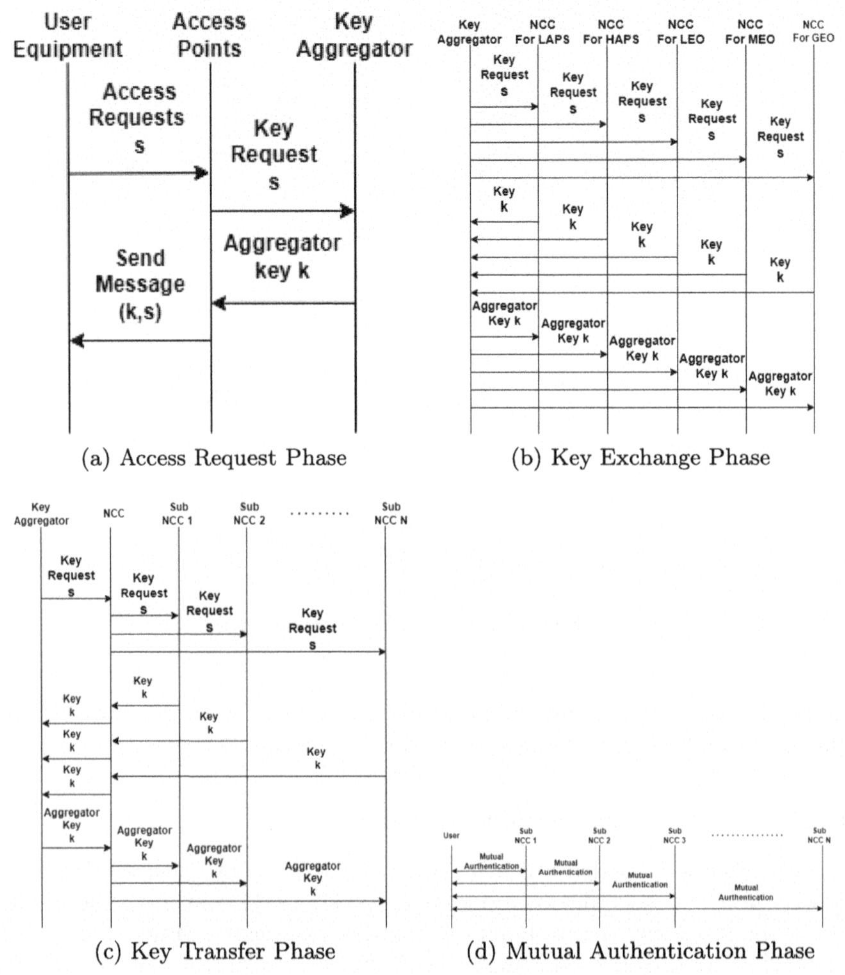

(a) Access Request Phase

(b) Key Exchange Phase

(c) Key Transfer Phase

(d) Mutual Authentication Phase

Fig. 4. Phases of the Proposed Authentication Protocol

SUBNCC(a) and each of its associated NCC as follows:

Step 1: NCC submits its identity ID_{ncc} to SUBNCC(a).

Step 2: SUBNCC(a) selects a random integer N_{ncc} and computes the pseudonym:

$RID_{\text{ncc}} = E_{k_{\text{ncca}}}(ID_{\text{ncc}} \parallel ID_a \parallel N_{\text{ncc}})$ for NCC. Then, SUBNCC(a) sends $\{ID_a, RID_{\text{ncc}}, N_{\text{ncc}}\}$ to NCC. Subsequently, SUBNCC(a) stores the record: $h(ID_{\text{ncc}} \parallel K_a)$ into a list of all associated NCCs.

Step 3: Finally, NCC computes: $K_{\text{ncca}} = h(ID_{\text{ncc}} \parallel N_{\text{ncc}})$ and saves the tuple $\{ID_a, RID_{\text{ncc}}, K_{\text{ncca}}\}$ into its database, as shown in Fig. 9 (Figs. 5 and 6).

NCC and each of its associated KA as follows:

Step 1: KA submits its identity ID_{ka} to NCC.

Table 1. Notations used

Notation	Description
$\mathcal{A}$	Adversary
ID_{ue}	Identity of UE
ID_{ka}	Identity of KA
ID_{ncc}	Identity of NCC
ID_a	Identity of SUBNCC(a)
PW_{uea}	Password of UE
TID_{ia}	Dynamic pseudonym of UE, $0 \leq i \leq n$
FID_{ka}	Dynamic pseudonym of KA
RID_{ncc}	Dynamic pseudonym of NCC
R_{ia}	Random number, $0 \leq i \leq n$
$N_a, N_{1a}, N_{2a}, N_{fa}$	Random integers
$R_{ia}^*, N_{ka}^*, N_{ncc}^*, N_1^*, N_2^*$	Random integers
k_a	Built-in secret key of SUBNCC(a)
PID_{ue}	Registration verifier of UE
V_a	Login verifier of UE
V_{ja}	Authentication verifier, $j = 1, 2, ..., 8$
K_{uea}	Dynamic key of MU
K_{kancc}	Dynamic key of KA
K_{ncca}	Dynamic key of NCC
SK	Session key
T_{mk}	Current timestamps ,$K = 1, 2, ..., 8$
Δ_{m1}	Valid time interval to receive a message from UE
Δ_{m2}	Valid time interval to receive a message from KA
$h(.)$	Collision-resistant one-way hash function
$E_{\text{key}}(.)/D_{\text{key}}(.)$	Symmetric Encryption/Decryption operation with key "key"
$\oplus$	Bit-wise Ex-OR operation
$\|$	Concatenation operation
$\text{Adv}_{\mathcal{A}}^{\text{INDCCA}}(t)$	Advantage of $\mathcal{A}$ in breaking INDCCA security in time t
$\text{Adv}_{\mathcal{A}}^{H}(t)$	Advantage of $\mathcal{A}$ in finding a hash collision in time t
$\text{Adv}_{\mathcal{A}}^{\text{MAKE}}(t)$	Advantage of $\mathcal{A}$ in breaking semantic security of SK in time t

Step 2: NCC selects a random integer N_{ka} and computes the pseudonym: $FID_{\text{ka}} = E_{k_{\text{kancc}}}(ID_{\text{ka}} \parallel ID_{\text{ncc}} \parallel N_{\text{ka}})$ for KA. Then, NCC sends $\{ID_{\text{ncc}}, FID_{\text{ka}}, N_{\text{ka}}\}$ to KA. Subsequently, NCC stores the record:$h(ID_{\text{ka}} \parallel K_{\text{ncc}})$ into a list of all associated KAs.

Step 3: Finally, KA computes: $K_{\text{kancc}} = h(ID_{\text{ka}} \parallel N_{\text{ka}})$ and saves the tuple $\{ID_{\text{ncc}}, FID_{\text{ka}}, K_{\text{kancc}}\}$ into its database, as shown in Fig. 8.

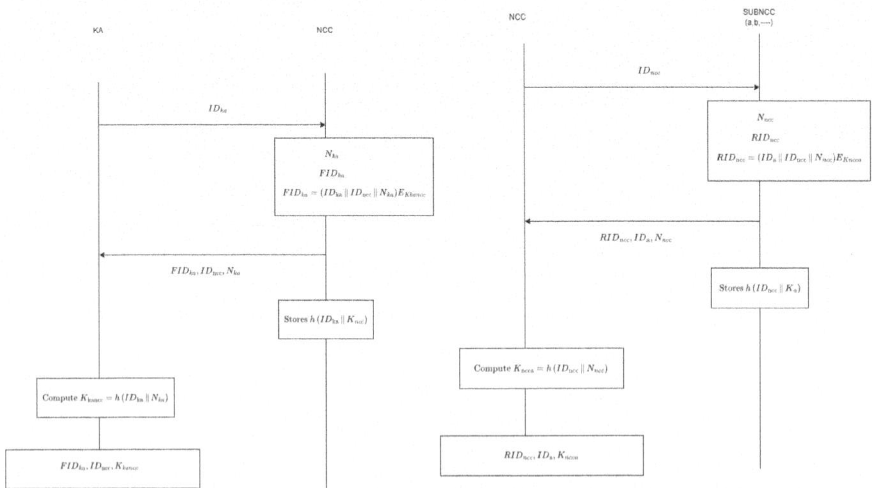

Fig. 5. Initialization between KA and NCC

Fig. 6. Initialization between NCC and SUBNCC

User Registration Phase. Before accessing the desired services, each UE must go through a registration process with SUBNCC(a). During this process, UE submits its registration details to SUBNCC(a) and, in return, obtains an SC via a secure channel as follows:

Step 1: Initially, UE generates a random integer N_a and computes its registration verifier: $PID_{\text{ue}} = h(ID_{\text{ue}} \parallel N_a)$ Then, UE submits the registration request $\{ID_{\text{ue}}, PID_{\text{ue}}\}$ to SUBNCC(a).

Step 2: SUBNCC(a) generates a set of (TID_{ia}, R_{ia}) pairs, where each pseudonym is computed as: $TID_{ia} = E_{k_{\text{uea}}}(ID_{\text{ue}} \parallel ID_a \parallel R_{ia})$, such that $0 < i < n$ and the random integers satisfy $R_0 \neq R_1 \neq \cdots \neq R_{n-1}$. Then, SUBNCC($a$) assigns an SC to UE consisting of the parameters: $ID_a, (TID_{ia}, R_{ia})$. Finally, SUBNCC(a) inserts the record $\{h(ID_{\text{ue}} \parallel K_a), PID_{\text{ue}}\}$ into a list of all registered UEs.

Step 3: Upon receipt, UE selects its password PW_{uea} and computes (Fig. 7):

$$N_a^* = N_a \oplus h\big(ID_{\text{ue}} \parallel PW_{\text{uea}}\big),$$
$$V_a = h\big(ID_{\text{ue}} \parallel ID_a \parallel N_a\big).$$

Subsequently, UE inserts the parameters $\{N_a^*, V_a\}$ into its SC. Finally, SC consists of the parameters:
$\{ID_a, (TID_{ia}, R_{ia}), N_a^*, V_a\}$. The structure is illustrated in Fig. 10.

MAKE Phase. In this phase, MU places its SC into a suitable UE and provides $\{ID_{\text{ue}}, PW_{\text{uea}}\}$ as inputs to initiate the MAKE phase. On successful execution, a fresh and unique session key is established between UE and KA to

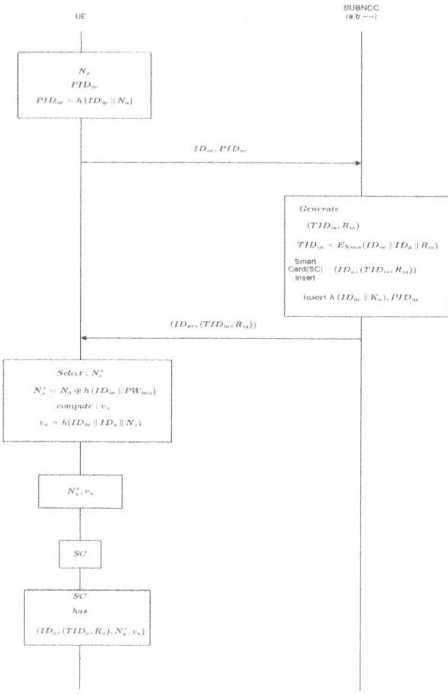

Fig. 7. Registration between UE and SUBNCC

encrypt/decrypt further messages. The entire process is demonstrated in Fig. 11 and explained as follows:

Step 1: SC evaluates: $N_{\mathrm{a}} = N_{\mathrm{a}}^{*} \oplus h(ID_{\mathrm{ue}} \parallel PW_{\mathrm{uea}})$ based on the input parameters $\{ID_{\mathrm{ue}}, PW_{\mathrm{uea}}\}$ and verifies whether: $V_{\mathrm{a}} = h(ID_{\mathrm{ue}} \parallel ID_{\mathrm{a}} \parallel N_{\mathrm{a}})$ holds. This check verifies the authenticity of UE locally. If there is any typo in the input credentials, SC can detect it while verifying V_{a} and take appropriate actions (e.g., prompting for re-entry or blocking after exceeding the maximum limit). Upon successful verification, SC randomly selects a pair $(TID_{\mathrm{ia}}, R_{\mathrm{ia}})$ from the set obtained from SUBNCC(a). SC then generates a random integer $N_{1\mathrm{a}}$ and computes:

$$\begin{aligned} PID_{ue} &= h\big(ID_{ue} \parallel N_{a}\big), \\ K_{uea} &= h\big(ID_{ue} \parallel R_{ia}\big), \\ N_{xa} &= K_{uea} \oplus N_{1a}, \\ V_{1a} &= h\big(PID_{ue} \parallel K_{uea} \parallel N_{1a} \parallel T_{m1}\big). \end{aligned}$$

Finally, SC sends the service access request: $M_1 = \{TID_{\mathrm{ia}}, N_{x\mathrm{a}}, V_{1\mathrm{a}}, ID_{\mathrm{a}}, Tm_1\}$ to KA.

Step 2: KA generates a random integer $N_{2\mathrm{a}}$ and computes: $N_{y\mathrm{a}} = K_{\mathrm{kaa}} \oplus N_{2\mathrm{a}}$ $V_{2\mathrm{a}} = h(ID_{\mathrm{ka}} \parallel K_{\mathrm{kancc}} \parallel N_{2\mathrm{a}} \parallel Tm_2)$ Then, KA sends the authentication request:
$M_2 = \{TID_{\mathrm{ia}}, FID_{\mathrm{ka}}, N_{x\mathrm{a}}, N_{y\mathrm{a}}, V_{1\mathrm{a}}, V_{2\mathrm{a}}, Tm_{\mathrm{a}}, Tm_2\}$ to NCC.
Authentication and Verification Process:

Step 3: NCC verifies Tm_2 and retrieves $\{ID_{\text{ka}}, ID_{\text{ncc}}, N_{\text{ka}}\}$ by decrypting the pseudonym FID_{ka} using its secret key K_{ncc}. NCC then checks whether the entries $h(ID_{\text{ka}} \parallel k_{\text{ncc}})$ exist in the corresponding lists. It verifies V_{2a}, generates a random integer N_{3a}, and computes:

$$N_{y\,ncc} = K_{ncc-a} \oplus N_{3a},$$
$$V_{3a} = h\big(ID_{ncc} \parallel K_{ncc-a} \parallel N_{3a} \parallel Tm_3\big).$$

Then, NCC sends the authentication request:

$$M_3 = \{TID_{ia}, RID_{\text{ncc}}, N_{xa}, N_{y_{\text{ncc-a}}}, V_{1a}, V_{3a}, Tm_1, Tm_3\}$$

to $SUBNCC(a)$.

Step 4: SUBNCC(a) verifies whether: $(Tm_4 - Tm_1) \leq \delta_{ue}$, $(Tm_4 - Tm_3) \leq \delta_{\text{ncc}}$. If any verification fails, $SUBNCC(a)$ assumes that the received message is a replay. Otherwise, it retrieves $\{ID_{\text{ue}}, ID_a, R_{ia}\}$ and $\{ID_{\text{ncc}}, ID_a, N_a\}$ by decrypting TID_{ia} and RID_{ncc} using its secret key K_a. $SUBNCC(a)$ then checks whether $h(ID_{\text{ue}} \parallel K_a)$ and $h(ID_{\text{ue}} \parallel R_{ia})$ exist in the corresponding lists. If found, it evaluates:

$$K_{ua} = h\big(ID_{ue} \parallel R_{ia}\big)$$
$$K_{ncc-a} = h\big(ID_{ncc} \parallel N_{xa}\big)$$
$$N_{1a} = K_{ua} \oplus N_{xa}$$
$$N_{3a} = K_{ncc-a} \oplus N_{yncc}$$

$SUBNCC(a)$ retrieves PID_{ue} and verifies:

$$V_{1a} = h\big(PID_{uea} \parallel K_{uea} \parallel N_{1a} \parallel Tm_1\big),$$
$$V_{3a} = h\big(ID_{ncc} \parallel K_{ncca} \parallel N_{3a} \parallel Tm_3\big).$$

Once verified, $SUBNCC(a)$ generates two new random integers R_{ia} and N^*_{ncc} and computes:

$$N_{xya} = h\big(K_{uea}\big) \oplus N_{3a},$$
$$N_{yxncc} = h\big(K_{ncca}\big) \oplus N_{1a},$$
$$TID^*_{ia} = E_{K_{uea}}\big(ID_{ue} \parallel ID_a \parallel R^*_{ia}\big) \oplus h\big(K_{uea}\big)$$
$$RID^* = E_{K_{ncca}}\big(ID_a \parallel ID_{ncc} \parallel N^*_{ncc}\big) \oplus h\big(K_{ncc\text{-}a}\big)$$
$$N_{xx_a} = h\big(K_{uea}\big) \oplus R^*_{ia}$$
$$N_{yy} = h\big(K_{ncca}\big) \oplus N^*_{ncc}$$

Finally, $SUBNCC(a)$ generates verifiers:

$$V_{4a} = h\big(TID^*_{ia} \parallel N_{1a} \parallel R^*_{ia} \parallel Tm_4\big),$$
$$V_{5a} = h\big(RID^* \parallel N_{1a} \parallel N_{3a} \parallel N^*_{ncc} \parallel Tm_4\big).$$

It then sends the authentication response:
$M_4 = \{TID^*_{ia}, RID^*, N_{xy_a}, N_{yx_{\text{ncc}}}, N_{xx_a}, N_{yy_{\text{ncc}}}, V_{4a}, V_{5a}, Tm_4\}$ to NCC.

Security Considerations: Any unauthorized attempt to modify timestamps will require computing valid verifiers:

$$V_{1a} = h\big(PID_{uea} \parallel K_{uea} \parallel N_{1a} \parallel Tm_1\big)$$
$$V_{3a} = h\big(ID_{ncc} \parallel K_{ncca} \parallel N_{3a} \parallel Tm_3\big).$$

These computations are infeasible due to the use of secret parameters $\{PID_{\text{ue}}, K_{ue-a}, N_{1a}, ID_{\text{ncc}}, K_{\text{ncc-a}}, N_{3a}\}$ in generating V_{1a} and V_{3a}. Since these parameters cannot be extracted from publicly available parameters V_{1a} and V_{3a} due to hash-based security, tampering with timestamps Tm_1 and Tm_3 would be detected by $SUBNCC(a)$ while verifying the values. Even if NCC tries to modify UE's timestamp Tm_1, the use of secret parameters $\{PID_{\text{ue}}, K_{uea}, N_{1a}\}$ in computing V_{1a} makes it infeasible.

NCC Authentication Response:

Step 5: NCC evaluates the parameters:$N_{1a} = h(K_{\text{ncca}}) \oplus N_{\text{yxncc}}$, $N^*_{\text{ncc}} = h(K_{\text{ncca}}) \oplus N_{\text{yya}}$. Then, NCC verifies whether: $V_{5a} = h(RID^* \parallel N_{1a} \parallel N_{3a} \parallel N^*_{\text{ncc}} \parallel T_{m4})$ for SUBNCC. Now, NCC sends the authentication response:
$M_{5a} = \{TID^*_{ia}, FID^*_{\text{ncc}}, N_{\text{xya}}, N_{\text{yx-ka}}, N_{\text{xx-a}}, N_{\text{yy-ncc}}, V_{6a}, V_{4a}, T_{m5}\}$ to KA. NCC again evaluates:

$$N_{1a} = h\big(K_{ncca}\big) \oplus N_{yxncc}$$
$$N^*_{ncc} = h\big(K_{ncca}\big) \oplus N_{yncc}$$

Then, NCC verifies: $V_{5a} = h(RID^* \parallel N_{1a} \parallel N_{3a} \parallel N^*_{\text{ncc}} \parallel T_{m4})$ and ensures T_{m4} holds to confirm the authenticity of SUBNCC(a). Once verified, NCC generates two new random integers FID^*_{NCC} and N^*_{ka} and computes:

$$N_{yxka} = h\big(K_{kancc}\big) \oplus N_{1a}$$
$$FID^*_{ncc} = E_{K_{kancc}}\big(ID_{ka} \parallel ID_{ncc} \parallel N^*_{ka}\big) \oplus h\big(K_{kancc}\big),$$
$$N_{yyncc} = h\big(K_{kancc}\big) \oplus N^*_{ka}.$$

Subsequently, NCC generates the verifier: $V_{6a} = h(FID^* \parallel N_{1a} \parallel N_{2a} \parallel N^*_{\text{ka}} \parallel T_{m5}) \parallel)$ for KA. Finally, NCC sends the authentication response: $M_5 = \{TID^*_{ia}, FID^*_a, N_{\text{xya}}, N_{\text{yxka}}, N_{\text{xxa}}, N_{\text{yyncc}}, V_{4a}, V_{6a}, T_{m5}\}$ to KA.

Evaluation and Verification by KA:

Step 6: KA evaluates the parameters:

$N_{1a} = h(K_{\text{kancc}}) \oplus N_{\text{yxka}}$ $N^*_{\text{ka}} = h(K_{\text{kancc}}) \oplus N_{\text{yyncc}}$Then, KA verifies whether the following holds: $V_{6a} = h(FID^* \parallel N_{1a} \parallel N_{2a} \parallel N^*_{\text{ka}} \parallel T_{m5})$ If the verification is successful, KA computes the session key: $SK = h(N_{1a})$ and: $V_{7a} = h(SK \parallel N_{1a} \parallel T_{m6})$ Subsequently, KA updates its stored parameters as:

$$FID_{ncc} = FID^*_{ncc} \oplus h\big(K_{kancc}\big),$$
$$K_{kancc} = h\big(ID_{ka} \parallel N^*_{ka}\big).$$

Finally, KA sends the access response: $M_6 = \{TID^*_{\text{ia}}, N_{\text{xya}}, N_{\text{xx-a}}, V_{7a}, V_{4a}, T_{m6}\}$ to MU and sends the session key encrypted with the shared secret key to NCC.

SC Evaluation and Session Key Computation:

Step 7: SC evaluates the parameter: $R^*_{ia} = h(K_{uea}) \oplus N_{xxa}$ Then, SC verifies whether: $V_{4a} = h(TID^*_{ia} \parallel N1a \parallel R^*_{ia} \parallel T_{m4})$ holds. On successful verification, SC computes the session key: $SK = h(N_{1a})$ Then, SC checks whether: $V_{7a} = h(SK \parallel N_{1a} \parallel T_{m6})$ and if T_{m6} holds in order to decide the acceptance of SK. Subsequently, SC replaces the selected pair (TID_{ia}, R_{ia}) with a newly generated pair $(TID_{\text{new}ia}, R_{\text{inew}*a})$.

Session Key Handling

Step 8: On the other hand, after receiving the session key encrypted with the shared secret key from KA, NCC decrypts it using the shared secret key between KA and NCC and learns the session key. NCC then sends the session key, encrypted with the shared secret key, to SUBNCC(a). After receiving the session key, encrypted with the shared secret key, from NCC, SUBNCC(a) decrypts it with the shared secret key between NCC and SUBNCC(a) and learns the session key.

Mutual Authentication and Session Key Agreement

Step 9: Now, SUBNCC(a) generates a random number N with timestamp T_{m9}, encrypted with the session key, and forwards it to UE via NCC and KA. Then, UE decrypts it with the session key it computed. UE acknowledges the message N^{-1}, T_{m9} encrypted with the session key SK to SUBNCC(a) via KA and NCC. SUBNCC(a) decrypts it upon successful verification. Thus, mutual authentication among all participants involved in the communication process (UE, KA, NCC, and SUBNCC(a)) has been preserved. Subsequently, both UE and KA agree on a fresh and unique session key SK after authenticating each other. Dynamic updates of the pseudonyms $\{TID_i, FID_{\text{ncc}}, RID_a\}$ and the shared keys $\{K_{\text{uea}}, K_{\text{kancc}}, K_{\text{ncc}}\}$ are also preserved upon successful completion of the entire MAKE phase.

Password Changing Phase. It is important to change the password of UE after a certain interval of time due to security reasons [13,22]. In our approach, UE can independently change its current password PW_{uea} (whenever required) as follows:

Step 1: Firstly, MU places its SC into the UE, and inputs ID_{ue}, PW_{uea}, and the newly chosen password $PW_{\text{uea-new}}$.

Step 2: Subsequently, SC evaluates the verifier and $N_a = N^*_a \oplus h(ID_{\text{ue}} \parallel PW_{\text{uea}})$, and verifies whether: $V_a = h(ID_{\text{ue}} \parallel ID_a \parallel N_a)$ holds. SC accepts $PW_{\text{uea-new}}$ if the verification remains successful. Finally, SC updates its existing parameter as: $N_a = N^*_a \oplus h(ID_{\text{ue}} \parallel PW_{\text{uea-new}})$

Session Key Update Phase. The current session key SK needs to be updated periodically whenever UE and KA participate in a long session [45]. This is crucial in preventing disclosure of the current SK. Our proposed protocol enables

UE and KA to mutually update their current SK and establish a new session key SK_{new} during an ongoing session as follows:

Step 1: MU generates a fresh random number N_1^*, and computes: $X = N_1^* \oplus h(SK)$, $V_{8a} = h(SK \parallel N_1^*)$ Subsequently, UE sends the session key update request $\{X, V_{8a}\}$ to KA.

Step 2: KA evaluates: $N_1^* = X \oplus h(SK)$, Verify: $V_{8a} = h(SK \parallel N_1^*)$ If verified, KA generates a fresh random number N_2^*, and computes the new session key: $SK_{\text{new}} = h(N_1^* \parallel N_2^*)$, $Y = N^2 \oplus h(SK)$, $V_{9a} = h(SK_{\text{new}} \parallel N_1^* \parallel N_2^*)$ Finally, KA sends the session key update response $\{Y, V_{9a}\}$ to UE.

Step 3: MU computes: $N^2 = Y \oplus h(SK)$ $SK_{\text{new}} = h(N_1^* \parallel N_2^*)$ Finally, UE accepts SK_{new} if: $V_{9a} = h(SK_{\text{new}} \parallel N_1^* \parallel N_2^*)$

5 AVISPA Simulation

The AVISPA tool automates security analysis of internet protocols, detecting vulnerabilities such as replay, impersonation, and man-in-the-middle attacks [3,32]. Protocols are modeled in HLPSL and verified using back-ends like OFMC and CL-AtSe [33,35], with intruders modeled under the Dolev-Yao threat model

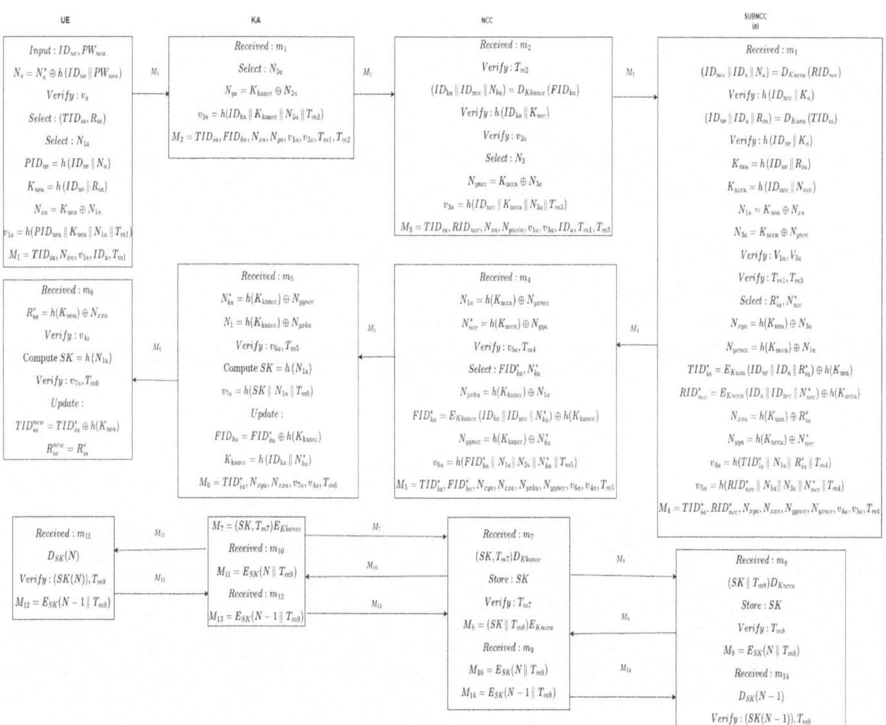

Fig. 8. Protocol

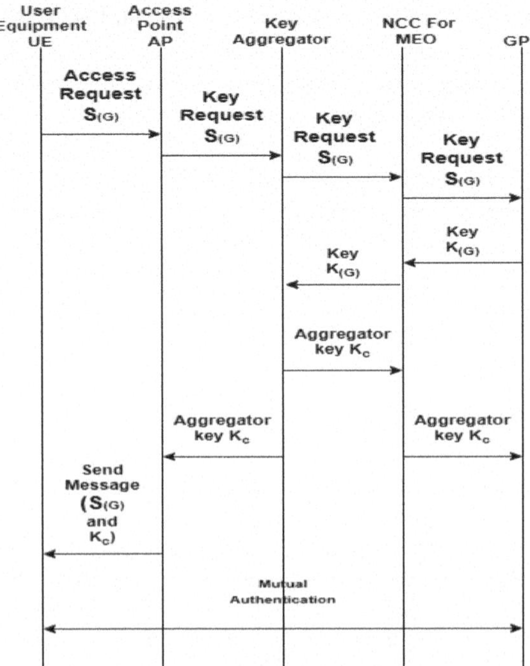

Fig. 9. Network Senario

[9]. Simulations check secrecy (parameter confidentiality) and authentication goals; violations indicate insecurity, while misconfigurations yield inconclusive results [32].

Our model includes four roles UE, Key Aggregator (KA), NCC, and SubNCC operating in two phases: registration (UE and SubNCC over secure channels using pre-shared symmetric keys) and mutual authentication and key exchange (MAKE) over insecure Dolev-Yao channels. The intruder can impersonate any participant and controls all public communications.

Secrecy goals cover key identifiers and nonces (e.g., $ID_{ue}, ID_{ka}, ID_{ncc}, N_a, N_{1a}, N_{2a}, N_{3a}$). Authentication goals verify mutual trust among UE, KA, NCC, and SubNCC during protocol exchanges.

Our HLPSL implementation accurately models these goals, confirming protocol correctness in AVISPA. Insider and stolen smart card attacks are excluded, as they occur outside public channels.

Figures 10 and 11 present simulation results using OFMC and CL-AtSe on a 32-bit Ubuntu system, showing the protocol satisfies all secrecy and authentication goals under the Dolev-Yao model.

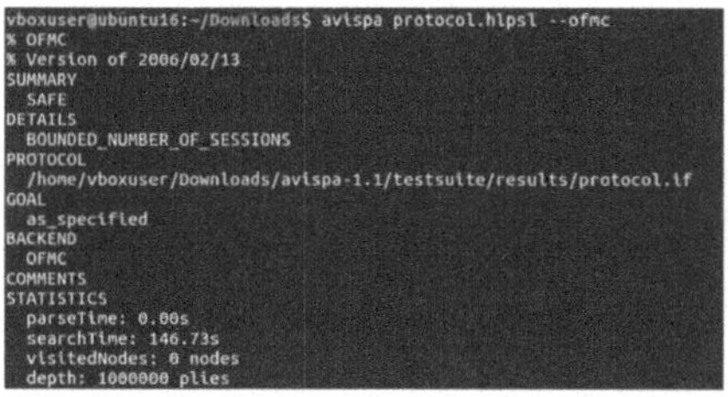

Fig. 10. AVISPA simulation result with OFMC

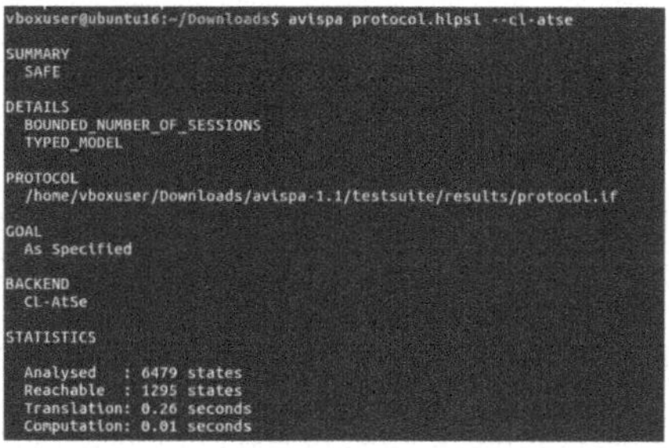

Fig. 11. AVISPA simulation result with CL-AtSe

6 Conclusion

This study presents a secure and efficient authentication protocol that ensures user privacy and mitigates threats such as replay and impersonation attacks. Formal verification using AVISPA confirms its robustness, while performance evaluations demonstrate minimal computational overhead, rendering it suitable for resource-constrained environments. Future research will focus on integration with emerging technologies and enhancement of security features to address evolving threats. This work contributes to the advancement of reliable authentication mechanisms in modern networked systems.

Disclosure of Interests. The authors have no competing interests to declare that are relevant to the content of this article.

References

1. Adams, W., Rider, L.: Circular polar constellations providing continuous single or multiple coverage above a specified latitude. J. Astronaut. Sci. **35**, 155–192 (1987)
2. Ali, A., Ali, A.: Securing space-based satellite networks: challenges and solutions. Technical report, EasyChair (2024)
3. Armando, A., et al.: The AVISPA tool for the automated validation of internet security protocols and applications. In: Etessami, K., Rajamani, S.K. (eds.) CAV 2005. LNCS, vol. 3576, pp. 281–285. Springer, Heidelberg (2005). https://doi.org/10.1007/11513988_27
4. Atzori, L., Iera, A., Morabito, G.: The internet of things: a survey. Comput. Netw. **54**(15), 2787–2805 (2010)
5. Chaoub, A., et al.: 6G for bridging the digital divide: wireless connectivity to remote areas. IEEE Wirel. Commun. **29**(1), 160–168 (2021)
6. Chen, T.H., Lee, W.B., Chen, H.B.: A self-verification authentication mechanism for mobile satellite communication systems. Comput. Electr. Eng. **35**(1), 41–48 (2009)
7. Chini, P., Giambene, G., Kota, S.: A survey on mobile satellite systems. Int. J. Satell. Commun. Netw. **28**(1), 29–57 (2010)

8. Coyle, A., et al.: The rise of the global internet satellite mega-constellation sector: opportunities and challenges for a sustainable transition. Master's thesis (2021)
9. Dolev, D., Yao, A.: On the security of public key protocols. IEEE Trans. Inf. Theor. **29**(2), 198–208 (1983)
10. Gaur, K., Kalla, A., Grover, J., Borhani, M., Gurtov, A., Liyanage, M.: A survey of virtual private LAN services (VPLS): past, present and future. Comput. Netw. **196**, 108245 (2021)
11. Geraci, G., López-Pérez, D., Benzaghta, M., Chatzinotas, S.: Integrating terrestrial and non-terrestrial networks: 3D opportunities and challenges. IEEE Commun. Mag. **61**(4), 42–48 (2022)
12. Giordani, M., Polese, M., Mezzavilla, M., Rangan, S., Zorzi, M.: Toward 6G networks: use cases and technologies. IEEE Commun. Mag. **58**(3), 55–61 (2020)
13. Gupta, M., Chaudhari, N.S.: Anonymous two factor authentication protocol for roaming service in global mobility network with security beyond traditional limit. Ad Hoc Netw. **84**, 56–67 (2019)
14. Huang, Y., Feng, B., Dong, P., Tian, A., Yu, S.: A multi-objective based inter-layer link allocation scheme for MEO/LEO satellite networks. In: 2022 IEEE Wireless Communications and Networking Conference (WCNC), pp. 1301–1306. IEEE (2022)
15. Kallatsa, M.: Strategies for network segmentation: a systematic literature review (2024)
16. Kang, H., Liu, G., Wang, Q., Meng, L., Liu, J.: Theory and application of zero trust security: a brief survey. Entropy **25**(12), 1595 (2023)
17. Kurt, G.K., et al.: A vision and framework for the high altitude platform station (HAPS) networks of the future. IEEE Commun. Surv. Tut. **23**(2), 729–779 (2021)
18. Lai, C., Chen, Z.: Group-based handover authentication for space-air-ground integrated vehicular networks. In: IEEE International Conference on Communications, ICC 2021, pp. 1–6. IEEE (2021)
19. Lasc, I., Dojen, R., Coffey, T.: Countering jamming attacks against an authentication and key agreement protocol for mobile satellite communications. Comput. Electr. Eng. **37**(2), 160–168 (2011)
20. Lee, C.C., Li, C.T., Chang, R.X.: A simple and efficient authentication scheme for mobile satellite communication systems. Int. J. Satell. Commun. Netw. **30**(1), 29–38 (2012)
21. Liu, R., Sheng, M., Lui, K.S., Wang, X., Zhou, D., Wang, Y.: Capacity analysis of two-layered LEO/MEO satellite networks. In: 2015 IEEE 81st Vehicular Technology Conference (VTC Spring), pp. 1–5. IEEE (2015)
22. Lu, Y., Xu, G., Li, L., Yang, Y.: Robust privacy-preserving mutual authenticated key agreement scheme in roaming service for global mobility networks. IEEE Syst. J. **13**(2), 1454–1465 (2019)
23. Ma, R., Cao, J., Feng, D., Li, H.: LAA: lattice-based access authentication scheme for IoT in space information networks. IEEE Internet Things J. **7**(4), 2791–2805 (2019)
24. Manulis, M., Bridges, C.P., Harrison, R., Sekar, V., Davis, A.: Cyber security in new space: analysis of threats, key enabling technologies and challenges. Int. J. Inf. Secur. **20**, 287–311 (2021)
25. Meng, W., Xue, K., Xu, J., Hong, J., Yu, N.: Low-latency authentication against satellite compromising for space information network. In: 2018 IEEE 15th International Conference on Mobile Ad Hoc and Sensor Systems (MASS), pp. 237–244. IEEE (2018)

26. Mohammed, A., Mehmood, A., Pavlidou, F.N., Mohorcic, M.: The role of high-altitude platforms (HAPS) in the global wireless connectivity. Proc. IEEE **99**(11), 1939–1953 (2011)
27. Qi, M., Chen, J.: An enhanced authentication with key agreement scheme for satellite communication systems. Int. J. Satell. Commun. Netw. **36**(3), 296–304 (2018)
28. Qiu, J., Grace, D., Ding, G., Zakaria, M.D., Wu, Q.: Air-ground heterogeneous networks for 5G and beyond via integrating high and low altitude platforms. IEEE Wirel. Commun. **26**(6), 140–148 (2019)
29. Singh Uppal, A.: Multi-factor authentication in network security (2021)
30. Srivastava, A., Jain, U.: Securing the future of IoT: a comprehensive framework for real-time attack detection and mitigation in IoT networks. In: 2023 14th International Conference on Computing Communication and Networking Technologies (ICCCNT), pp. 1–6. IEEE (2023)
31. Strinati, E.C., et al.: 6G in the sky: on-demand intelligence at the edge of 3D networks. arXiv preprint arXiv:2010.09463 (2020)
32. The AVISP Team: AVISPA v1.1 user manual. Information Society Technology Program, June 2006. **62**, 112 (2006). http://avispa-project.org
33. Turuani, M.: The CL-ATse protocol analyser. In: Pfenning, F. (ed.) International Conference on Rewriting Techniques and Applications. LNCS, vol. 4098, pp. 277–286. Springer, Heidelberg (2006). https://doi.org/10.1007/11805618_21
34. Vermesan, O., Friess, P.: Internet of Things Applications-from Research and Innovation to Market Deployment. Taylor & Francis (2014)
35. Von Oheimb, D.: The high-level protocol specification language HLPSL developed in the EU project AVISPA. In: Proceedings of APPSEM 2005 Workshop, APPSEM'05, Tallinn, Estonia, pp. 1–17 (2005)
36. Xue, K., Meng, W., Li, S., Wei, D.S., Zhou, H., Yu, N.: A secure and efficient access and handover authentication protocol for internet of things in space information networks. IEEE Internet Things J. **6**(3), 5485–5499 (2019)
37. Yang, Q., Xue, K., Xu, J., Wang, J., Li, F., Yu, N.: AnFRA: anonymous and fast roaming authentication for space information network. IEEE Trans. Inf. Forensics Secur. **14**(2), 486–497 (2018)
38. Yi, Z., Du, X., Liao, Y., Lu, X.: An access authentication algorithm based on a hierarchical identity-based signature over lattice for the space-ground integrated network. In: 2019 International Conference on Advanced Communication Technologies and Networking (CommNet), pp. 1–9. IEEE (2019)
39. Yoon, E.J., Yoo, K.Y., Hong, J.W., Yoon, S.Y., Park, D.I., Choi, M.J.: An efficient and secure anonymous authentication scheme for mobile satellite communication systems. EURASIP J. Wirel. Commun. Netw. **2011**, 1–10 (2011)
40. Zhang, Y., Wu, Q., Lai, Z., Li, H.: Enabling low-latency-capable satellite-ground topology for emerging LEO satellite networks. In: IEEE Conference on Computer Communications, IEEE INFOCOM 2022, pp. 1329–1338. IEEE (2022)
41. Zhang, Y., Deng, R.H., Bertino, E., Zheng, D.: Robust and universal seamless handover authentication in 5G HetNets. IEEE Trans. Dependable Secure Comput. **18**(2), 858–874 (2019)
42. Zhang, Y., Chen, J., Huang, B.: Security analysis of an authentication and key agreement protocol for satellite communications. Int. J. Commun Syst **27**(12), 4300–4306 (2014)
43. Zhang, Y., Chen, J., Huang, B.: An improved authentication scheme for mobile satellite communication systems. Int. J. Satell. Commun. Netw. **33**(2), 135–146 (2015)

44. Zhao, B., Liu, P., Wang, X., You, I.: Toward efficient authentication for space-air-ground integrated internet of things. Int. J. Distrib. Sens. Netw. **15**(7), 1550147719860390 (2019)
45. Zhao, D., Peng, H., Li, L., Yang, Y.: A secure and effective anonymous authentication scheme for roaming service in global mobility networks. Wirel. Pers. Commun. **78**, 247–269 (2014)

STRIKE: A Structured Taxonomy of Cybercrime for Risk, Impact, Knowledge, and Evolution

Melissa Pappy, Linh Nguyen, Suman Kumar(✉), Byungkwan Jung, and Bernard Chen

Troy University, Troy, AL 36082, USA
{mvenable150308,lnguyen93238,skumar,bjung,chenb}@troy.edu

Abstract. Cybercrime has grown exponentially in both scale and sophistication, posing significant threats. As attack methods evolve rapidly, traditional classification schemes often fail to capture the complexity and diversity of modern threats. To address this gap, we introduce STRIKE, a Structured Taxonomy for Risk, Impact, Knowledge, and Emerging Threats, which provides a unified, multi-dimensional framework for categorizing cybercrimes. STRIKE spans both conventional and emerging domains, including ransomware, phishing, network intrusion, child sexual abuse material (CSAM), cryptojacking, deepfakes, and supply chain attacks. It organizes threats using criteria such as attack vectors, adversarial tactics, societal impact, detection techniques, and mitigation strategies. Alongside the taxonomy, we review recent advances in detection methodologies and present a response workflow to assist practitioners under active threat conditions. This work offers researchers, security professionals, and policymakers a practical foundation for threat analysis, comparative evaluation, and adaptive cyber defense.

Keywords: Cybercrime Taxonomy · Threat Classification · Cybersecurity · Intrusion Detection · Ransomware · Cryptojacking · Deepfakes · CSAM · Supply Chain Attacks · Cyber Threat Intelligence

1 Introduction

Cyberspace is the new horizon, a globally interconnected domain controlled by machines for the exchange of information and communication among human beings. Crimes committed in cyberspace are collectively known as *cybercrimes*, broadly defined as criminal acts perpetrated using computers, networks, or the Internet itself [79]. A *cybercriminal* is an individual who engages in or facilitates such activities. There are many kinds of cybercriminals [88], including crackers, hackers, pranksters, career criminals, cyber terrorists, cyberbullies, and salami attackers, among others.

Given the widespread adoption of computer technology across all sectors of society from business operations to critical infrastructure the problem of information protection has become more urgent than ever. Early attacks relied on

I. Roy et al. (Eds.): CAINE 2025, CCIS 2740, pp. 189–211, 2026.
https://doi.org/10.1007/978-3-032-14893-3_14

distributing malicious code via floppy disks and removable media, infecting systems one at a time. However, the Internet quickly evolved into the primary vector for mass infection and exploitation. Whether the mechanism is a computer virus, trojan, worm, or ransomware, attackers can now compromise thousands of systems within hours. High-profile incidents such as the *WannaCry* ransomware outbreak (detected more than 213,000 times in 112 countries) [6] have underscored the scale and speed at which cybercrime can disrupt economic activity, endanger democratic processes, and damage social trust. Reports consistently show that cybercrime has been growing exponentially, with Verizon's Data Breach Investigations Reports documenting a continuous increase in both frequency and impact [5].

In recent years, the threat landscape has further diversified. According to Europol's 2023 Internet Organised Crime Threat Assessment [44], *ransomware remains the most significant cyber threat*, fueled by the rise of *Ransomware-as-a-Service (RaaS)* marketplaces and double extortion strategies, where attackers not only encrypt data but also threaten to leak sensitive information publicly. Simultaneously, phishing campaigns have become more sophisticated, often leveraging *synthetic media* such as deepfake audio and video to deceive both individuals and organizations [24]. The COVID-19 pandemic has dramatically expanded the attack surface by accelerating remote work, increasing cloud dependence, and normalizing digital transactions [73]. Moreover, *supply chain compromises*, such as the SolarWinds attack, have demonstrated that even well-defended networks remain vulnerable through trusted third-party vendors [4].

Cybercrime today targets virtually every facet of society: individuals, small businesses, multinational corporations, and national governments. Criminals employ a wide array of tools, including AI-powered social engineering, cryptojacking malware, and advanced persistent threats (APTs) [14,36]. The World Economic Forum consistently ranks cybercrime among the most pressing global risks, citing its potential to undermine economic stability and national security [47].

In response to the expanding landscape of cyber threats targeting individuals, businesses, and critical infrastructure, building on prior work [29,41], this paper introduces STRIKE: a Structured Taxonomy for Risk, Impact, Knowledge, and Emerging threats. The STRIKE framework offers a comprehensive classification of modern cybercrimes by integrating traditional domains (e.g., phishing, intrusion, ransomware) with rapidly evolving threats such as deepfake-enabled fraud, cryptojacking, and supply chain attacks. Each threat category is examined through a multi-dimensional lens that is its delivery method, impact profile, attacker motivation, and defensive countermeasures. By aligning taxonomy development with current detection, mitigation, and policy needs, STRIKE serves as both an analytical model and a practical guide for cybersecurity practitioners, researchers, and policy-makers. Its extensible structure also enables continuous adaptation to new attack vectors and evolving threat actor behaviors. By consolidating insights from technical, legal, and societal perspectives, this taxonomy aims to support researchers, practitioners, and policymakers in developing more resilient and informed cybersecurity strategies.

The remainder of the paper is organized as follows: Sect. 2 reviews key recent contributions in the field of cybercrime research and outlines how our approach differs from existing taxonomies and frameworks. Section 3 describes the methodology followed to develop STRIKE. Section 4 introduces the proposed STRIKE taxonomy, detailing various cybercrime categories along with their impacts and associated detection techniques. Furthermore, we examine commonly adopted mitigation strategies and discuss broader issues related to internet governance and cybersecurity policy in the same section. Finally, Sect. 5 concludes the paper and outlines potential future research directions aimed at strengthening cyber resilience in the face of evolving digital threats.

2 Existing Work

Over the past two decades, researchers have proposed a variety of taxonomies, ontologies, and detection frameworks to classify cybercrime and support defensive strategies. This section summarizes the most notable contributions and highlights how the proposed work differs from existing literature.

Development of a theory-based taxonomy for cybercrime and the need for clear definitions and standards for measurement and management are presented in [29]. Chandra and Snowe describe the building blocks of their taxonomy, including mutual exclusivity, structure, exhaustiveness, and well-defined categories, which form a robust theoretical foundation. In addition, their work discusses potential implications in the realms of management, reporting, disclosure, governance, regulation, and the judiciary. However, while conceptually rigorous, this approach primarily emphasizes definitional clarity rather than the integration of contemporary detection methodologies or practical response workflows.

The need for a universally agreed-upon classification scheme to enhance understanding of cybercrime and to facilitate legal and policy responses is tackled in [40]. Donalds and Osei-Bryson introduce a cybercrime ontology that incorporates multiple perspectives and aims to offer a more holistic viewpoint for classification. Their work advances the field by acknowledging the diversity of stakeholder perspectives and the need for a unified conceptual framework. Nonetheless, it predates the recent shifts in the threat landscape, such as ransomware-as-a-service and deepfake-enabled phishing, and does not explicitly link taxonomy categories to detection and mitigation strategies.

Al-Khater et al. [15] conducted a comprehensive review of cybercrime detection and prevention techniques, exploring various types of cybercrimes, their impacts (e.g., privacy breaches, security violations, business losses, financial fraud), and the strategies commonly employed by attackers. While this review provides valuable insights into detection methods and recommendations for more effective models, it does not propose a unified taxonomy that links threat categories to countermeasures or response workflows.

More recent surveys illustrate the growing complexity of the field. Conti et al. [36] examined the evolution of ransomware, highlighting the rise of double extortion and the increasing importance of behavior-based detection approaches. Nisioti et al. [14] reviewed machine learning techniques for intrusion detection, emphasizing both their potential and their limitations, including challenges with false positives and dataset biases. Kshetri [73] analyzed the surge in cybercrime driven by the COVID-19 pandemic and the rapid expansion of remote work. Brundage et al. [24] discussed the malicious use of artificial intelligence, including deepfake-enabled social engineering and automated reconnaissance. Recently, there has been a rapid evolution of ransomware evasion strategies [45], deep learning-based intrusion detection [68], and the critical impact of deepfake-enabled social engineering [42]. Ali and Zhang [16] have documented the surge in AI-powered phishing attacks, while Chen and Gupta [95] have classified emerging supply chain threats in IoT environments. Rodriguez and Ahmed [25] provide a global overview of new cybercrime categories, motivations, and countermeasures.

Several international reports also contribute important perspectives. The Europol Internet Organised Crime Threat Assessment (IOCTA) [44] and the World Economic Forum's Global Risks Report [47] document the escalation of ransomware-as-a-service marketplaces, cryptojacking, supply chain compromises, and synthetic media attacks.

Despite the progress made by these contributions, existing work exhibits several limitations. Taxonomies and ontologies are often static and may not adequately capture emerging threats such as deepfake fraud, cryptojacking, and IoT supply chain compromises. Many frameworks stop at conceptual categorization and do not integrate detection methodologies, response workflows, or mitigation strategies in a practical manner. There is limited focus on accessibility for multidisciplinary stakeholders, including policymakers, educators, and non-technical people. While prior works have offered taxonomies focused on cyberattack vectors or technical vulnerabilities, they often emphasize technical classification or system-layer attacks alone. STRIKE differs from existing efforts by explicitly incorporating impact analysis, adversary behavior, and mitigation strategies as integral classification dimensions. For example, unlike Chandra and Snowe's theory-driven model [28] or Donalds and Osei-Bryson's ontology [40], STRIKE integrates new and underrepresented cybercrimes such as CSAM, synthetic identity fraud, and deepfake-based impersonation—within a unified framework. Moreover, STRIKE bridges academic insight with operational relevance by structuring attack profiles alongside their associated risks, social impact, and current defense practices, filling an important gap between theoretical taxonomy and applied cybersecurity defense.

3 Methodology

This study adopts a multi-phase methodology to construct and validate the proposed STRIKE (Structured Taxonomy for Risk, Impact, Knowledge, and Emerging Threats) framework for cybercrime classification. The methodology comprises four key stages:

3.1 Literature Survey and Source Selection

A structured literature review was conducted to capture the current landscape of cybercrime detection, mitigation, and classification. Peer-reviewed journals, conference proceedings, industry whitepapers, technical reports, and governmental publications were surveyed using databases such as IEEE Xplore, ACM Digital Library, SpringerLink, and arXiv. Search terms included ransomware detection, phishing attacks, cybercrime taxonomy, deepfake detection, and child safety online. Sources were selected based on relevance, recency (primarily 2017–2024), citation strength, and practical significance.

The survey methodology followed a targeted thematic approach. Literature was categorized by cybercrime domain, e.g., ransomware, phishing, CSAM, intrusion, cryptojacking, and analyzed across consistent dimensions: attack vectors, adversarial techniques, targets, impact, detection methods, and mitigation strategies. Preference was given to comprehensive surveys, technical frameworks, and meta-analytical studies that addressed emerging threats and novel defense strategies. Representative works include [36], Donalds & Osei-Bryson [40], and Chandra & Snowe [28]. This process enabled a comparative synthesis of threats and informed the construction of the STRIKE taxonomy.

3.2 STRIKE Construction

Cybercrime characteristics were systematically extracted and organized across several dimensions, including attack techniques and delivery vectors, target entities and adversary profiles, impact domains spanning technical, financial, psychological, and societal consequences, detection and mitigation strategies, and relevant legal, policy, and governance contexts. These dimensions were synthesized into two comprehensive taxonomy tables: one that consolidates detection, mitigation, and prevention strategies across cybercrime categories, and another that provides detailed profiles of each attack type, highlighting methods, vectors, targets, and threat actors.

3.3 Mapping and Validation

To ensure completeness and operational alignment, the STRIKE taxonomy was mapped against established cybersecurity frameworks such as the NIST Cybersecurity Framework, MITRE ATT&CK, and Europol IOCTA reports. Threat categories were also cross-validated with industry threat intelligence, including the Verizon DBIR and GAO's SolarWinds incident report.

3.4 Analysis and Profiling

Real-world attack scenarios were integrated to contextualize each cybercrime category. For each category, typical adversary behaviors, technical indicators, and known response workflows were examined. Where applicable, illustrative examples and empirical findings from prior literature were used to demonstrate the taxonomy's practical relevance and extensibility.

4 STRIKE: A Structured Taxonomy of Cybercrime

Figure 1 and Table 1 provide a unified view of the cybercrime landscape. The figure outlines attack profiles by connecting techniques, vectors, attackers, targets, and impacts, while the table complements this by detailing detection, mitigation, and prevention strategies for each category. Together, they highlight both shared patterns and unique features of threats ranging from traditional attacks like ransomware and phishing to emerging risks such as cryptojacking and deepfake-enabled fraud, offering a practical foundation for threat modeling and defense planning.

4.1 Ransomware

Ransomware is malicious software that encrypts or locks a victim's files and systems, demanding payment—typically in cryptocurrency in exchange for the password that can decrypt or unlock the system [36]. Historically, ransomware emerged as scareware or locker ransomware targeting individuals. Over the past decade, it has evolved into a sophisticated ecosystem with extensive underground infrastructure, giving rise to *Ransomware-as-a-Service* (RaaS) operations [45]. In these models, ransomware developers lease their tools to others who conduct attacks and then split ransom proceeds among themselves. This Software-as-a-Service model has made it significantly easier for cybercriminals to launch attacks, fueling exponential growth in attacks.

Ransomware can be delivered through email attachments or malicious links that trick the user into downloading the ransomware software [80]. Often, ransomware is present on compromised websites, also known as drive-by downloads, which infect users when they visit these sites containing hidden iframes or exploit unpatched vulnerabilities in the system. Once installed, the malware executes and takes the system hostage [66].

Ransomware is of two types: *Locker Ransomware* and *Crypto Ransomware* [85]. Locker ransomware locks the desktop or the user interface, preventing access but leaving files intact. This form is generally less destructive, since systems can be restored after removing the malware. In contrast, crypto ransomware silently searches for valuable files and encrypts them without the user's knowledge, rendering them inaccessible and forcing the user to pay a ransom [85].

In modern ransomware attacks, attackers exfiltrate data before encryption and threaten public release if payment is not made [36]. This tactic increases leverage over victims and has contributed to ransomware becoming one of the most financially damaging forms of cybercrime [44]. Often, the victim does not report the attack to the public or even to authorities, fearing it might damage their reputation and cause more harm than simply paying the ransom.

Impact. Ransomware attacks play on the fear of victims, who are often so anxious about losing critical data or facing reputational harm that they pay without verifying whether recovery is possible. Payments are usually demanded

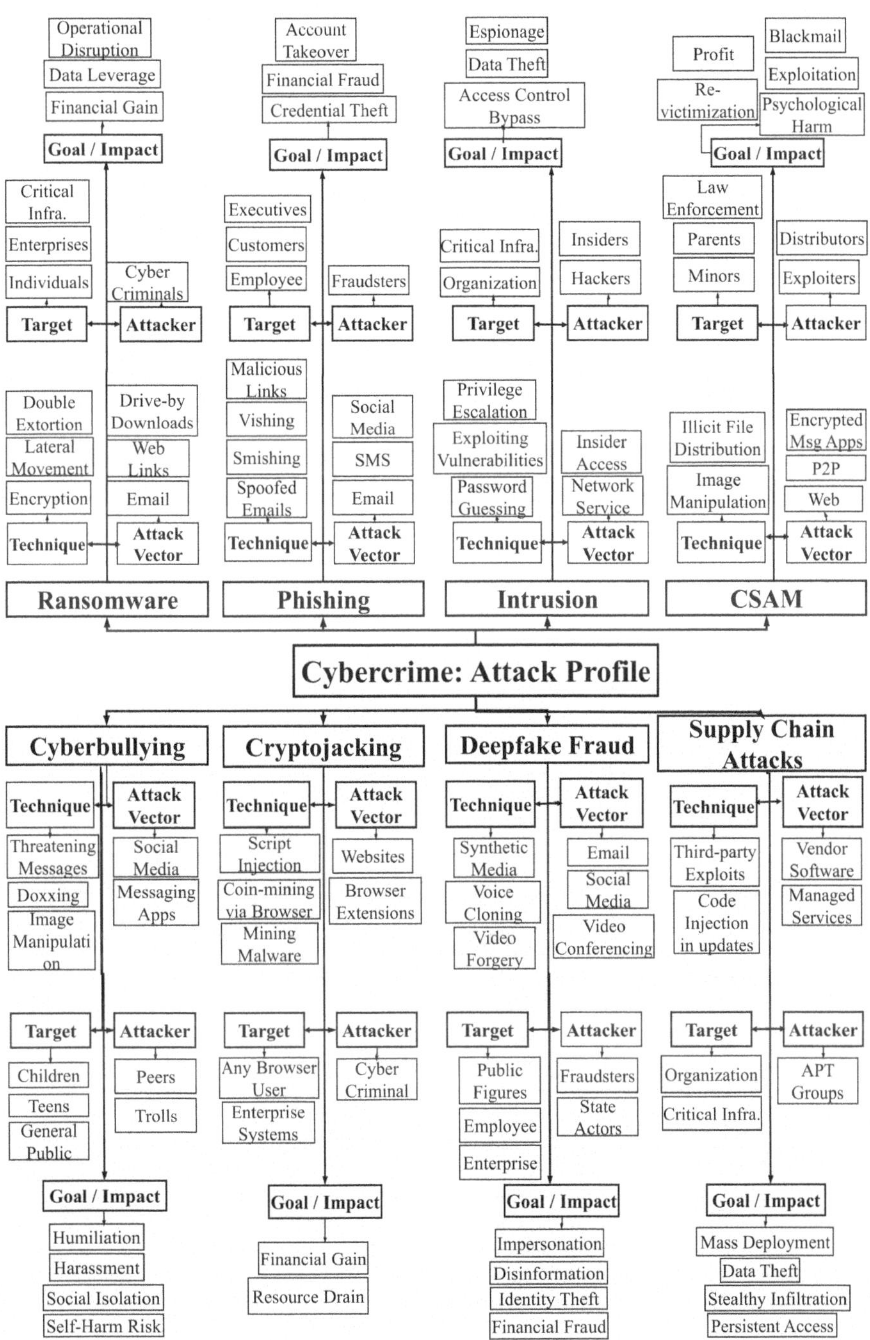

Fig. 1. Cybercrime: Attack Profile.

Table 1. Cybercrime: Detection, Mitigation, and Prevention

Category	Detection	Mitigation	Prevention	Key Ref.
Ransomware	File system monitoring, memory forensics, entropy and I/O pattern analysis, honeypots, anomaly detection	System isolation, decryption tools, restoring from clean backups, threat response teams	Offline and cloud backups, software patching, access control, user training	[36,45,66,68, 80,95]
Phishing	Email header analysis, link inspection, NLP classifiers, blacklist and reputation scoring, browser plugins	Blacklisting, warning systems, credential resets, alerting security teams	Anti-phishing training, spam filters, multi-factor authentication (MFA), email protocol enforcement (DMARC, SPF, DKIM)	[16,39,42,57, 82]
Network Intrusions	Signature/anomaly-based IDS, flow analysis, log correlation, kill chain modeling	Quarantine endpoints, patch vulnerabilities, credential revocation, incident response	Segmentation, least privilege, frequent audits, zero-trust architecture	[19,58,63,74, 89]
Child Sexual Abuse Material (CSAM)	Perceptual hashing (PhotoDNA), CNN-based image recognition, facial landmark estimation, keyword filtering	Image takedown, URL/IP blacklisting, law enforcement reporting, suspect tracing	Platform moderation, legal frameworks, child protection education, collaboration with NGOs	[42,60,61,77, 100]
Cyberbullying	Sentiment analysis, NLP-based toxicity detection, deepfake detection, complaint systems	Post removal, account suspension, psychological support, auto-moderation	Awareness campaigns, digital literacy education, active monitoring, community standards	[75,90,104]
Cryptojacking	Resource usage anomalies, web mining signature analysis, endpoint alerts, script auditing	Process termination, cleanup tools, block mining pools/IPs, extension removal	Browser hardening, JS filtering, no-script policies, endpoint protection platforms	[22,43,71]
Deepfake-Enabled Fraud	Video/audio forensics, liveness detection, ML anomaly detection in media, watermark tracing	Content takedown, account review, source verification, forensic review	Blockchain-based media validation, public awareness, regulation, detection tools for journalists	[33,78,96,98]
Supply Chain Attacks	Behavior profiling, SBOM diffing, code signing, vendor intelligence	Disconnect vulnerable links, revoke certificates, block updates, notify partners	Secure development lifecycle (SDLC), vendor audits, inventory monitoring, endpoint protection	[4,23,37,95]

in untraceable cryptocurrencies such as Bitcoin [80]. In many cases, even after payment, the files remain unrecoverable, resulting in permanent data loss. Studies show that approximately 2.9% of victims ultimately pay the ransom [80].

Attackers typically demand amounts ranging from $50 to $200, and with sufficient scale, can earn hundreds of thousands of dollars monthly. For example, one estimate suggests that an attacker could make up to $394,400 in a single month, factoring in money laundering and conversion fees [53].

Beyond financial losses, ransomware has profound psychological impacts, including anxiety, depression, and social alienation [25,47,73]. Incidents affecting hospitals, schools, and critical infrastructure have demonstrated that ransomware can also pose serious risks to public health and safety.

Detection, Mitigation and Prevention. Detecting ransomware remains a challenging task due to continuous innovation in evasion techniques. Various detection strategies have been proposed, including file system and I/O monitoring, which identify ransomware by extracting distinctive access patterns characterized by tuples of process, file, operation type, and entropy [65,66]. Locker ransomware can also be detected through desktop monitoring techniques that capture and analyze screenshots for ransom messages. Graph-based approaches construct file relationship graphs using metrics like Jaccard similarity and locality-sensitive hashing to classify malicious files [94]. Memory and process analysis assist in detecting in-memory encryption activities [45], while machine learning methods analyze system call sequences, entropy, and behavioral features to detect anomalies [68]. Honeypots and deception-based systems deploy decoy files or environments to observe and detect early ransomware activity [36]. Mitigation and prevention demand a layered strategy including endpoint protection, user awareness training, behavioral monitoring, frequent backups, and incident response planning. Additional measures, such as timely patching, restricting administrative privileges, and enforcing multi-factor authentication, help minimize exposure to ransomware attacks.

Emerging Challenges. Ransomware continues to adapt through obfuscation, polymorphism, and in-memory execution, which hinder detection and forensics [36,45]. It exploits trusted tools like PowerShell and targets supply chains, as seen in the Kaseya breach [95]. Double extortion tactics and Ransomware-as-a-Service have accelerated growth and impact [44,45] since one does not require deep technical skills. Attacks on critical infrastructure underscore the need for layered defenses, proactive monitoring, and robust response plans [73]. These approaches rely on supervised learning and therefore, lacks the ability to detect new types of Ransomwares.

4.2 Computer and Network Intrusions

Intrusion is an act by a user of an information system who is not legally authorized to take a particular action. This user is referred to as an intruder. Intrusions can be performed by external attackers exploiting vulnerabilities in network services or by insiders exceeding their legitimate privileges [72]. Such attacks range from simple password guessing to sophisticated advanced persistent threats (APTs) that remain undetected for extended periods [91].

Impact. Computer files, databases, networks, and Internet-based applications have gradually become critical assets for organizations. When these assets are attacked or compromised, data integrity and privacy are threatened, business operations are disrupted, and customer trust may be lost—sometimes resulting in severe financial and reputational damage [59]. In critical infrastructure contexts, intrusions can interrupt essential services such as energy grids, healthcare systems, and transportation networks [67].

Detection, Mitigation and Prevention. Intrusion Detection (ID) is a type of security management system for computers and networks. An IDS gathers and analyzes information from multiple sources to detect security breaches. Typically, IDSs are classified by their monitoring location and detection strategy:

- Host-Based IDS (HIDS) monitors the activities of specific machines, collecting data such as audit logs, configuration files, and system calls [19].
- Network-Based IDS (NIDS) captures and analyzes packets traversing the network, often using deep packet inspection or flow analysis [99].
- Hybrid IDS combines host and network data and may take proactive measures (e.g., blocking traffic) or simply alert administrators.
- Advanced Persistent Threats (APTs) involve prolonged, targeted campaigns against specific organizations. Detection often relies on correlating events over time, combining multiple data sources, and using kill-chain analysis frameworks [58,63]. Graph-based anomaly detection and threat intelligence integration are increasingly important in uncovering APT activity.

There are two main IDS techniques. Signature-based IDS searches for known patterns or byte sequences indicative of attacks [89]. Anomaly detection establishes a baseline of normal system or network behavior and flags deviations [27]. Modern systems increasingly combine both approaches in hybrid and ensemble methods [74].

The effectiveness of mitigation approaches depends on multiple layers of protection. Network segmentation limits lateral movement, while multi-factor authentication reduces the risk of credential compromise. Other important measures include regular software patching, integrating threat intelligence, and enforcing the principle of least privilege to prevent unauthorized access.

Emerging Challenges. Proliferation of end-to-end encryption reduces packet visibility, creating a major challenge in intrusion detection [51]. Attackers nowadays deceptively use the combination of malicious actions with legitimate processes (living-off-the-land techniques) [21,81]. Modern intrusion attacks include AI, making it adaptive to detection techniques [93]. The widespread use of untested and vulnerable IoT devices significantly expands the attack surface, especially when there are many autonomous systems combined to participate in the ecosystem, providing some services.

4.3 Phishing and Social Engineering

Phishing is a form of online identity theft in which attackers impersonate legitimate entities to trick victims into revealing sensitive information, installing malware, or transferring funds [57]. Also known as "carding" or "brand spoofing," phishing attacks frequently target individuals and organizations through email, SMS (smishing), and voice calls (vishing) [3]. Modern campaigns increasingly leverage AI-generated content and deepfake audio or video to enhance credibility [42]. There are four main types of phishing attacks: spear phishing, which targets specific individuals or organizations; whaling, which aims at senior executives or high-profile figures; clone phishing, which duplicates legitimate emails but replaces links or attachments with malicious content; and vishing and smishing, which rely on voice and SMS-based campaigns.

A typical phishing attack proceeds in four stages [32]. First, the attacker creates a fraudulent website that closely mimics a legitimate service, complete with realistic domain names, web pages, and branding. Second, large volumes of deceptive emails or messages are distributed to potential victims. Third, unsuspecting users are lured to the fake site and prompted to enter credentials, payment details, or other sensitive information. Finally, attackers use the stolen data to commit fraud or gain unauthorized access to accounts. Techniques such as URL obfuscation, compromised legitimate accounts, and malware-laden attachments are commonly employed to improve success rates [16].

Impact. Studies have shown that over 15,000 phishing websites are created daily, with 95% automatically generated by attacker infrastructure [10]. The impact of phishing extends beyond privacy breaches to include credential theft, unauthorized access, and identity fraud. These attacks often lead to fraudulent transactions [26], financial losses, malware infections such as ransomware, and significant reputational damage. Phishing remains the most common initial access vector for ransomware and costs organizations billions of dollars annually [16].

Detection, Mitigation and Prevention. Detection strategies fall into two main categories [34]. Server-based detection includes brand monitoring to identify fraudulent websites, track abnormal activity, and aggregate logs to detect coordinated attacks. Detected phishing domains are added to centralized blacklists to block access. Client-based detection focuses on user devices, where browsers and email clients load URL blacklists, analyze message content with machine learning [39], and monitor information flow for suspicious behavior [38]. Mitigation involves updating blacklists, disabling compromised accounts, and removing malicious content. Whereas prevention relies on user education, as awareness programs significantly reduce click-through rates and improve resilience against phishing [82].

Emerging Challenges. AI-generated emails and deepfake audio are increasingly convincing [42]. There is an increase in Multi-channel campaigns targeting users across email, SMS, and social media [62,103]. There have been rise of malware-as-a-service offerings [86].

4.4 Child Sexual Abuse Material (CSAM)

Child Sexual Abuse Material (CSAM) refers to any visual representation depicting the sexual exploitation of minors, including photographs, videos, digital images, and other media [44]. CSAM includes both real and simulated explicit conduct involving a minor, as well as any depiction of a minor's sexual parts for primarily sexual purposes [9,54]. Even images that do not depict sexual activity, such as nudity presented in a sexualized context, may be classified as illegal. Visual material can include photographs, videos, digital downloads, undeveloped film, and electronic files.

Impact. The impact of CSAM on victims is severe and enduring. Long-term psychological trauma, social stigmatization, re-victimization each time the material is shared or accessed, and barriers to recovery due to the permanence of online distribution are just a few of the consequences. From a societal perspective, CSAM fuels a market for child exploitation, normalizes predatory behavior, and contributes to broader criminal networks [44]. Research suggests that the consumption of such material can have a corrosive effect on offenders. Viewing CSAM may prime individuals to escalate from passive consumption to hands-on offending, normalize deviant sexual interests, and increase desensitization to harm [18].

Detection, Mitigation and Prevention. Combating CSAM requires a combination of technical safeguards, legal frameworks, and international cooperation. Detection relies on hash databases maintained by organizations such as NCMEC and Interpol, with tools like PhotoDNA generating perceptual hashes to identify altered images [77]. Deep learning models, including convolutional neural networks, can classify content based on facial features, body proportions, and contextual clues [61]. Some systems estimate age using facial landmarks [2,31], while two-tier representations improve robustness against manipulation [100].

Mitigation involves platform monitoring and reporting workflows that automate scanning, flag content, and coordinate removal. Cross-border law enforcement collaboration helps identify offenders and rescue victims [60]. Preventative measures include proactive blocking of uploads, strong platform policies, and user reporting mechanisms. Rigorous laws and severe penalties—such as imprisonment, sex offender registration, and fines—serve as deterrents [54].

Emerging Challenges. The proliferation of online platforms, encrypted messaging applications, peer-to-peer networks, and anonymizing technologies has increased circulation and complicated enforcement [60]. Offenders increasingly rely on hidden services and dark web markets. Deepfake technology enables the creation of synthetic CSAM, complicating legal and technical responses [42], making it extremely challenging to tackle since it is now easy to create. It is worth noting that machine learning based detection accuracy varies widely (between 70% and 85%) and false positives remain a challenge [100].

4.5 Cyberbullying

Cyberbullying is the use of electronic communication technologies to intimidate, harass, or harm individuals, typically through repeated aggressive behavior [75]. The U.S. Department of Health and Human Services defines cyberbullying as willful and repeated harm inflicted using electronic devices and communication tools such as social networks, text messaging, and websites [7,84]. While the term generally applies to minors, similar behaviors targeting adults are referred to as cyberharassment or cyberstalking [8,75]. Cyberbullying is most prevalent among children and adolescents ages 6–18 [1,49], although adults increasingly experience online abuse on social media and messaging platforms. Cyberbullying includes sending threatening or humiliating messages, spreading rumors or manipulated images, impersonation to damage reputation, and orchestrating campaigns to isolate individuals. The proliferation of smartphones, instant messaging apps, and social networking platforms has amplified the reach and persistence of online harassment [20].

Impact. Cyberbullying has severe psychological, emotional, and social consequences. Anxiety, depression, and decreased self-esteem can lead to disengagement from regular or essential activities. It often causes relationship challenges and increases the risk of self-harm and suicidal tendencies [55,69]. Victims are twice as likely to attempt suicide compared to non-victims [55]. Emotional consequences frequently include feelings of frustration, anger, sadness, helplessness, and vengefulness [83].

Detection, Mitigation, and Prevention. Detection of cyberbullying relies on a combination of manual and automated approaches. Natural language processing models automatically flag abusive or harassing text by analyzing sentiment, toxicity, and linguistic cues [90]. Computer vision techniques detect manipulated or explicit imagery used to embarrass or threaten victims, including deepfakes [104]. Platforms also monitor user behavior to identify patterns consistent with harassment, such as repeated targeting or coordinated attacks. Most social media services and messaging apps provide tools to report abuse, though effectiveness depends on enforcement and moderation workflows [56].

Mitigation includes support services such as counseling and helplines to address psychological harm. Clear platform policies and community standards should be consistently enforced [75]. Prevention requires collaboration among parents, educators, and platform providers. Discouraging bullying, implementing enforcement policies, and integrating anti-bullying programs into schools can help reduce incidents. At the family level, promoting safe online behavior, monitoring online activities when appropriate, and maintaining open communication are essential. Coordinated efforts across communities, technology platforms, and policymakers, combined with early intervention and education, can significantly reduce the likelihood and impact of cyberbullying [30].

Emerging Challenges. Surveys estimate that up to 35% of adolescents have experienced online harassment [20]. The COVID-19 pandemic further increased incidents due to greater reliance on digital communication [84]. Deepfake technologies are enabling more sophisticated forms of defamation and harassment [104]. Cross-border jurisdictional issues often limit the enforcement of protective laws. Cyberbullying is exacerbated by the ease of anonymity, widespread use of end-to-end encryption, and the difficulty of removing harmful content once it has been shared [7].

4.6 Supply Chain Attacks

Supply chain attacks are intended to create disruption across the end-to-end delivery of goods and services. The supply chain includes suppliers, providers, third-party service vendors, and their supporting infrastructures. This type of attack exploits the trust ecosystem among multiple entities in the chain. It can target hardware or software and may involve anything from malicious software updates to backdoor insertion at any point in the supply chain infrastructure [23].

Impact of Supply Chain Attacks. Malicious code insertion can disrupt operations and spread rapidly across the many organizations involved [4]. Trojan horses are often installed under the guise of legitimate software updates, affecting large numbers of users [37]. Ransomware attacks can impact systems ranging from accounting platforms to critical databases [50]. Major consequences include compromise of government and enterprise networks, theft of intellectual property and sensitive data, operational disruption, reputational damage, and, in some cases, significant financial losses and social unrest.

Detection and Mitigation. Detection and mitigation techniques include code signing, authentication, and strict control of software changes [23], as well as comprehensive software update tracking [23]. End-to-end monitoring and anomaly detection help identify suspicious activities across the supply chain [95]. Because modern supply chains are highly complex and global in scope, securing them requires cross-industry collaboration, regulatory frameworks, and continuous oversight. Given the number of stakeholders involved, mitigation strategies increasingly rely on autonomous trust mechanisms among entities. Blockchain has emerged as a promising solution to facilitate secure, verifiable transactions and improve supply chain transparency [13,102].

4.7 Cryptojacking

Cryptojacking is one of the most modern forms of cybercrime, in which attackers use victimś computing resources to mine cryptocurrency without their knowledge [43]. A report by Symantec revealed that cryptojacking incidents grew by 8,500% between 2017 and 2018 as attackers shifted from ransomware to stealthy mining [22]. Attackers embed mining scripts in websites, exploit software vulnerabilities, or distribute malicious browser extensions [97], remaining hidden to maximize profit over time. Typically, this crime involves three entities: the third-party website, the user who visits it, and the attacker. In browser-based mining, attackers inject JavaScript code into web pages, causing visitors' browsers to mine cryptocurrency while they browse [43]. Cryptojacking can also be achieved by installing malicious software directly on the victim's machine [36]. Additionally, attackers may compromise cloud environments to deploy mining software and conduct large-scale mining operations [22].

Impact. Cryptojacking increases CPU and GPU usage, which reduces performance, disrupts normal device operation, and shortens hardware lifespan [76]. In cloud environments, it can lead to significantly higher electricity costs, increased resource consumption, and rising maintenance expenses. In addition, there are ramifications in a variety of critical sectors such as banking and finance [70]. Recent study [11] found that over half the website surveyed remained malicious, hosting mining scripts like CoinHive and Webminepool. The study revealed that cryptojacking infrastructure often overlaps with other threats, including phishing and malware distribution.

Detection and Mitigation. Detection strategies for cryptojacking include monitoring resource utilization, where unusual CPU or GPU activity, especially on idle systems, is a common indicator. Machine learning algorithms are increasingly useful for detecting such anomalies [71]. Monitoring network traffic helps identify outbound connections to mining pools and suspicious DNS requests [97]. Signature-based detection and heuristic analysis of known mining behaviors also play a role. For browser-based attacks, blocking JavaScript mining scripts can be effective, although this approach is limited: attackers can obfuscate code, and some websites may not function properly if JavaScript is disabled. With the rise of AI improving attackers' capabilities, cryptojacking remains a persistent threat that requires adaptive defenses [22]. Digital Forensics and Incident Response (DFIR) can be an effective approach for detection and mitigation. Study [11] suggests that attackers leverage shared hosting, domain obfuscation, and outdated software to persist over time. These findings underscore the need for continuous monitoring, better transparency in domain registration, and cross-layer detection techniques to counter the evolving infrastructure behind cryptojacking campaigns

Emerging Challenges. Cryptojacking has evolved beyond simple browser-based attacks to sophisticated techniques that exploit cloud infrastructure, fileless malware, and memory-resident scripts. Attackers increasingly use stealthy methods such as PowerShell, WMI, and obfuscated JavaScript to evade detection, while mining-as-a-service platforms and third-party software vulnerabilities complicate defenses. Recent research [35] has shown that even advanced detection tools can be bypassed by tactics like function renaming, loop unrolling, and modifying WebSocket behaviors, highlighting how quickly attackers adapt. As a result, cryptojacking often goes unnoticed for long periods.

4.8 Deepfake-Enabled Fraud

Deepfakes-enabled frauds are Generative AI created content to impersonate individuals or fabricate evidence involving images, audios, or videos [33]. In recent years, deepfakes have been increasingly used to cause identity fraud, social engineering attacks, mass public deception, and public image tarnishing [78]. Deep fake can be active or passive, active deep fake includes Face and voice swapping in audio and videos to impersonate a person in live video calls [12], causing business-related authentication to be broken than relying on audio and video of a person. The passive attacks include distribution of impersonated images to cause social reputation harm or public deception, such as to cause policy changes and election manipulation [87], see many such in recent election years, as well as celebrity or public personabilities in porn or other places, say memes [46,92]. Attackers often combine deepfakes with traditional settings to improve the impact of the attack.

Impact. Deepfake-enabled fraud impacts the social reputation of the person, resulting in damages that span from monetary [33], image tarnishing, to the ability to be elected or hold a position of importance. Effectively, it creates a major distrust among the public about digital communication, where the public can view legitimate media as false.

Detection, Mitigation and Prevention. Media forensics, verification, control, and independent trust infrastructure are major techniques to detect, mitigate, and prevent such frauds. Deepfake detection methods utilize various inconsistencies in artifacts [96]. Some modern techniques include frequency domain analysis and deep learning [48,52]. There are domain-specific techniques that also exist, for example, voice biometric systems targeted to distinguish synthetic speech from real audio [101]. Modern trust models include the use of Blockchain and digital watermarking to perform the authenticity of media [98]. Generative AI and the proliferation of ready-to-use open source tools are helping this attack evolve, making it hard to detect

Emerging Challenges. The increasing realism and accessibility of deepfake tools make fraudulent impersonation more convincing and widely available.

Attackers are also blending deepfakes with phishing, vishing, and social engineering to maximize impact, creating complex multi-vector threats [52]. At the same time, gaps in legal and policy frameworks hinder effective regulation, and the erosion of trust in digital media enables the "liar's dividend," where authentic evidence can be dismissed as fake [17,64]. These factors collectively position deepfake-enabled fraud as one of the most dynamic and difficult cybercrime categories to address.

Table 2. Cybercrime: Summary of Emerging Challenges

Cybercrime Type	Emerging Challenges
Ransomware	Increasing sophistication of encryption techniques; use of double/triple extortion (data theft + DDoS + leaks); targeting of critical infrastructure; Ransomware-as-a-Service (RaaS) ecosystems lowering entry barriers.
Phishing	Highly personalized spear-phishing using AI; multi-vector attacks combining email, SMS, and social media; use of deepfake audio/video in vishing; evasion of spam filters via adversarial content generation.
Intrusions / APTs	Advanced persistent threats with stealthy lateral movement; supply-chain infiltration; "living off the land" techniques to avoid detection; exploitation of zero-day vulnerabilities.
CSAM	Increasing use of encrypted platforms; AI-generated synthetic CSAM; distribution via decentralized networks; legal and jurisdictional barriers to enforcement.
Cyberbullying	Use of ephemeral messaging apps to evade detection; AI-generated harassment content (text-to-image/video abuse); psychological impact magnified by algorithm-driven amplification; cross-platform targeting.
Cryptojacking	Targeting of cloud/IoT environments for large-scale mining; polymorphic malware that adapts to avoid detection; economic incentives driving constant innovation; integration with other malware campaigns.
Deepfake-Enabled Fraud	Realism; accessibility of tools; arms race with detection methods; combination with phishing/vishing/social engineering; legal/regulatory gaps; erosion of trust via "liar's dividend."
Supply Chain Attacks	Growing exploitation of third-party vendors and software; difficulty in verifying software integrity; attacks on CI/CD pipelines; widespread downstream impact due to single compromise.

4.9 Summary

We developed a taxonomy of cybercrime encompassing traditional threats such as ransomware, phishing, network intrusions, and child sexual abuse material, as well as emerging challenges like cryptojacking, deepfake-enabled fraud, and supply chain attacks. Each category has distinct characteristics, impacts, and detection strategies. Table 2 consolidates emerging challenges highlighting how each threat type is evolving, and the systemic difficulties that complicate detection, mitigation, and prevention. In summary, defending against modern cybercrime requires a combination of technical safeguards, policy frameworks, and multidisciplinary collaboration. As threats continue to evolve, continuous research and adaptive security strategies remain critical.

5 Conclusion

This paper presents a structured Taxonomy of major cybercrime called STRIKE. The Taxonomy chart classifies the cybercrimes based on the type of attack vectors, operational impact, informational impact, defense and targets. This classification scheme will aid a defender in protecting their network by providing vital attack information.

There is an ongoing new attack manifestation, therefore this taxonomy could be extended to include new categories within each classification. It will provide a defender with the appropriate information to make an educated decision in defending against cyber attacks. Creative approaches to defending attacks will become available, and providing an extensible taxonomy able to capture new defenses is imperative to defense. We believe this taxonomy provides a foundation for the cybersecurity community and provides the ability to continuously grow as attacks and defenses become more sophisticated. In future work, to build a better Defense System, more research can be done to see the applicability of this taxonomy in determining the action space of the attackers.

References

1. Bullying definition. https://www.stopbullying.gov/bullying/what-is-bullying
2. A new way to detect digital child pornography. http://abcnews.go.com/Technology/detect-digital-child-pornography/story?id=10362616
3. Report on phishing. https://www.justice.gov/sites/default/files/opa/legacy/2006/11/21/report_on_phishing.pdf
4. The solarwinds cyberattack. https://www.gao.gov/blog/solarwinds-cyberattack-demands-significant-federal-and-private-sector-response-infographic, Accessed 12 July 2024
5. Verizon: 2017 data breach investigations report, https://www.itgovernance.co.uk/blog/verizons-2017-data-breach-investigations-report
6. Wannacry ransomware attack. Wannacry ransomware attack. https://blog.avast.com/ransomware-that-infected-telefonica-and-nhs-hospitals-is-spreading-aggressively-with-over-50000-attacks-so-far-today, Accessed 12 July 2024

7. What is cyberbullying. https://www.stopbullying.gov/cyberbullying/what-is-it/index.html
8. What is cyberbullying, exactly?. http://www.stopcyberbullying.org/what_is_cyberbullying_exactly.html
9. What legally makes it child pornography?. https://www.hg.org/article.asp?id=38082
10. Phishing: An analysis of a growing problem. SANS Institute InfoSec Reading Room (2007). https://www.sans.org/reading-room/whitepapers/threats/phishing-analysis-growing-problem-1417
11. Adeniran, A., Human, K., Mohaisen, D.: Dissecting the infrastructure used in web-based cryptojacking: a measurement perspective. In: International Conference on Information Security Applications, pp. 270–283. Springer (2024)
12. Agarwal, S., Farid, H., Gu, Y., He, M., Nagano, K., Li, H.: Detecting deep-fake videos from appearance and behavior. In: Proceedings of the ACM Workshop on Information Hiding and Multimedia Security, pp. 81–87. ACM (2020)
13. Agarwal, U., et al.: Blockchain technology for secure supply chain management: a comprehensive review. IEEE Access **10**, 85493–85517 (2022)
14. Ahmad, Z., Shahid Khan, A., Wai Shiang, C., Abdullah, J., Ahmad, F.: Network intrusion detection system: a systematic study of machine learning and deep learning approaches. Trans. Emerging Telecommun. Technol. **32**(1), e4150 (2021)
15. Al-Khater, W.A., Al-Maadeed, S., Ahmed, A.A., Sadiq, A.S., Khan, M.K.: Comprehensive review of cybercrime detection techniques. IEEE Access **8**, 137293–137311 (2020). https://doi.org/10.1109/ACCESS.2020.3011259
16. Ali, F., Zhang, W.: Phishing attacks in 2025: evolution, trends, and countermeasures. J. Inf. Secur. Appl. **78**, 103529 (2025)
17. Amerini, I., et al.: Deepfake media forensics: status and future challenges. J. Imaging **11**(3), 73 (2025)
18. Babchishin, K.M., Hanson, R.K.H.: Child pornography. Child Pornography (2011). http://research.universalessays.com/criminal-justice-research-paper/child-abuse-research-paper/child-pornography-research-paper/
19. Bace, R.: Intrusion Detection. Macmillan Technical Publishing (2000)
20. Barlett, C.P.: Cyberbullying among youth: a comprehensive review of current international research and its implications and application to policy and practice. Aggress. Violent. Beh. **45**, 93–101 (2019)
21. Barr-Smith, F., Ugarte-Pedrero, X., Graziano, M., Spolaor, R., Martinovic, I.: Survivalism: systematic analysis of windows malware living-off-the-land. In: 2021 IEEE Symposium on Security and Privacy (SP), pp. 1557–1574. IEEE (2021)
22. Blancaflor, E., Dionisio, P.J., Juacalla, D.J., Oraya, A., Yabut, J.A.: Cryptojacking: a comprehensive review of attack techniques, victims, and effective mitigation strategies. In: Proceedings of the 2024 7th International Conference on Information Science and Systems, pp. 42–48 (2024)
23. Boyens, J., Paulsen, C., Moorthy, R., Bartol, N.: Key practices in cyber supply chain risk management: observations from industry. Tech. Rep. NISTIR 8276, NIST (2020). https://doi.org/10.6028/NIST.IR.8276
24. Brundage, M., Avin, S., Clark, J., Toner, H., Eckersley, P., Garfinkel, B.: The malicious use of artificial intelligence: forecasting, prevention, and mitigation. arXiv preprint arXiv:1802.07228 (2018)
25. Buçaj, E., Idrizaj, K.: The need for cybercrime regulation on a global scale by the international law and cyber convention. Multidis. Rev. **8**(1), 2025024–2025024 (2025)

26. Caputo, D.D., Pfleeger, S.L., Freeman, J.D., Johnson, M.E.: Going spear phishing: exploring embedded training and awareness. IEEE Secur. Privacy **12**(1), 28–38 (2014)
27. Chaabouni, N., Mosbah, M., Zemmari, A., Sauvignac, C., Faruki, P.: Network intrusion detection for IoT security based on learning techniques. IEEE Commun. Surv. Tutorials **21**(3), 2671–2701 (2019)
28. Chandra, A., Snowe, M.J.: A taxonomy of cybercrime: theory and design. Int. J. Account. Inf. Syst. **38**, 100467 (2020), 2019 UW CISA Symposium
29. Chandra, A., Snowe, M.J.: A taxonomy of cybercrime: theory and design. Int. J. Account. Inf. Syst. **38**, 100467 (2020)
30. Charles E. Notar, Sharon Padgett, J.R.: Cyberbullying: resources for intervention and prevention (2013). https://doi.org/10.13189/ujer.2013.010301, http://www.hrpub.org
31. Chatzis, V., Panagiotopoulos, F., Mardiris, V.: Face to iris area ratio as a feature for children detection in digital forensics applications. In: 2016 Digital Media Industry Academic Forum (DMIAF), pp. 121–124 (2016). https://doi.org/10.1109/DMIAF.2016.7574915
32. Chawla, M., Chouhan, S.S.: A survey of phishing attack techniques. Int. J. Comput. Appl. **93**(3), 0975 – 8887 (2014)
33. Chesney, R., Citron, D.K.: Deepfakes and the new disinformation war. Foreign Aff. **98**(1), 147–155 (2019)
34. Chhikara, J., Dahiya, R., Garg, N., Rani, M.: Phishing and anti-phishing techniques: case study. Int. J. Adv. Res. Comput. Sci. Softw. Eng. **3**(5) (2013)
35. Chmiel, K., Rajba, P.: How to evade modern web cryptojacking detection tools? a review of practical findings. In: Proceedings of the 19th International Conference on Availability, Reliability and Security, pp. 1–10 (2024)
36. Conti, M., Cozzo, D., Gangwal, S., Kumar, N.: A survey on ransomware detection techniques. ACM Comput. Surv. **55**(3), 1–38 (2022)
37. Cordey, S.: Software supply chain attacks: an illustrated typological review. Tech. rep, ETH Zurich (2023)
38. Davidson, R.: The fight against malware as a service. Netw. Secur. **2021**(8), 7–11 (2021)
39. Do, N.Q., Selamat, A., Krejcar, O., Herrera-Viedma, E., Fujita, H.: Deep learning for phishing detection: taxonomy, current challenges and future directions. IEEE Access **10**, 36429–36463 (2022)
40. Donalds, C., Osei-Bryson, K.M.: Toward a cybercrime classification ontology: a knowledge-based approach. Comput. Hum. Behav. **92**, 403–418 (2019)
41. Donalds, C., Osei-Bryson, K.M.: Toward a cybercrime classification ontology: a knowledge-based approach. Comput. Hum. Behav. **92**, 403–418 (2019)
42. Dsouza, D.S., Hajjar, A.E., Jahankhani, H.: Deepfakes in social engineering attacks, pp. 153–183 (2024)
43. Eskandari, S., Leoutsarakos, A., Mursch, T., Clark, J.: A first look at browser-based cryptojacking. arXiv preprint arXiv:1803.02887 (2018)
44. Europol: Internet organised crime threat assessment (IOCTA) 2023 (2023). https://www.europol.europa.eu/publications-events/main-reports/iocta-report
45. Ferdous, J., Islam, R., Mahboubi, A., Islam, M.Z.: AI-based ransomware detection: a comprehensive review. IEEE Access (2024)
46. Fido, D., Rao, J., Harper, C.A.: Celebrity status, sex, and variation in psychopathy predict judgements of and proclivity to generate and distribute deepfake pornography. Comput. Hum. Behav. **129**, 107141 (2022)

47. Forum, W.E.: Global risks report 2023. World Economic Forum Reports (2023). https://www.weforum.org/reports/global-risks-report-2023
48. Frank, J., Eisenhofer, T., Schönherr, L., Fischer, A., Kolossa, D., Holz, T.: Leveraging frequency analysis for deep fake image recognition. In: International Conference on Machine Learning, pp. 3247–3258. PMLR (2020)
49. Gina S. Smith, Maria A. Minor, H.M.B.: Cyberbullying in higher education: implications and solutions (2014). https://doi.org/10.5590/JERAP.2014.04.1.05
50. Greenberg, A.: The untold story of notpetya, the most devastating cyberattack in history. WIRED (2018). https://www.wired.com/story/notpetya-cyberattack-ukraine-russia-code-crashed-the-world
51. Heidari, A., Jabraeil Jamali, M.A.: Internet of things intrusion detection systems: a comprehensive review and future directions. Clust. Comput. **26**(6), 3753–3780 (2023)
52. Heidari, A., Jafari Navimipour, N., Dag, H., Unal, M.: Deepfake detection using deep learning methods: a systematic and comprehensive review. Wiley Interdis. Rev. Data Mining Knowl. Discov. **14**(2), e1520 (2024)
53. Hernandez-Castro, Julio, Cartwright, Edward, Stepanova, A.: Economic analysis of ransomware (2017)
54. Hessick, C.B.: Refining child pornography law. University of Michigan Press (2016)
55. Hinduja, S., Patchin, J.W.: Bullying, cyberbullying, and suicide. Arch. Suicide Res. **14**(3), 206–221 (2010)
56. Hinduja, S., Patchin, J.W.: Cyberbullying: identification. Prevention and Response, Cyberbullying Research Center (2014)
57. Hong, J.: The state of phishing attacks. Commun. ACM **55**(1), 74–81 (2012)
58. Hutchins, E.M., et al.: Intelligence-driven computer network defense informed by analysis of adversary campaigns and intrusion kill chains. Leading Iss. Inf. Warfare Secur. Res. **1**(1), 80–106 (2011)
59. Institute, P.: Cost of a data breach report 2023 (2023). https://www.ibm.com/reports/data-breach
60. Interpol: Interpol: Crimes against children (2021). https://www.interpol.int
61. Islam, M., Mahmood, A.N., Watters, P., Alazab, M.: Forensic detection of child exploitation material using deep learning, pp. 211–219 (2019)
62. Kaur, K., Jain, A.K.: A survey on phishing attack taxonomy, detection techniques, datasets, and security measures. J. Appl. Secur. Res. 1–52 (2025)
63. Khalid, A., Zainal, A., Maarof, M.A., Ghaleb, F.A.: Advanced persistent threat detection: a survey, 1–6 (2021)
64. Khan, N., Nguyen, T., Bermak, A., Khalil, I.: Unmasking synthetic realities in generative AI: a comprehensive review of adversarially robust deepfake detection systems. arXiv preprint arXiv:2507.21157 (2025)
65. Kharraz, A., Robertson, W., Arshad, S., Kirda, E., Mulliner, C.: Unveil: a large-scale, automated approach to detecting ransomware. In: Proceedings of the 25th USENIX Security Symposium. usenix (2016)
66. Kharraz, A., Robertson, W., Balzarotti, D., Bilge, L., Kirda, E.: Cutting the gordian knot: a look under the hood of ransomware attacks. Detection of Intrusions and Malware, and Vulnerability Assessment Lecture Notes in Computer Science **9148**, 3–24 (2015)
67. Khraisat, A., Gondal, I., Vamplew, P., Kamruzzaman, J.: Survey of intrusion detection systems: techniques, datasets and challenges. Cybersecurity **2**(1), 1–22 (2019). https://doi.org/10.1186/s42400-019-0038-7

68. Kocher, G., Kumar, G.: Machine learning and deep learning methods for intrusion detection systems: recent developments and challenges. Soft. Comput. **25**(15), 9731–9763 (2021). https://doi.org/10.1007/s00500-021-05893-0
69. Kowalski, R.M., et al.: Bullying in the digital age: a critical review and meta-analysis of cyberbullying research among youth. Psychol. Bull. **140**(4), 1073–1137 (2014)
70. Kshetri, N., Rahman, M.M., Sayeed, S.A., Sultana, I.: Cryptoran: a review on cryptojacking and ransomware attacks in the banking industry—threats, challenges, and problems. In: 2024 2nd International Conference on Advancement in Computation and Computer Technologies (InCACCT), pp. 523–528. IEEE (2024)
71. Kshetri, N., Rahman, M.M., Sayeed, S.A., Sultana, I.: Cryptoran: a review on cryptojacking and ransomware attacks wrt banking industry-threats, challenges, and problems, 523–528 (2024)
72. Kumar, S., Gupta, S., Arora, S.: Research trends in network-based intrusion detection systems: a review. IEEE Access **9**, 157761–157779 (2021)
73. Lallie, H.S., et al.: Cyber security in the age of covid-19: a timeline and analysis of cyber-crime and cyber-attacks during the pandemic. Comput. Secur. **105**, 102248 (2021)
74. Liao, H.J., Lin, C.H.R., Lin, Y.C., Tung, K.Y.: Intrusion detection system: a comprehensive review. J. Netw. Comput. Appl. **36**(1), 16–24 (2013)
75. Lizut, J.: Cyberbullying: Educators and Parents. IGI Global (2019)
76. Meland, P.H., Johansen, B.H., Sindre, G.: An experimental analysis of cryptojacking attacks. In: Nordic Conference on Secure IT Systems, pp. 155–170. Springer (2019)
77. Microsoft: Photodna technology (2019). https://www.microsoft.com/en-us/photodna
78. Mirsky, Y., Lee, W.: The creation and detection of deepfakes: a survey. ACM Comput. Surv. **54**(1), 1–41 (2021)
79. N, R.: Evolution of cyber crimes (2008)
80. O'Gorman, G., McDonald, G.: Ransomware: a growing menace (2012)
81. Ongun, T., et al.: Living-off-the-land command detection using active learning. In: Proceedings of the 24th International Symposium on Research in Attacks, Intrusions and Defenses, pp. 442–455 (2021)
82. Parsons, K., McCormac, A., Pattinson, M., Butavicius, M., Jerram, C.: Phishing for the truth: a scenario-based experiment of users' behavioural response to emails. Behav. Inf. Technol. **34**(3), 247–258 (2015)
83. Patchin, J., Hinduja, S.: Offline consequences of online victimization: school violence and delinquency (2007)
84. Patchin, J.W., Hinduja, S.: Cyberbullying prevention and response: expert perspectives. Routledge (2012)
85. Pathak, P., Nanded, Y.M.: A dangerous trend of cybercrime: ransomware growing challenge. Int. J. Adv. Res. Comput. Eng. Technol. **5**(2), 371–373 (2016)
86. Patsakis, C., Arroyo, D., Casino, F.: The malware as a service ecosystem. In: Malware: Handbook of Prevention and Detection, pp. 371–394. Springer (2024)
87. Pawelec, M.: Deepfakes and democracy (theory): how synthetic audio-visual media for disinformation and hate speech threaten core democratic functions. Digital Soc. **1**(2), 19 (2022)
88. Sabillon, R., Cano, J.J., Serra-Ruiz, J., et al.: Cybercrime and cybercriminals: a comprehensive study. Int. J. Comput. Netw. Commun. Secur. 2016, **4**(6) (2016)
89. Scarfone, K., Mell, P.: Guide to intrusion detection and prevention systems (IDPS). NIST Special Publication **800**(94) (2007)

90. Schmidt, A., Wiegand, M.: A survey on hate speech detection using natural language processing, 1–10 (2017)
91. Sharma, A., Gupta, B.B., Singh, A.K., Saraswat, V.: Advanced persistent threats (apt): evolution, anatomy, attribution and countermeasures. J. Ambient. Intell. Humaniz. Comput. **14**(7), 9355–9381 (2023)
92. Sharma, R., Bhatnagar, M., Ozen, E.: The role of intellectual capital in combatting deepfake threats to corporate reputation and trust. In: Mastering Deepfake Technology: Strategies for Ethical Management and Security, pp. 1–11. River Publishers (2025)
93. Song, L., Mittal, P.: Systematic evaluation of privacy risks of machine learning models, 2615–2632 (2021)
94. Tamersoy, A., Roundy, K., Chau, D.H.: Guilt by association: Large scale malware detection by mining file-relation graphs. In: Proceedings of the 20th ACM SIGKDD International Conference on Knowledge Discovery and Data Mining, pp. 1524–1533. KDD '14, ACM, New York (2014). https://doi.org/10.1145/2623330.2623342
95. Tan, Z., Parambath, S.P., Anagnostopoulos, C., Singer, J., Marnerides, A.K.: Advanced persistent threats based on supply chain vulnerabilities: challenges, solutions and future directions. IEEE Internet Things J. (2025)
96. Tolosana, R., Vera-Rodriguez, R., Fierrez, J., Morales, A., Ortega-Garcia, J.: Deepfakes and beyond: a survey of face manipulation and fake detection. Inf. Fusion **64**, 131–148 (2020)
97. Tuttle, H.: Cryptojacking: how hackers steal resources to mine digital gold. Risk Manage. **65**(7), 22–28 (2018)
98. Verdoliva, L.: Media forensics and deepfakes: an overview. IEEE J. Selected Topics Sign. Process. **14**(5), 910–932 (2020)
99. Vigna, G., Kemmmerer, R.A.: Netstat: a network-based intrusion detection approach. NetSTAT: a network-based intrusion detection approach
100. Vitorino, P., Avila, S., Rocha, A.: A two-tier image representation approach to detecting child pornography. In: Proceedings of Workshop de Visão Computacional (WVC'16) (2016)
101. Wang, X., et al.: Asvspoof 2019: a large-scale public database of synthetic and spoofed speech. In: IEEE International Conference on Acoustics, Speech and Signal Processing (ICASSP), pp. 6846–6850. IEEE (2020)
102. Xu, P., Lee, J., Barth, J.R., Richey, R.G.: Blockchain as supply chain technology: Considering transparency and security. Int. J. Phys. Distrib. Logist. Manag. **51**(3), 305–324 (2021)
103. Yang, P., Zhao, G., Zeng, P.: Phishing website detection based on multidimensional features driven by deep learning. IEEE Access **7**, 15196–15209 (2019)
104. Zhang, T.: Deepfake generation and detection, a survey. Multimedia Tools Appl. **81**(5), 6259–6276 (2022)

A New Pseudo-Random Binary Generator Based on the Combination of Self-Shrinking Generators and Modified Self-Shrinking Generators Using Boolean Functions

Narayan Debnath[1], Andrés Francisco Farías[2](✉), Andrés Alejandro Farías[2], Germán Antonio Montejano[3], Ana Gabriela Garis[3], and Daniel Riesco[3]

[1] School of Computing and Information Technology, Eastern International University, Binh Duong, Vietnam
narayan.debnath@eiu.edu.vn

[2] Academic Department of Physical, Mathematical and Natural Sciences, National University of La Rioja, La Rioja, Argentina
afarias665@yahoo.com.ar

[3] Department of Computer Science, Faculty of Physical-Mathematical and Natural Sciences, National University of San Luis, San Luis, Argentina
{gmonte,agaris,driesco}@unsl.edu.ar

Abstract. Random or pseudo-random binary sequences are essential to various branches of science, such as cryptography, simulation, and mathematics. These values are obtained through the use of random or pseudo-random binary generators. These random strings must have a high period and high linear complexity, and must pass statistical randomness tests, to ensure their effectiveness. The development of a generator must take all of the above factors into account, and each step of the process must be meticulously controlled to ensure a satisfactory final result. Combining cryptographic components can result in an unsuitable generator. This project describes the methodology for developing a reliable binary generator. The basic components of the generator proposed in this presentation are Self-Shrinking Generators and Modified Self-Shrinking Generators, which are combined using four-variable Boolean functions.

Keywords: Self-Shrinking Generators · Modified Self-Shrinking Generators · LFSR · cipher · key · Boolean function

1 Introduction

The design of a pseudorandom binary generator is presented based on the combination of eight Self-Shrinking Generators (SSGs) and eight Modified Self-Shrinking Generators (MSSGs) of different lengths, with their respective primitive connection polynomials, using four-variable Boolean functions [1, 2].

The construction procedure for a pseudorandom generator with these characteristics requires several steps, as follows:

I. Roy et al. (Eds.): CAINE 2025, CCIS 2740, pp. 212–222, 2026.
https://doi.org/10.1007/978-3-032-14893-3_15

- Generator design
  - o Characteristics of the Self-Shrinking Generators (SSGs)
  - o Characteristics of the Modified Self-Shrinking Generators (MSSGs)
  - o Schematic definition of the Generator.
  - o Election of the different LFSR that make up each Self-Shrinking Generators (SSGs) and Modified Self-Shrinking Generators (MSSGs)
  - o Boolean Function Selection
  - o Key and procedure to generate the initial states of the LFSR.
  - o Composition of the Generator with the elements already selected.
- Tests of randomness
  - o Choice of Statistical Tests.
- Tests on the Generator
  - o Results Analysis.
  - o Proportion of samples that pass tests.
  - o Uniform distribution of P-values
  - o Final analysis.

2 Generator Design

2.1 Characteristics of the Self-shrinking Generators (SSGs)

The proposed SSG shrinks two bits of an LFSR sequence to one output bit. For the binary sequence generated by an LFSR $(x) = x_0x_1x_2...$, let $(y) = y_0y_1...$ be the corresponding self-shrunken sequence. Consider two consecutive bits of (x), (x_{2i}, x_{2i+1}). If x_{2i} is 1, then x_{2i+1} is output. If x_{2i} is 0, then is discarded [3, 4].

The truth table for this generator, in Table 1.

Table 1. The truth table for this SSGs

x_{2i}	x_{2i+1}	y_i
0	0	No output
0	1	No output
1	0	0
1	1	1

2.2 Characteristics of the Modified Self-shrinking Generators (MSSGs)

The proposed MSSG shrinks two bits of an LFSR sequence to one output bit. For the binary sequence generated by an LFSR $(x) = x_0x_1x_2...$, let $(y) = y_0y_1...$ be the corresponding self-shrunken sequence. Consider three consecutive bits of (x), $(x_{3i}, x_{3i+1}, x_{3i+2})$. If $x_{3i} \oplus x_{3i+1}$ is 1, then x_{3i+2} is output. If $x_{3i} \oplus x_{3i+1}$ is 0, then is discarded [3, 4].

The truth table for this generator, in Table 2.

Table 2. The truth table for this MSSGs

x_{3i}	x_{3i+1}	x_{3i+2}	y_i
0	0	0	No output
0	0	1	No output
0	1	0	0
0	1	1	1
1	0	0	0
1	0	1	1
1	1	0	No output
1	1	1	No output

2.3 Schematic Definition of the Generator

The Fig. 1 shows the proposed Pseudo-Random Binary Generator Based on the Combination of Self-Shrinking Generators (SSGs) and Modified Self-Shrinking Generators (MSSGs) using Boolean Functions:

2.4 Election of the Different LFSR that make up each Self-shrinking Generators (SSGs) and Modified Self-shrinking Generators (MSSGs)

The lengths and primitive polynomials [5–7] of the LFSR that make up each block are those indicated in Table 3:

Table 3. LFSR, lengths and primitive polynomials

LFSR	Lengths	Primitive polynomials
1	61	$P(x)_1 = x^{61} + x^{59} + x^{52} + x^{47} + x^{38} + x^{33} + 1$
2	37	$P(x)_2 = x^{37} + x^{33} + x^{31} + x^{30} + x^{21} + x^{3} + 1$
3	59	$P(x)_3 = x^{59} + x^{53} + x^{45} + x^{34} + x^{22} + x^{21} + 1$
4	53	$P(x)_4 = x^{53} + x^{51} + x^{37} + x^{29} + x^{18} + x^{4} + 1$

(*continued*)

Table 3. (*continued*)

LFSR	Lengths	Primitive polynomials
5	47	$P(x)_5 = x^{47} + x^{42} + x^{32} + x^{19} + x^{17} + x^5 + 1$
6	31	$P(x)_6 = x^{31} + x^{16} + x^{14} + x^{10} + x^8 + x^1 + 1$
7	43	$P(x)_7 = x^{43} + x^{35} + x^{32} + x^{30} + x^{25} + x^8 + 1$
8	41	$P(x)_8 = x^{41} + x^{40} + x^{32} + x^{20} + x^{12} + x^{11} + 1$

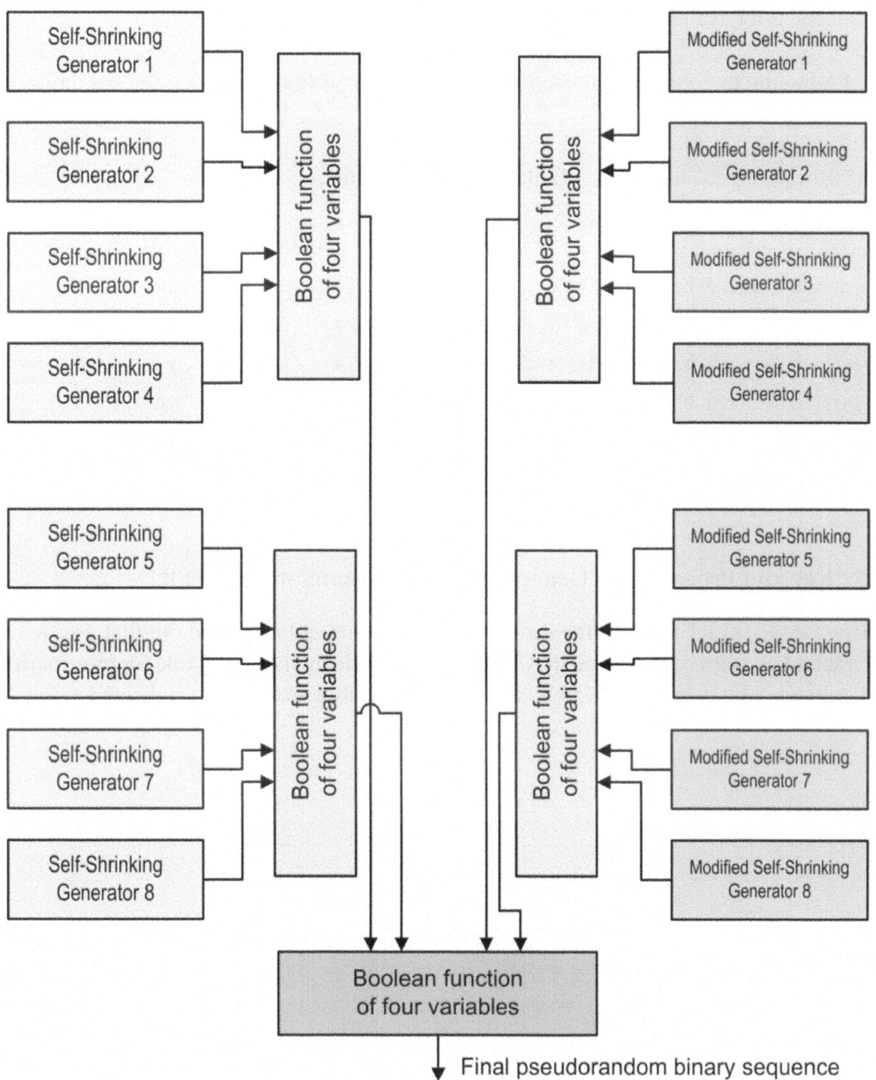

Fig. 1. Proposed Pseudo-Random Binary Generator scheme

2.5 Boolean Function Selection

For the selection, some of the desirable cryptographic properties for these functions are taken into account. Boolean functions of four variables are adopted, both for those that perform non-linear filtering and for the one that fulfills the combination task.

Desirable Cryptographic Properties. Some of the most cryptographically important properties adopted for this project are presented [8–10]:

- Balanced Function
- High non-linearity
- Meets strict avalanche criteria (SAC)

Following the criteria indicated above, the accepted Boolean functions are shown in Table 4:

Table 4. Boolean functions

f_{NAF}	Balanced	Non-linearity	SAC Compliant
$f_{1316} = b \oplus a \cdot b \oplus b \cdot c \oplus d \oplus b \cdot d$	yes	4	yes
$f_{1585} = b \oplus a \cdot b \oplus a \cdot c \oplus d \oplus a \cdot d$	yes	4	yes
$f_{2365} = b \oplus a \cdot c \oplus b \cdot c \oplus a \cdot d \oplus b \cdot d$	yes	4	yes
$f_{2402} = b \oplus a \cdot c \oplus b \cdot c \oplus d \oplus c \cdot d$	yes	4	yes
$f_{2678} = b \oplus c \oplus a \cdot c \oplus b \cdot c \oplus c \cdot d$	yes	4	yes

2.6 Key and Procedure to Generate the Initial States of the LFSR

To create the initial states of the different LFSR, a process is carried out that uses a 32-character key, which, expressed in ASCII code (American Standard Code for Information Interchange), has a length of 256 bits. The cryptographic procedure is indicated in Fig. 2.

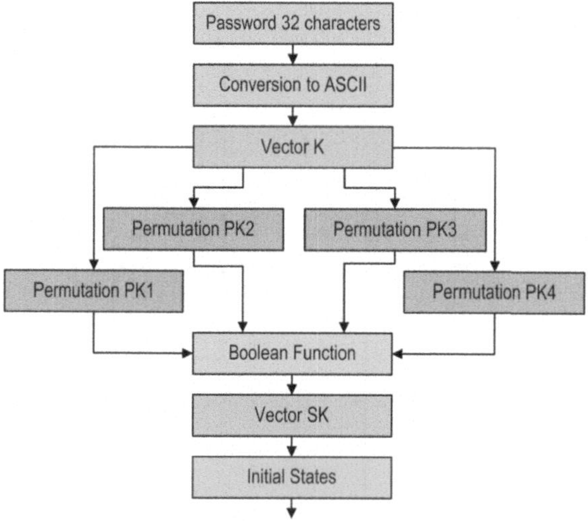

Fig. 2. Key to the generator

Permutation. Permutations are determined by a multiplicative congruent generator [11]. The generator has the following expression:

$$x_{i+1} = (a_x \cdot x_i) mod m_x$$

$$(a_x = multiplier; m_x = module; x_0 = seed)$$

Table 5 shows the values of the vectors, modules, multipliers and seeds:

Table 5. Vectors, modules, multipliers and seeds

Vector	module	multiplier	seed
PK1	1048573	2281	1553
PK2	1048573	2281	1559
PK3	1048573	2281	1567
PK4	1048573	2281	1571

The operation results in a 256-bit vector SK[j], which will provide the initial states of the LFSR, sequentially.

2.7 Composition of the Generator with the Elements Already Selected

With the previously selected components, the structure of the pseudorandom binary generator is completed, Fig. 3.

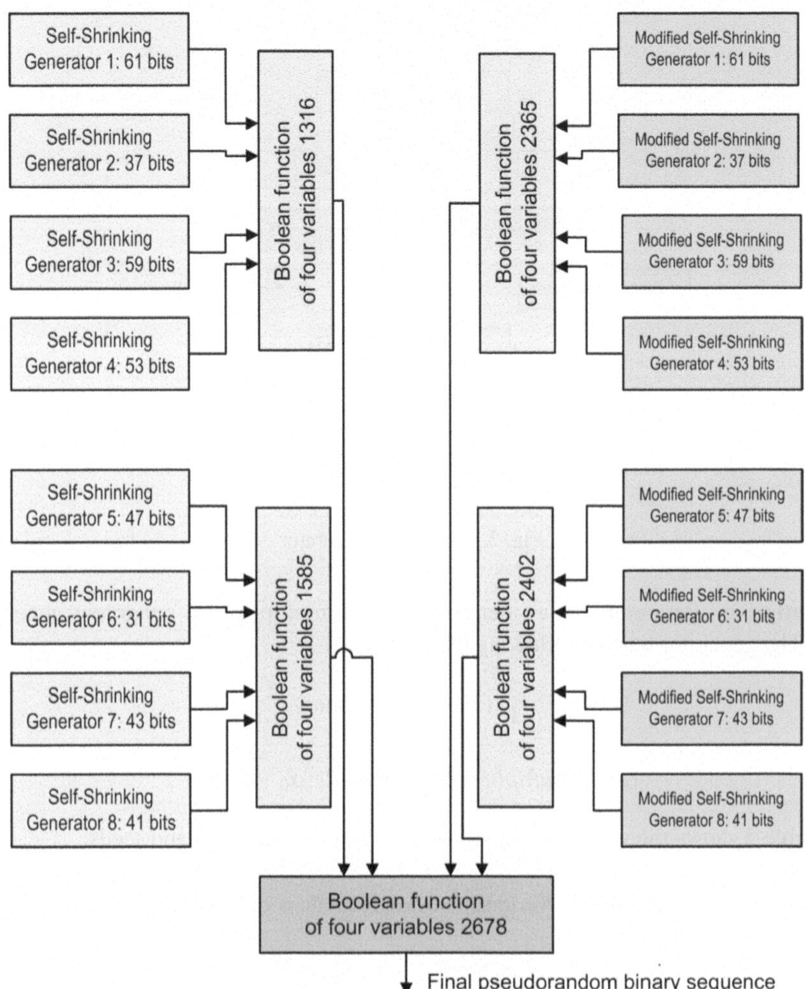

Fig. 3. Composition of the Generator with the elements already selected

3 Tests of Randomness

3.1 Choice of Statistical Tests

The statistical test suite for random and pseudorandom number generators for cryptographic applications was selected from the National Institute of Standards and Technology (NIST) Special Publication 800–22 revision 1a, from the work of Rukhin (et al.) [12]. Table 6 shows the statistical tests for random and pseudo-random numbers that make up the package.

Table 6. Statistical Tests for Random and Pseudorandom Number

	Statistical Tests for Random and Pseudorandom Number
1	Frequency (Monobit)
2	Frequency Test within a Block
3	Approximate Entropy Test
4	Cumulative Sums Test
5	Runs Test
6	Serial Test
7	Maurer's "Universal Statistical" Test
8	Non-overlapping Template Matching Test
9	Linear Complexity Test
10	Discrete Fourier Transform (Spectral) Test

4 Tests on the Generator

One hundred binary sequences of 1,000,000 bits were analyzed, obtained from the generator from one hundred different keys.

The significance level adopted for the statistical tests is: $\alpha = 0.01$

The null hypothesis is: $H_0 \rightarrow p\_value > 0.01$

4.1 Results Analysis

Following the directives of NIST 800–22, having the results, two processes are carried out to interpret them:

- Proportion of samples that pass tests.
- Test for Uniformity of p-value

4.2 Proportion of Samples that Pass Tests

To analyze the results, the proportion of samples that pass the tests is determined, and with this data a point graph is constructed, where it must be met that all points are within the upper and lower limits, to accept that the tests were successful.

$$LS, LI = (1 - \alpha) \pm 3 \cdot \sqrt{\frac{\alpha(1 - \alpha)}{k}}$$

In our case: $k = 100$, and the chosen significance level is: $\alpha = 0.01$.

All tests are considered and the results are indicated in Table 7:

Table 7. Proportion of Samples That Pass Tests

	Statistical Tests for Random and Pseudorandom Number	Total	Pass	Ratio	Upper	Lower
1	Frequency (Monobit)	100	99	0.99	1.02	0.96
2	Frequency Test within a Block	100	99	0.99	1.02	0.96
3	Approximate Entropy Test	100	98	0.98	1.02	0.96
4	Cumulative Sums Test	100	99	0.99	1.02	0.96
5	Runs Test	100	100	1.00	1.02	0.96
6	Serial Test	100	99	0.99	1.02	0.96
7	Maurer's "Universal Statistical" Test	100	100	1.00	1.02	0.96
8	Non-overlapping Template Matching Test	100	100	1.00	1.02	0.96
9	Linear Complexity Test	100	99	0.99	1.02	0.96
10	Discrete Fourier Transform (Spectral) Test	100	98	0.98	1.02	0.96

The result can be seen in the graph. The points are within the acceptance limits, finally the sequences delivered by the generator pass the randomness tests in Fig. 4:

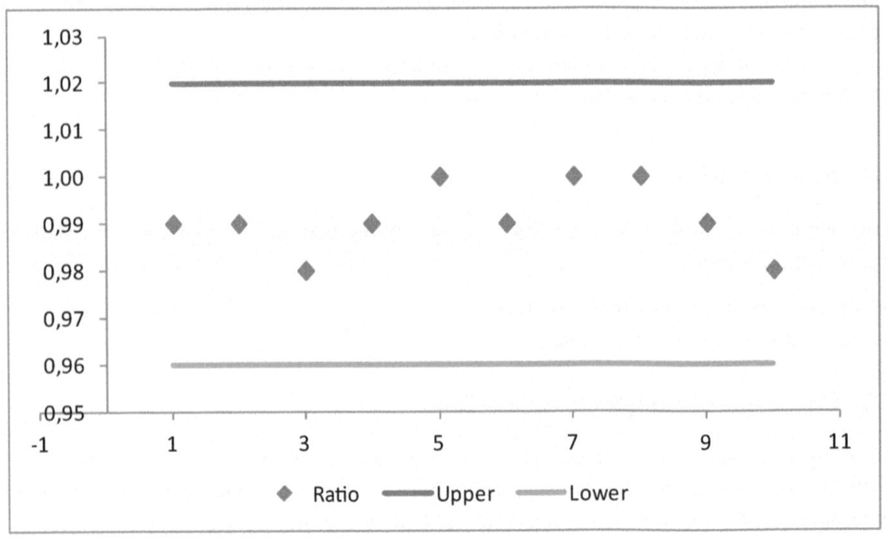

Fig. 4. Dot plot of Proportion of Samples that Pass Tests

4.3 Uniform Distribution of P-Values

Goodness-of-fit tests are performed. This control is carried out for each test on the hundred samples. With the results of the p-value frequencies obtained. Table 8 indicates the satisfactory results obtained:

Table 8. Uniform distribution of P-values

	Tests	p-value	p-value limit	Pass
1	Frequency (Monobit)	0.0428	0.0001	Yes
2	Frequency Test within a Block	0.5544	0.0001	Yes
3	Approximate Entropy Test	0.3041	0.0001	Yes
4	Cumulative Sums Test	0.3191	0.0001	Yes
5	Runs Test	0.9781	0.0001	Yes
6	Serial Test	0.6371	0.0001	Yes
7	Maurer's "Universal Statistical" Test	0.9781	0.0001	Yes
8	Non-overlapping Template Matching Test	0.4012	0.0001	Yes
9	Linear Complexity Test	0.7197	0.0001	Yes
10	Discrete Fourier Transform (Spectral) Test	0.5955	0.0001	Yes

The result can be seen in the graph; the points are within the acceptance limits. Finally, the sequences returned by the generator pass the goodness-of-fit tests, as shown in Fig. 5:

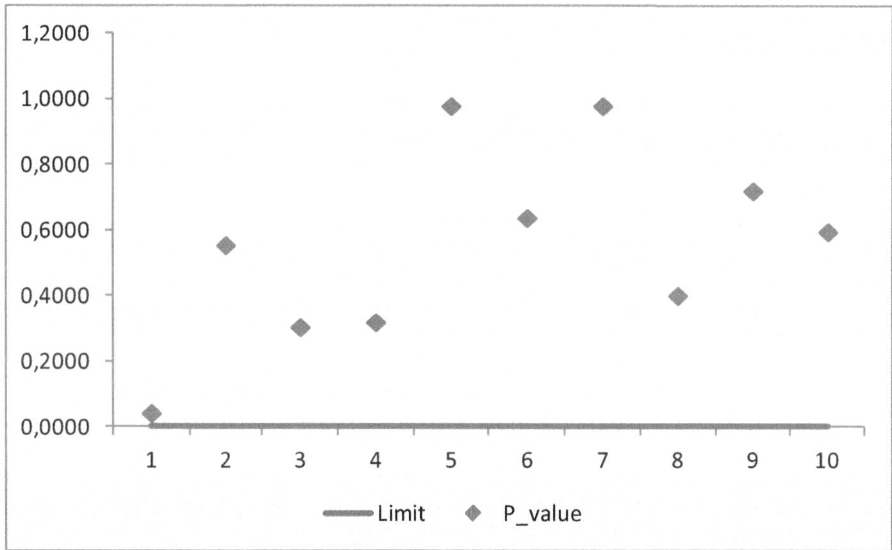

Fig. 5. Dot plot of Uniform Distribution of P-values

4.4 Final Analysis

In short the sequences delivered by the cipher pass both methods of interpreting results and are therefore pseudo-random.

5 Conclusions and Future Works

The generator nonlinearly combines the pseudorandom binary sequences generated by eight Self-Reducing Generators (SSGs) and eight Modified Self-Reducing Generators (MSSGs) using four-variable Boolean functions to obtain a final pseudorandom binary string.

The Self-Reducing Generators (SSGs) and Modified Self-Reducing Generators (MSSGs) that comprise the generator have primitive connection polynomials that guarantee the maximum period of the resulting sequences.

After completing the selection process, the devices were integrated into the generator for subsequent operation with different key values and the generation of the respective binary sequences.

After running statistical randomness tests, the results were interpreted. The values obtained were satisfactory, so the presented generator is considered valid for producing pseudorandom binary sequences of good cryptographic quality.

Future developments will include more Self-Shrinking Generators (SSGs) and Modified Self-Shrinking Generators (MSSGs) with coupled Boolean functions for nonlinear filtering which will be four and five variables. In the fusion sector it is standard practice to combine all the resulting sequences with four or more devices. Ultimately the concluding combination will be executed through the utilisation of a Boolean function whose number of variables will correspond to the number of sequences received.

References

1. Masoodi, F., Alam, S., Bokhari, M.U.: An analysis of linear feedback shift registers in stream ciphers. Int. J. Comput. Appl. **46**(17), 46–49 (2012)
2. Menezes, A.J., Van Oorschot, P.C., Vanstone, S.A.: Handbook of Applied Cryptography. CRC Press (2018)
3. Kanso, A.: Modified self-shrinking generator. Comput. Electr. Eng. **36**(5), 993–1001 (2010)
4. Hu, Y., Xiao, G.: Generalized self-shrinking generator. IEEE Trans. Inf. Theory **50**(4), 714–719 (2004)
5. Paar, C., Pelzl, J.: Understanding Cryptography, vol. 1. Springer-Verlag Berlin Heidelberg (2010). https://doi.org/10.1007/978-3-662-69007-9
6. Stahnke, W.: Primitive binary polynomials. Math. Comput. **27**(124), 977–980 (1973)
7. Seroussi, G.: Table of Low-Weight Binary Irreducible Polynomials, pp. 98–135. Hewlett-Packard Laboratories, Palo Alto (1998)
8. Clark, J.A., Jacob, J.L., Maitra, S., Stănică, P.: Almost boolean functions: the design of boolean functions by spectral inversion. Comput. Intell. **20**(3), 450–462 (2004)
9. Preneel, B., Braeken, A.: Cryptographic properties of boolean functions and s-boxes. Departement elektrotechniek (ESAT), Leuven, Belgium (2006)
10. Elhosary, A.M., Hamdy, N., Farag, I.A., Rohiem, I.: State of the art in boolean functions cryptographic assessment. Int. J. Comput. Netw. Commun. Secur. **1**(3), 88–94 (2013)
11. Fishman, G.S.: Multiplicative congruential random number generators with modulus 2β: an exhaustive analysis for $\beta = 32$ and a partial analysis for $\beta = 48$. Math. Comput. **54**(189), 331–344 (1990)
12. Rukhin, A., et al.: A Statistical Test Suite for Random and Pseudorandom Number Generators for Cryptographic Applications, vol. 22, p. 1. US Department of Commerce. Technology Administration. National Institute of Standards and Technology, Gaithersburg (2001)

LiveTest: An Open-Source OMR Testing Software

Garrett Mathers(✉), Terry Griffin, and Catherine Stringfellow

Midwestern State University, Wichita Falls, TX 76308, USA
Garrett.Mathers@gmail.com, {Terry.Griffin, Catherine.Stringfellow}@msutexas.edu

Abstract. In education, the effortless analysis of test results is vital to teacher effectiveness. Instructors at all levels have relied on hardware-based OMR systems to meet this need. Although effective and requiring minimal instructor effort, this method imposes financial burdens. Costs begin immediately at the university level with the purchase and maintenance of the OMR scanner. Universities incur costs for purchasing and maintaining OMR scanners, while students must buy compatible answer sheets for each test. These expenses can be especially challenging for those with limited resources or immediate funding.

This paper presents LiveTest, a highly customizable and configurable testing solution. It integrates Pictron to generate flexible answer sheets, which are graded using the dynamic OMRGrader module. Instructors can upload images of completed student answer sheets via any web browser to a predefined test for immediate grading. Results and analytics are updated in real-time, reflecting live changes as new submissions are received. The paper also details the processes of extracting, standardizing, interpreting, grading, and archiving the answer sheets.

Keyword: Automated OMR Scanning

1 Introduction

Optical mark recognition (OMR) is the process of scanning paper to detect marks in predefined locations, commonly used by instructors to efficiently grade multiple-choice tests. Although a shift toward free-response questions has led to reduced emphasis on multiple-choice questions (MCQ) and increased focus on handwriting AI, recent advances in computer vision have brought renewed interest in software-based OMR solutions [15]. However, these tools often remain costly and inaccessible for the average educational institution [4].

For years, students have carefully filled in answer sheet bubbles and submitted them to instructors, who then scan and upload results using a hardware-based OMR system. One professor, discovering the OMR scanner was broken with no plan for repair, faced the prospect of grading by hand—a process that can take up to ten times longer than using software-based OMR [3]. Unwilling to do this manually, he attempted to use OpenCV to extract results, but faced challenges due to the sheet's inconsistent structure and irregular

I. Roy et al. (Eds.): CAINE 2025, CCIS 2740, pp. 223–243, 2026.
https://doi.org/10.1007/978-3-032-14893-3_16

bubble shading. Several existing public repositories implement OMR scanning, but none provide the level of automation seen in paid apps [5, 6, 12]. This led to the development of LiveTest to provide an intuitive, customizable, and free software-based OMR scanning system.

2 Related Work

Two established software-based OMR solutions, similar to LiveTest, are publicly available. One such option, Omrquiz.com, offers a robust, free toolset for instructors to create answer sheets with 3 to 5 choices and up to 200 questions. While the service allows users to generate and grade sheets at no cost, its functionality is limited to returning basic results without offering additional analysis or features [6]. ZipGrade is another widely available software-based OMR application that offers free grading for up to 100 answer sheets per month. This reliable, enterprise-level tool provides pre-designed answer sheets for 20, 50, and 100 questions, while also supporting custom sheet creation. For those needing more capacity, the full version—priced affordably at just $7 per year—removes the monthly grading limit [5]. While the cost of some OMR solutions is affordable and can save instructors significant time, the financial burden should not fall on individual educators. Schools should take the initiative to adopt software-based OMR systems. However, many available tools—like those previously mentioned—are designed for individual use rather than school-wide implementation. Comprehensive solutions built for institutional use, such as Remark Office, can cost around $1,300 per year per license, making them impractical for most instructors without institutional support [12].

The best open-source library on GitHub for scanning diverse OMR documents is OMR Checker by Udayraj123 [18]. It implements a very robust process for a general-purpose OMR. You can configure the "OMR layout" with the app and then feed images to it individually to receive the scores.

3 Approach

An initial version of the custom-built OMR LiveTest system using OpenCV achieved considerable success, however, it struggled to handle edge cases due to the open contours of letter answers. This limitation necessitated the design of a new custom answer sheet for this application. At the core of the proposed LiveTest system are two custom-built Python modules: Pictron and OMRGrader, which leverage well-known third-party libraries, including Pillow (a user-friendly fork of PIL), OpenCV, and NumPy. Pictron was developed in response to the lack of open-source tools for generating customizable answer sheets, while OMRGrader was created to accurately grade those newly designed sheets. This section describes the Pictron and OMRGrader modules.

3.1 Creating Answer Sheets

An instance of Pictron is created and an answer sheet is customized with information that contains layout and formatting settings for generating a multiple-choice answer sheet.

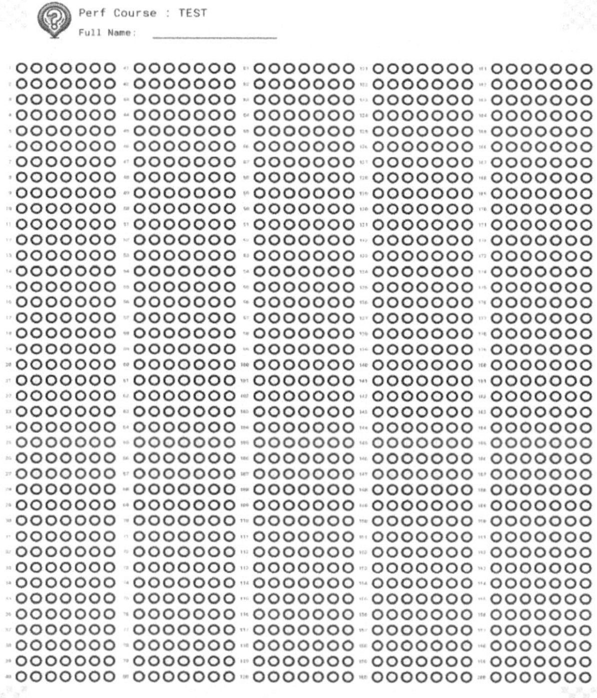

Fig. 1. Example of a Pictron resulting image.

It specifies page size, fonts, bubble appearance, spacing parameters, and the number of questions and answer options to be displayed. Figure 1 shows an example of a Pictron resulting image.

Dynamic configuration options allow educators to extensively customize their answer sheets, potentially creating new ones for every test or customizing them for students with disabilities. Various layouts of answer sheets produced by Pictron are shown in Fig. 2.

Figure 3 shows the two relevant code snippets for the Pictron method addAnswer-Bubbles, illustrating the dynamic process. The addAnswerBubbles method starts by initializing the layout parameters. Then, the answer bubbles are placed for all questions. For each question, the question number, the answer labels and bubbles are placed with the desired spacing. When a bubble is placed, it is filled in if it corresponds to the correct response.

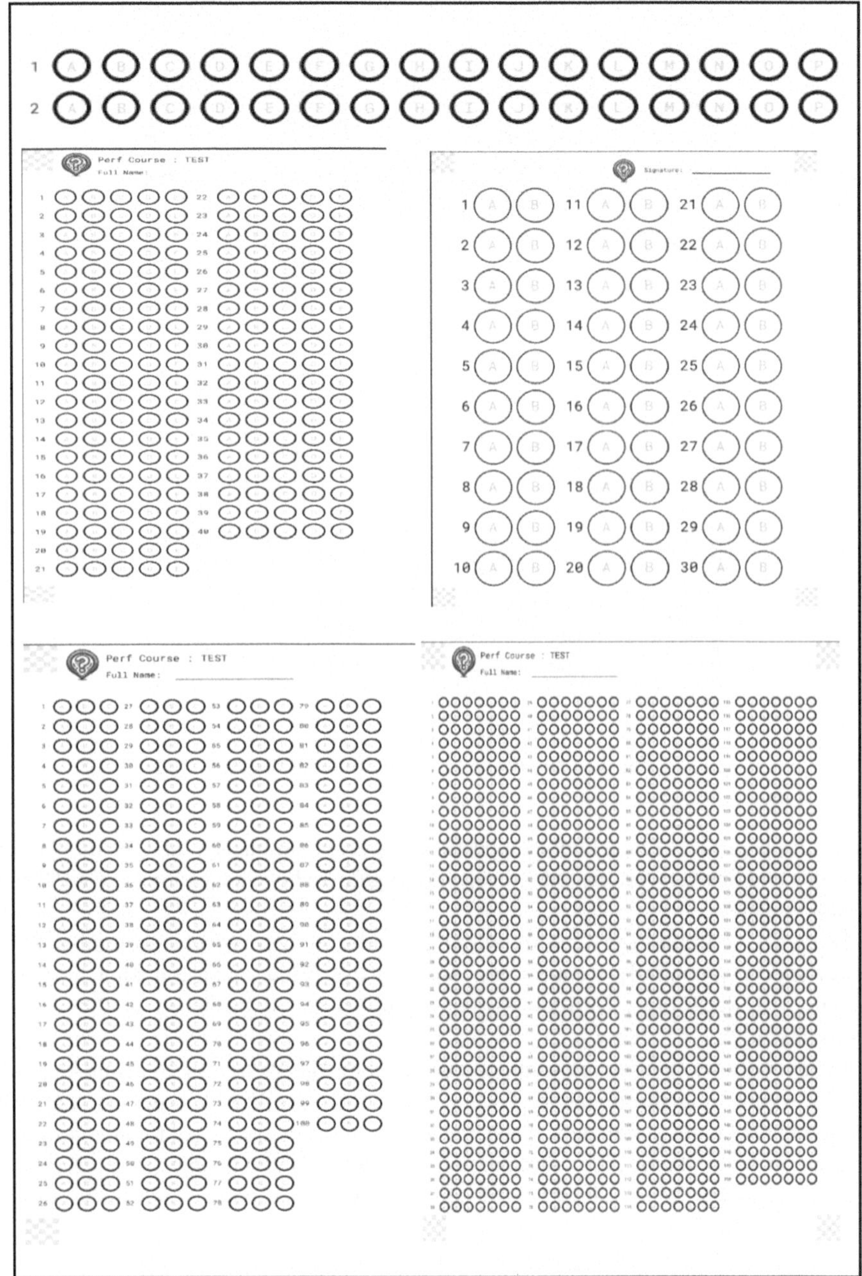

Fig. 2. More example of answer sheets produced by Pictron.

```
Initialize Layout parameters:
  Set the current (x, y) position to the starting coordinates.
  Set counters:
   i to track total answer options placed
   n to track the current question number
  Compute the width required for a full set of answer options and spacing.

Repeat for each answer choice until all questions are processed:
   If this is the first answer choice for a new question:
      If placing a new question would exceed the page height:
          Move to the next column by shifting x-coordinate.
          Reset the y-coordinate to the starting vertical position.

    Display the question number label at the current position.
    Add spacing after the label to begin answer bubbles.

   Determine the answer option label (e.g., A, B, C...) based on its index.
   Place the answer bubble and label at the current (x, y) location
      if it matches a specified correct answer index:
         Fill the bubble.
   Shift the x-coordinate to position the next answer bubble.
   If all answer choices for the current question have been placed:
     Reset x to the original horizontal start.
     Move y downward to the next row.
     Increment the question number n.
     Increment the total answer option counter i.
```

Fig. 3. Pictron.addAnswerBubbles algorithm to build answer sheets of varying sizes.

3.2 Dynamically Grading Answer Sheets

The OMRGrader module provides the core functionality for dynamically grading answer sheets generated by Pictron. It begins by constructing an instance of OMRGrader and configuring key test parameters such as the number of questions, the number of choices per question, and whether the input image is a synthetic Pictron output or a photo of a completed sheet. An optional setting allows the grading process to be visualized step-by-step using OpenCV. Once initialized, the grader processes an image and an accompanying answer key via the run method and returns the final grade, detailed grading results, and a dictionary of the selected choices.

The run method in OMRGrader is the helper function that runs the entire process of grading images of answer sheets produced by Pictron. It starts by isolating the document. To isolate the answer sheet from the image, the original image is loaded into a Python cv2 object and pre-processed. Figure 4 shows how it prepares an image for contour detection by converting it to grayscale and applying Gaussian blur to smooth it out. It then uses OTSU's method to perform thresholding, inverting the result to highlight the edges. Canny edge detection is applied to identify the edges in the image. Finally, the detected edges are dilated to enhance their prominence and reduce background noise [7], making the edges of the answer sheet more distinct for further processing. The dilation handles potential splotchy brightness surrounding the answer sheet in the image.

```
gray = cv2.cvtColor (image, cv2.COLOR_BGR2GRAY)
blur = cv2.GaussianBlur (gray, (5,5), 0)
thresh = cv2.threshold (blur, 0, 255, cv2.THRESH_BINARY_INV | cv2.THRESH_OTSU)
edges = cv2.Canny(thresh, 50, 150)
kernal = np.ones ((4,4), np.uint8)
dilated = cv2.dilate (edges, kernel, iterations=2)
return dilated
```

Fig. 4. Pre-processing of submitted image.

Once the image is pre-processed, the contours are detected in the pre-processed image, retrieving only the outermost contours, ignoring nested contours and compressing contour points by only storing the start and end points of each vertically, horizontally, and diagonally compressed segment. The detected contours are then sorted by area in descending order. (See Fig. 5.) Only the top three largest contours are retained.

```
contours, _ = cv2.findContours(image_proc, cv2.RETR_EXTERNAL,
                                                  cv2.CHAIN_APPOX_SIMPLE)
contours = sorted (contours, key=cv2.contourArea, reverse=True)
```

Fig. 5. findcontours - CHAIN_APPROX_SIMPLE for finding contours in distorted images.

The algorithm then loops over these large contours (see Fig. 6), starting by calculating the perimeter and using it to perform contour approximation with a neutral epsilon value that works well for simple shapes. The first contour with four vertices found will have a four-point transformation performed to obtain a bird's eye view of the image [8, 15], finalizing the document extraction. If the document is not found, it will raise an exception to notify the user that the image needs to be retaken.

After the perspective transformation is applied, answer bubbles are identified using the getAnswerBubbles method (see Fig. 7). Prior to this, it is essential to address low-contrast conditions that may arise from suboptimal image capture or limitations in camera hardware (e.g., mobile or webcam sensors). To enhance local contrast, the algorithm applies Contrast-Limited Adaptive Histogram Equalization (CLAHE) to the transformed image [9]. Unlike global histogram equalization, CLAHE operates on small image tiles and performs local equalization, thereby mitigating issues such as noise amplification and over-saturation in non-uniform illumination. The clip limit parameter constrains contrast enhancement to prevent over-amplification, ensuring stable and consistent contrast across the image.

```
Loop through each contour in the list of contours

  For each contour, calculate its perimeter using the arc length
     function.
  Approximate the contour to a polygon using the Douglas-Peucker
     algorithm with a precision of 2.5% of the perimeter.
  if the approximated polygon has exactly 4 points
     Print message: "Document found. Performing transformation."
     Apply a four-point perspective transformation to the image
        using the four-point polygon.
     Display the transformed image if the option to show the
        process is enabled.
     Return the transformed image as the result.

If no contour with 4 points is found after checking all contours
   Raise an error indicating: "Document could not be isolated."
```

Fig. 6. Document detection and transformation.

```
Enhance the input image:
  Apply histogram equalization to normalize brightness and contrast.
  Convert the image to grayscale.
  Apply Gaussian blur using a 5×5 kernel to reduce noise.

Apply OTSU's thresholding with a binary inverse flag to convert the blurred
image into a binary format where bubbles appear as white regions on a black
background.

Find external contours in the thresholded binary image using simple chain
approximation to simplify the contour shapes.

Initialize an empty list to store contours identified as question bubbles.

Iterate over each detected contour:
  Compute the bounding rectangle of the contour.
  If contour approximates a circular shape (use the is_circle method):
    Add it to the list of question contours.
    If process visualization is enabled:
          Draw the contour on original image in green with a line
          thickness of 2.
     Display the original image with the identified bubbles highlighted.

Return the list of contours identified as question bubbles.
```

Fig. 7. OMRGrader.getAnswerBubbles - Detect and Identify Question Bubbles in an Image.

Subsequently, the image is converted to grayscale and denoised using a Gaussian blur with a 5×5 kernel. OTSU's binarization is then applied to generate a binary inverse image where foreground regions (bubbles) are white and the background is black—facilitating edge and contour detection. Contours are extracted using cv2.findContours, focusing on external ones with simplified polygonal approximation. Each contour is evaluated based on geometric properties, and those approximating circular shapes are added to question_contours. If self.show_process is enabled, these contours are visually overlaid on the original image in green. The method returns the complete list of detected question contours for downstream processing. Figure 8 shows intermediate outputs from the OMRGrader pipeline, including the original, pre-processed, transformed, equalized, and re-thresholded images, along with the image where answer bubbles are highlighted in green.

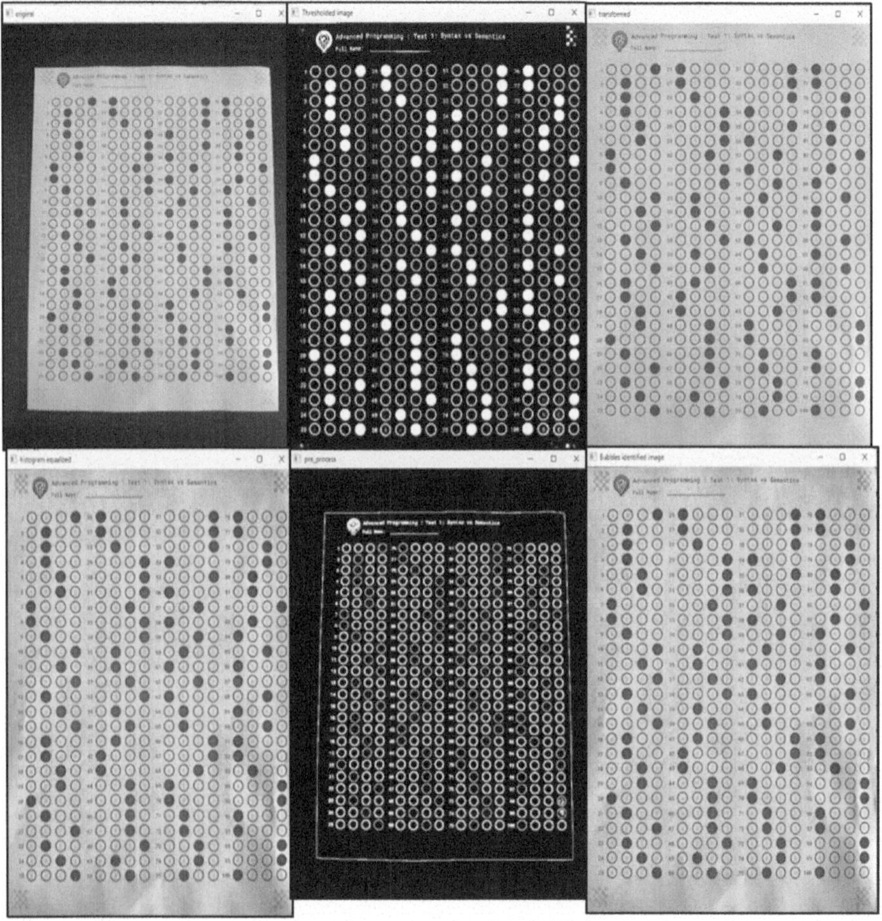

Fig. 8. Images of Original, Pre-Processed, 4-point transformation, histogram equalization, re-thresholded, and answer bubbles highlighted (in green). (Color figure online)

Next, the algorithm must associate each detected answer bubble with its corresponding question. This is done by grouping the bubble contours into rows based on the spatial proximity of their y-coordinates, leveraging vertical alignment as a heuristic. Once row-level grouping is complete, the sort_rows_to_questions method (detailed in Fig. 9) is applied to dynamically assign bubbles to their respective questions, accounting for layout variations and positional tolerances.

```
Define an empty dictionary called questions to store the results.
Initialize starting column index to 0.
Initialize current question number i to 0.

While i is less than the total number of questions:
  Assign the question at index i + 1 a list of bubble contours.
  Get the appropriate row by taking i modulo the number of rows.
  From that row, select the bubbles in columns starting at col_num and
      ending at col_num + number of choices.
        Increment i to move to the next question.
        If i is evenly divisible by the number of rows:
          Increment col_num by the number of choices to shift to next
              group of columns.
Return the questions dictionary.
```

Fig. 9. OMRGrader.sort_rows_to_questions.

After the contours have been grouped for each question, they are already in order from left to right and finally ready to be processed to determine each answer's choice using OMR. The questions are then run through identify_answer_choices. Figure 10 shows that the method first initializes choices. It then begins looping over each of the answers. First, a mask is created to isolate each contour, and the non-zero pixels produced from the threshold within the masked area are counted. The contour with the highest count for each question is identified as the chosen answer. The contour's total count, data, and choice letter are stored in the choices dictionary.

```
Create a dictionary, choices, to store the selected answer for each
question.
For each question and its corresponding contours list in the questions
dictionary:
  For each index and contour in the list of contours:
    Create a blank mask with the same shape as the thresholded image.
    Draw the current contour onto the mask, filling it completely.
    Apply a bitwise AND operation between the mask and the thresholded
      image to isolate the marked area.
    Count the number of non-zero pixels in the resulting mask.
    If this question is not already in choices or this contour has a
      higher total than the previously recorded one:
        Store this contour as the new best choice for the question,
        along with its total and its corresponding letter .
Return the choices dictionary.
```

Fig. 10. OMRGrader.identify_question_choices.

Finally, the grade_choices method is run with the original answer key, highlighting correct answers in green and wrong answers in red. Figure 11 shows the final grade placed in the top right corner.

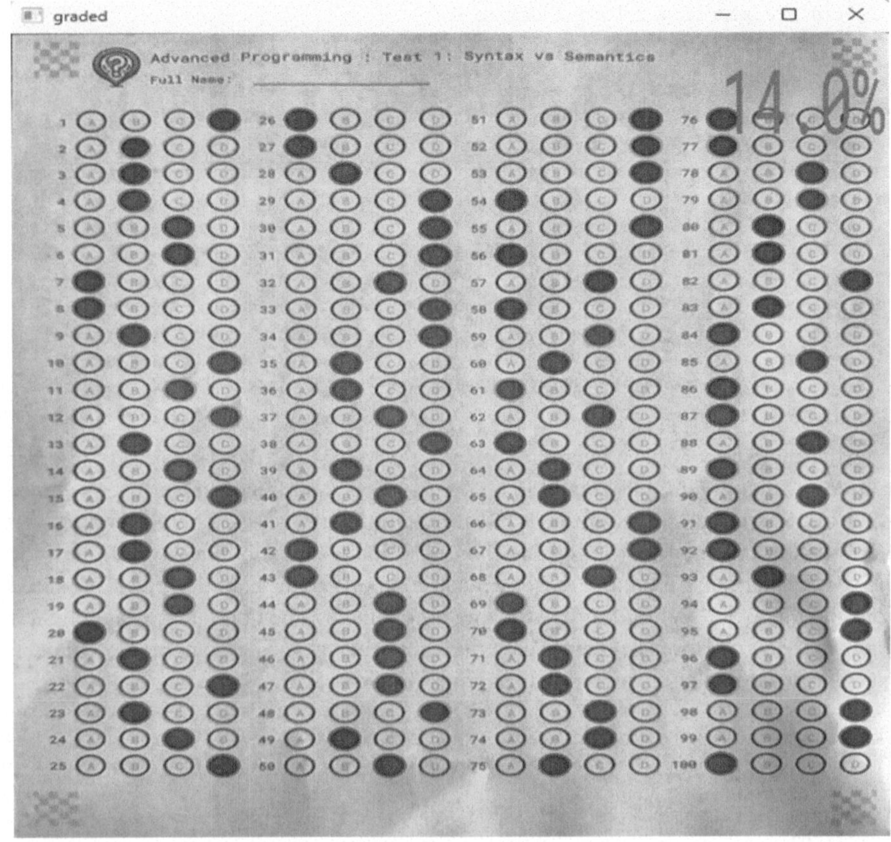

Fig. 11. Final graded image of OMRGrader.run.

3.3 Testing the System

Testing the Pictron and OMRGrader modules involved a rigorous evaluation process, significantly facilitated by a custom demo frontend built in React. An iPhone 14 was used to perform ten complete test sessions, each involving multiple submissions, with all data collection completed in under an hour. In total, over 3,000 answer bubbles were processed. Of these, only 14 were incorrectly graded, yielding an accuracy rate exceeding 99%—comparable to established OMR systems [1]. Nevertheless, each submission should still be reviewed manually, as the system does not guarantee perfect accuracy. Figures 12, 13, 14, 15 present detailed visualizations of key processing stages, demonstrating the dynamic capabilities of both OMRGrader and Pictron.

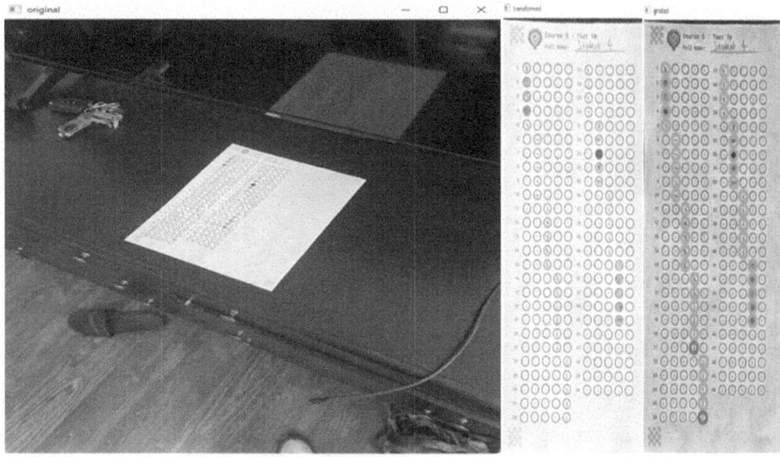

Fig. 12. Showcases ability to take a varied brightness image and grade it.

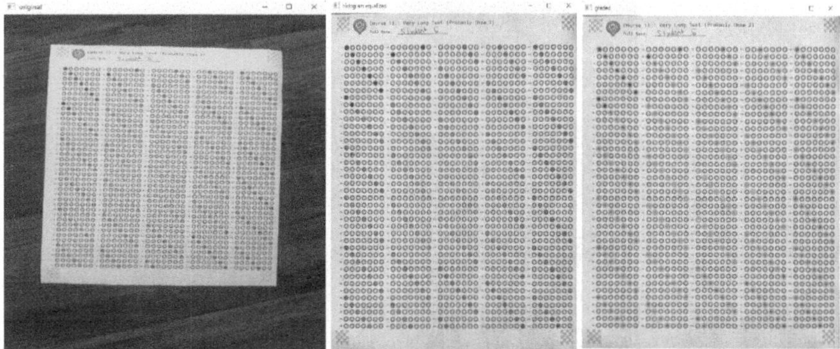

Fig. 13. 1400 individual question bubbles graded with 100% accuracy.

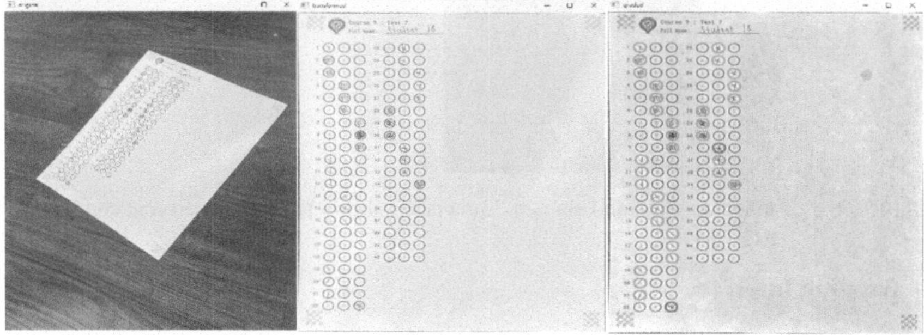

Fig. 14. Angled and tilted, splotchy lighting showcases the benefits of histogram equalization.

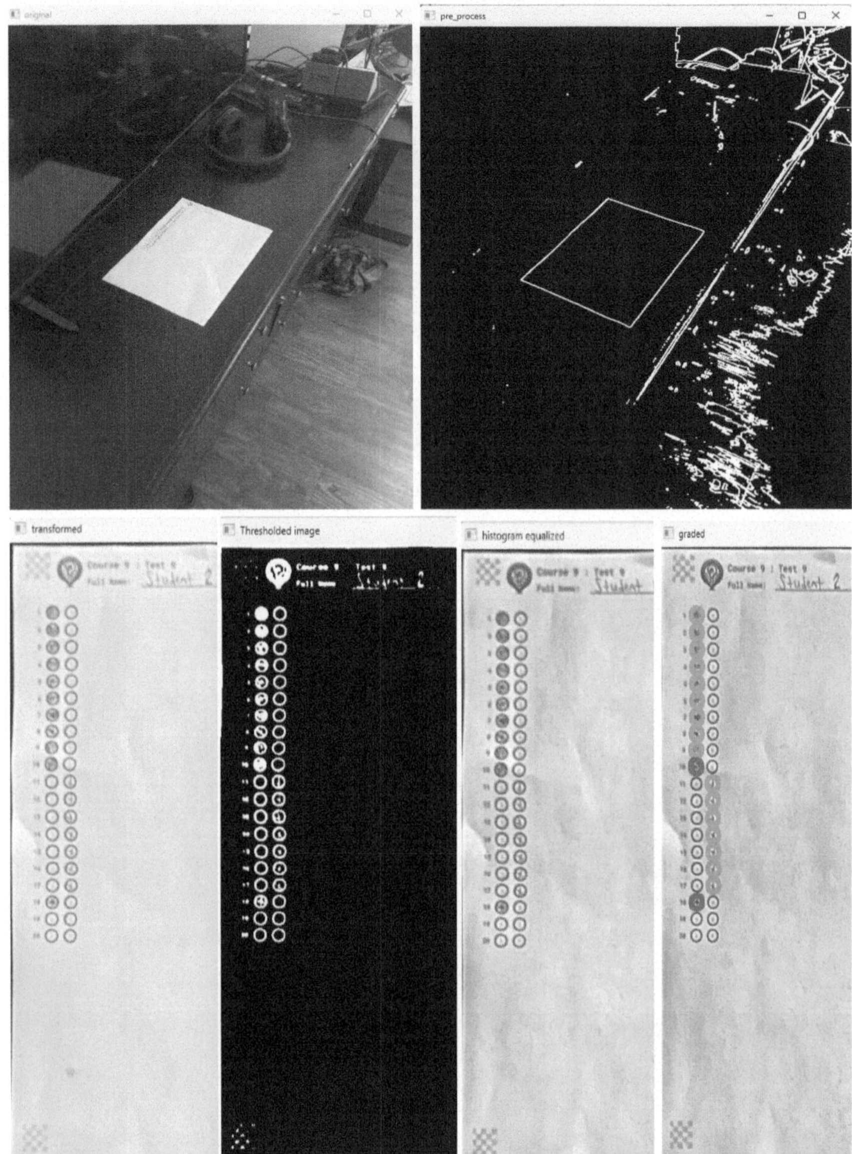

Fig. 15. Far distance, angled, and slanted. Showcases ability to grade in diverse conditions.

3.4 Grading Interface

LiveTest features an intuitive interface designed with faculty usability in mind. Its underlying database architecture operates on a school-specific basis, meaning each LiveTest instance maintains a distinct set of courses, students, instructors, tests, enrollments, and submissions. Upon logging in, both instructors and students are directed to a personalized dashboard displaying all of their associated courses, as illustrated in Fig. 16.

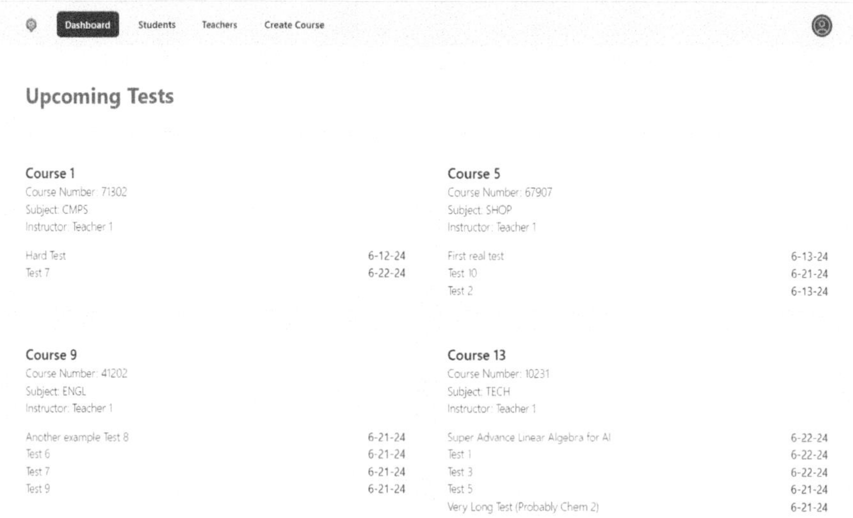

Fig. 16. LiveTest dashboard.

Selecting a course renders the corresponding course page, as shown in Fig. 17, providing access to detailed course information and administrative functionality. Instructors can create new tests, manage student enrollments, and review individual student submissions directly from this interface. Students can be added to a course easily via a unique enrollment QR code, which directs them to a login or registration page. Upon successful authentication or registration, the course is automatically added to their account. This streamlined process simplifies course administration and offers instructors greater flexibility in organizing and managing their classes.

Creating a test should be intuitive. With LiveTest, it is. The teacher simply gives the test a name, start/end time, number of questions/choices, and an answer key. Once all fields are filled out properly, the teacher clicks "Create Test". LiveTest then automatically creates the test artifact for the specific course (see Fig. 18).

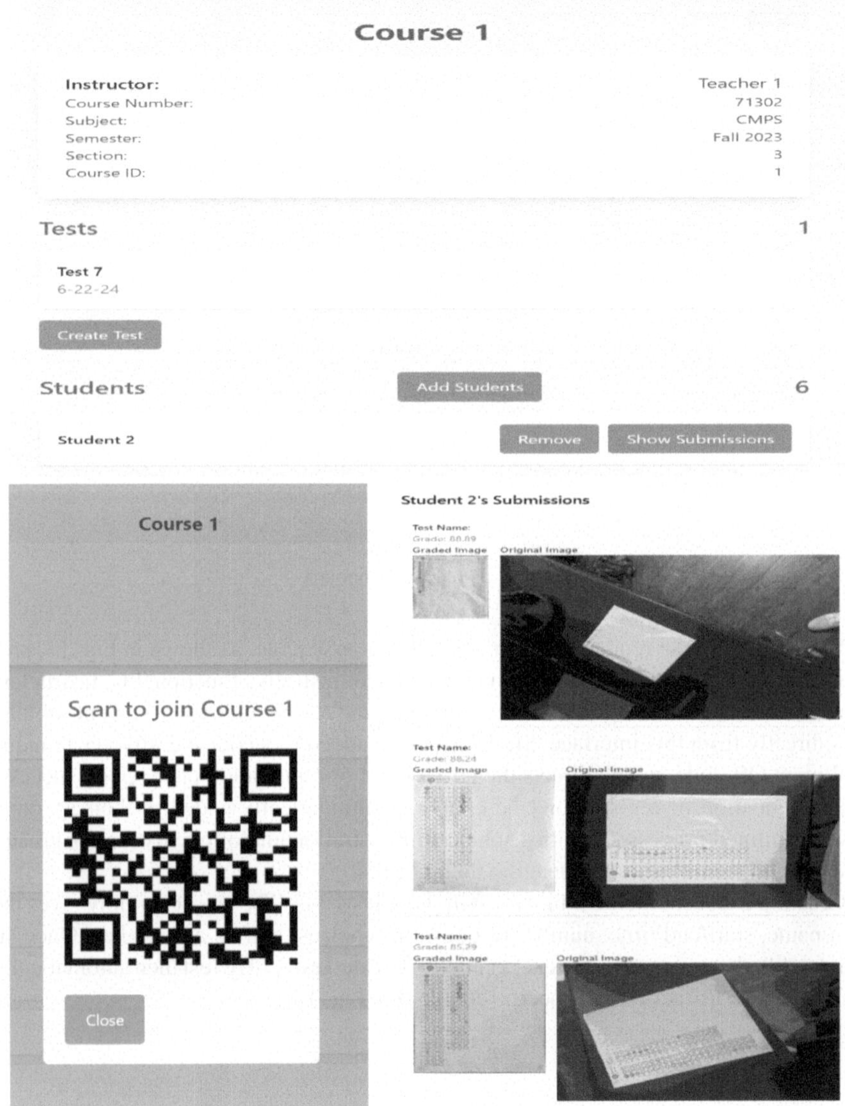

Fig. 17. Course Page, Course enrollment QR code, Student's Submissions.

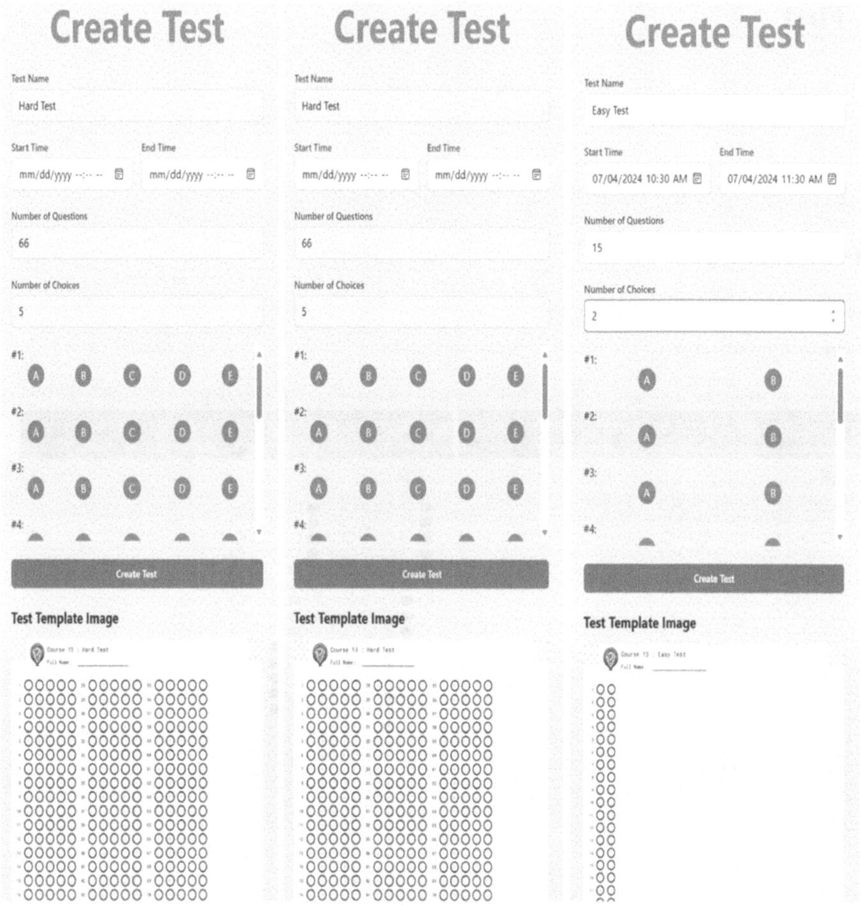

Fig. 18. Create Test - U.I. Examples showcasing dynamic capabilities of Pictron.

Once a test is created, users can access its dedicated Test Page, as shown in Fig. 19. Prior to the scheduled start time, the test remains in an "IN WAIT" state and does not accept submissions. From this page, instructors can modify the test's name, adjust the start and end times, delete the test, download blank answer sheets, and retrieve the corresponding answer key used for grading.

4 First

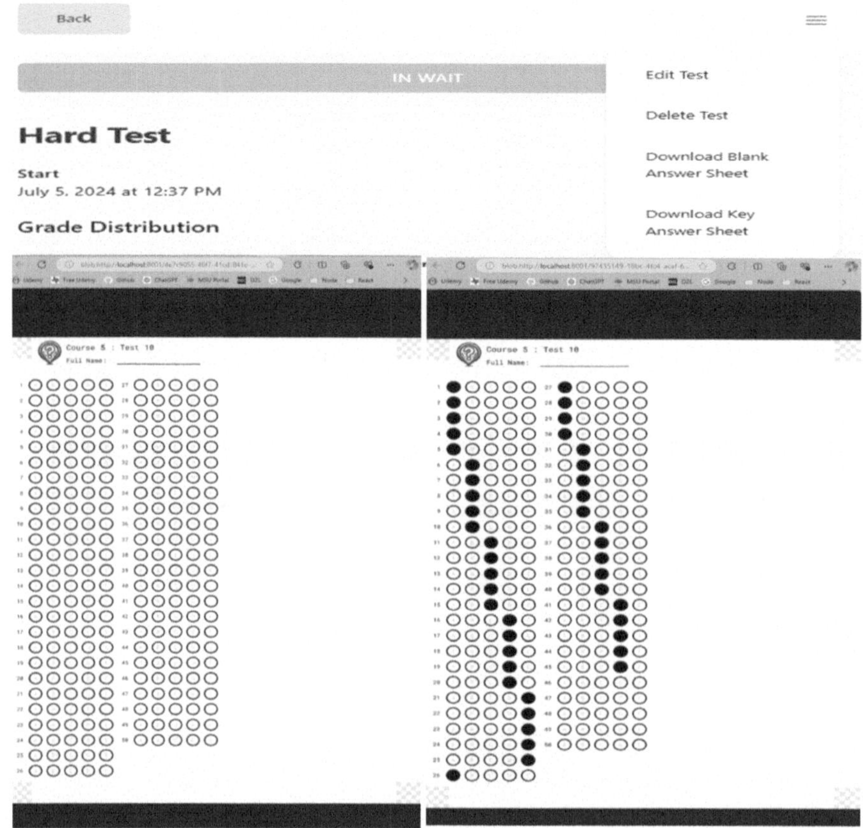

Fig. 19. Test Page "IN WAIT" along with downloaded blank/key answer sheets.

Figure 20 shows that once test grading begins, the status bar turns green and the test enters the "LIVE" state, indicating that it is now an active LiveTest. While the test is live, instructors can submit graded answer sheets by selecting the blue "Create a new submission" button, initiating the automated grading process.

Figure 21 highlights the scanner's current limitations by showing examples of both accepted and rejected markings. If students have difficulty following marking guidelines, instructors can reference these examples to clarify expectations. To submit a response, the instructor selects a student from the relevant course, captures an image of the test sheet using an iOS or Android device, or uploads one from a PC or Mac, and then initiates the grading process. Currently, if grading fails, the submission must be retried manually. This interface supports single-image submissions after a student is selected. While future updates aim to incorporate a more advanced, continuous video feed—accessible via web or mobile camera—that automatically detects and repeatedly attempts to grade the document, the current release provides a simplified Submission Page (see Fig. 22). Upon successful grading, results are immediately displayed, allowing the instructor to either retake the image or proceed to the next student.

Once submissions have been received for a test, the Analytics Page automatically generates visualizations such as grade distribution and most frequently missed questions, as shown in Fig. 23. While full analytics are displayed once a test is finalized, the system also updates these metrics in real time during the live testing period.

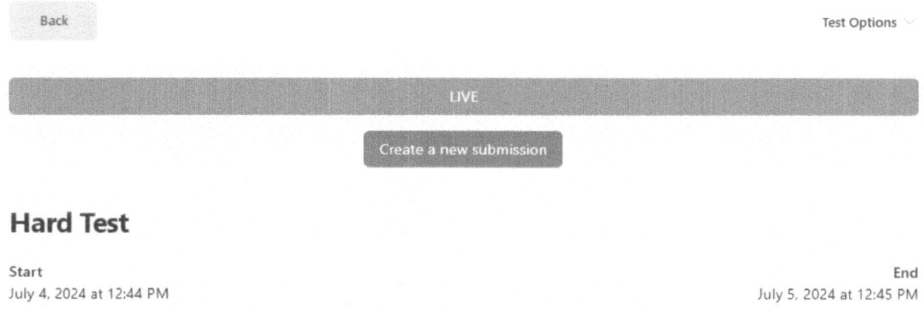

Fig. 20. Test Page "LIVE.

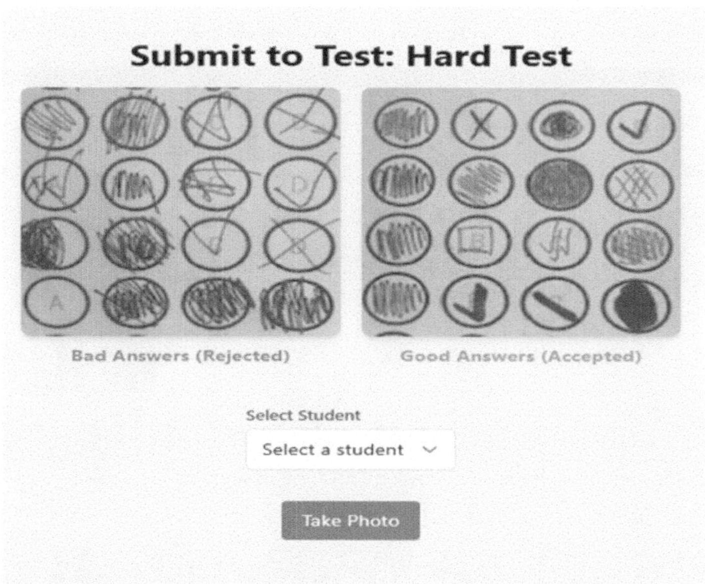

Fig. 21. Rejected and accepted markings.

Fig 22

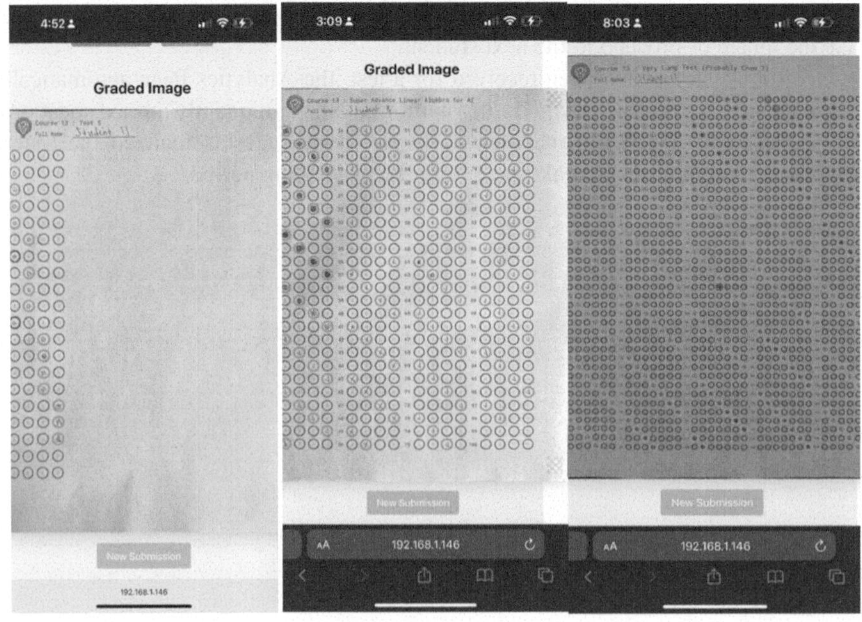

Fig. 23. Submission Page.

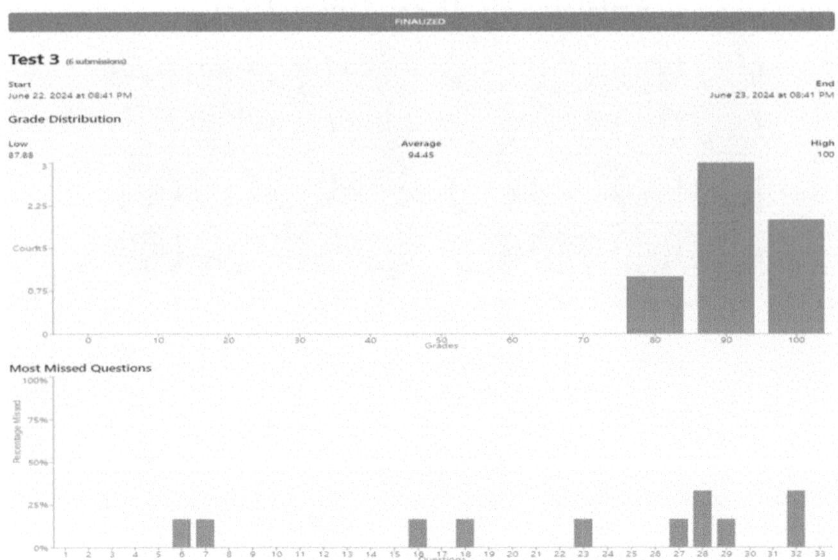

Fig. 24. Test Analytics Page – Finalized.

For grading purposes, all test submissions are listed beneath the Analytics Page, with the option to view individual student submissions as shown in Fig. 24. While the current demo provides useful functionality, the frontend remains limited compared to the robustness of the backend. A technically proficient user can effectively leverage LiveTest's existing capabilities to process and grade student assessments; however, the platform is not yet refined enough for general use or broad adoption in its present state (Fig. 25).

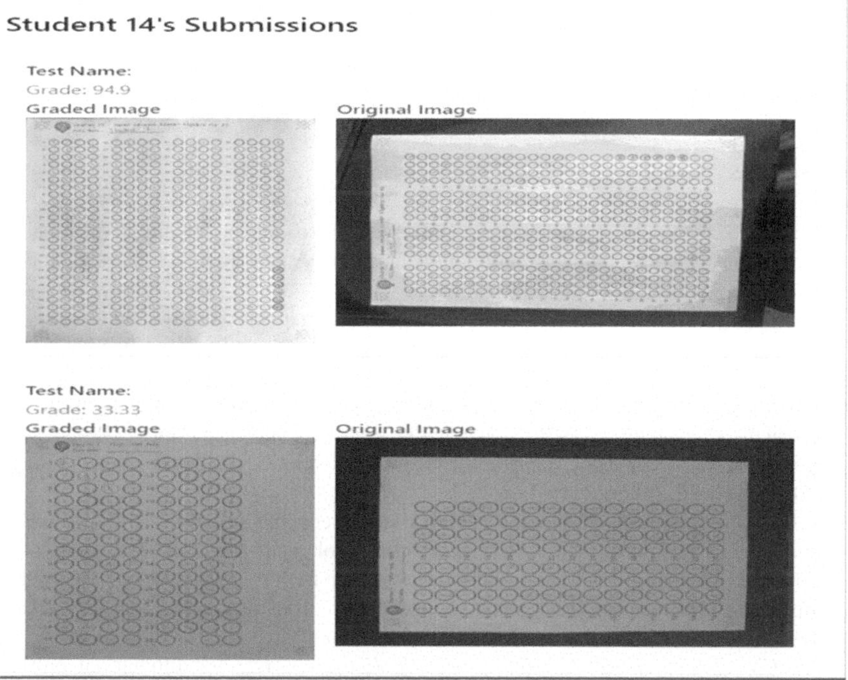

Fig. 25. Student's Submissions Page

5 Conclusion

Optical Mark Recognition (OMR) has long been a foundational tool in educational assessment, valued for its efficiency and accuracy in grading multiple-choice exams. Despite technological advancements, OMR remains prevalent and is expected to continue as a standard testing method. However, traditional hardware-based OMR systems pose financial and accessibility challenges for many institutions. LiveTest was developed to overcome these limitations by providing a cost-effective, software-based alternative that is both accessible and easy to use. Early trials demonstrated an accuracy rate of 99%, but further development is required to scale the platform for widespread institutional use.

Interesting to note, a convolutional neural network (CNN), specifically using a generative adversarial network (GAN), utilized sample data in this work and synthetic data to improve the accuracy of recognizing markings from 99% to 100% in nine real-world samples [11].

To foster continued enhancement and community engagement, LiveTest has been open-sourced. The GitHub repository (github.com/gramcracker40/LiveTest) offers comprehensive setup instructions, enabling users to deploy the demo frontend and backend using Docker with just two commands. Live testing is currently active at livetest.us, where educators can register and begin administering assessments.

Acknowledgments. This research was inspired by Adrian Rosebrock's blog post on pyimagesearch.com, Bubble sheet multiple-choice scanner, and grader using OpenCV [13].

Disclosure of Interests. The authors have no competing interests to declare that are relevant to the content of this article.

References

1. Agarwal, S., Varun, M., Prabakeran, S.: OMR reader info scanner. In: 9th International Conference on Advanced Computing and Communication Systems (ICACCS2023), Coimbatore, India, pp. 205–209 (2023)
2. Calaguas, N., Consunji, P.: Mobile optical mark recognition application as a non-inferior alternative to manual marking for lower and middle income countries. J. Adv. Stud. **5**, 10–11 (2022)
3. Ison, M.: Effectiveness of automation in evaluating test results using EvalBee as an alternative optical mark recognition (OMR): a quantitative-evaluative approach from a Philippine public School. Int. J. Theory Appl. Elementary Secondary School Educ. (2021). https://doi.org/10.31098/IJTAESE.V3I2.661
4. de Elias, E., Tasinaffo, P. Hirata, R.: Optical mark recognition: advances, difficulties, and limitations. SN COmput. Sci. **2**, 367 (2021). https://doi.org/10.1007/s42979-021-00760-z
5. ZipGrade LLC (n.d.): IPhone and Android grading app for formative assessment and quizzes. https://www.zipgrade.com/. Accessed 16 June 2025
6. Omr Quiz: OMR Quiz (n.d.). https://www.omrquiz.com/. Accessed 15 May 2024
7. OpenCV: eroding and dilating (n.d.). https://docs.opencv.org/3.4/db/df6/tutorial_erosion_dilatation.html. Accessed 15 May 2024
8. OpenCV: Geometric Transformations of Images (n.d.). https://docs.opencv.org/3.4/da/d6e/tutorial_py_geometric_transformations.html#:~:text=To%20find%20this%20transformation%20matrix%2C%20you%20need%204,Then%20apply%20cv.warpPerspective%20with%20this%203x3%20transformation%20matrix. Accessed 15 May 2024
9. OpenCV: Histograms - 2: Histogram Equalization (n.d.). https://docs.opencv.org/4.x/d5/daf/tutorial_py_histogram_equalization.html. Accessed 15 May 2024
10. OpenCV: Image thresholding. (n.d.). https://docs.opencv.org/4.x/d7/d4d/tutorial_py_thresholding.html. Accessed 15 May 2024
11. Pulliam, K., Griffin, T., Stringfellow, C.: Generating synthetic datasets to improve deep learning model training. In: Proceedings of International Conference on Computational Science and Computational Intelligence (CSCI2024). Springer Nature (2025)
12. Remark Software Online Store. Remark Software Store (n.d.). https://buy.remarksoftware.com/
13. Rosebrock, A.: 4 point OpenCV GetPerspective Transform Example. PyImageSearch. (2021). https://pyimagesearch.com/2014/08/25/4-point-opencv-getperspective-transform-example/. Accessed 17 June 2025

14. Rosebrock, A.: Bubble sheet multiple choice scanner and test grader using OMR, Python, and OpenCV. https://pyimagesearch.com/2016/10/03/bubble-sheet-multiple-choice-scanner-and-test-grader-using-omr-python-and-opencv/. Accessed 17 June 2025
15. Singh, A., Karayev, S., Gutowski, K., Abbeel, P.: Gradescope: a fast, flexible, and fair system for scalable assessment of handwritten work. In: 4th ACM Conference on Learning @ Scale (L@S 2017), pp. 81–88. Association for Computing Machinery, New York, NY, USA (2017)
16. Software Online Store. Remark Software Store (n.d.). https://buy.remarksoftware.com/. Accessed 15 May 2024
17. Stimper, V., Bauer, S., Ernstorfer, R., Schölkopf, B., Xian, R.: multidimensional contrast limited adaptive histogram equalization. IEEE Access **7**, 165437–165447 (2019)
18. Udayraj (n.d.): GitHub - Udayraj123/OMRChecker: Evaluate OMR sheets fast and accurately using a scanner or your phone. GitHub. https://github.com/Udayraj123/OMRChecker. Accessed 17 June 2025

Securing E-Commerce: A Framework for Risk Management and Compliance

Derek Taylor[1(✉)], Mehdi Barati[2], and Nick Rahimi[1]

[1] School of Computing Sciences and Computer Engineering, University of Southern Mississippi, Hattiesburg, MS, USA
{derek.taylor,nick.rahimi}@usm.edu

[2] School of Social Science and Global Studies, University of Southern Mississippi, Hattiesburg, MS, USA
mehdi.barati@usm.edu

Abstract. The rapid growth of e-commerce has transformed how businesses operate globally, creating new opportunities but also introducing significant security challenges. As online platforms process millions of transactions daily, they face increasingly sophisticated cyber threats that can compromise customer data and disrupt business operations.

Security management becomes particularly complex in the global marketplace. Organizations must navigate multiple regulations, such as the European Union's General Data Protection Regulation (GDPR) and California's Consumer Privacy Act (CCPA), which set strict standards for data protection and privacy. Additionally, supply chain security presents critical challenges, as demonstrated by recent breaches where attackers exploited vulnerabilities in vendor systems to access larger networks.

This research analyzes current e-commerce security challenges and potential solutions. We examine how organizations implement established frameworks like the Payment Card Industry Data Security Standard (PCI DSS), while also adopting newer technologies such as advanced encryption, blockchain, and AI-based threat detection. Our findings indicate that while technological solutions are essential, they must be complemented by strong human factors - particularly staff training and security awareness, as human error remains a significant factor in security breaches.

Based on empirical evidence and industry practices, we propose practical security recommendations. These include implementing Zero Trust Architecture to strengthen access controls, following international security standards for better compliance, and developing ethical data handling practices. Our research emphasizes the importance of regular security assessments to address emerging threats.

The findings provide organizations with practical approaches to protect sensitive data and maintain customer trust. By implementing these recommendations, businesses can strengthen their security measures while building a foundation for sustainable growth in digital commerce.

Contributing authors—M. Barati and N. Rahimi

I. Roy et al. (Eds.): CAINE 2025, CCIS 2740, pp. 244–261, 2026.
https://doi.org/10.1007/978-3-032-14893-3_17

Keywords: e-commerce security · cyber threats · data protection · regulatory compliance · AI-enabled detection · zero trust architecture

1 Introduction

E-commerce has fundamentally transformed modern business, creating a digital marketplace that never sleeps. Online shopping platforms now process millions of transactions daily, handling sensitive customer data including credit card numbers, personal information, and shopping habits [1,2]. This wealth of valuable data makes these platforms prime targets for cybercriminals, who employ increasingly sophisticated attack methods ranging from basic phishing to advanced ransomware [3,4].

The threat landscape continues to evolve as attackers adopt new technologies [5]. Artificial intelligence now powers automated attacks, while deepfake technology enables more convincing fraud attempts [6–8]. These technological advances force businesses to constantly update their security measures, creating an ongoing challenge for e-commerce platforms [9].

Beyond direct cyber threats, e-commerce businesses must navigate a complex web of international regulations. The European Union's General Data Protection Regulation (GDPR) and California's Consumer Privacy Act (CCPA) impose strict rules on how companies collect and handle customer data [10–13]. Breaking these rules can result in severe consequences - from heavy fines to loss of customer trust and business disruption [14].

Supply chain security adds another layer of complexity [15]. Most e-commerce platforms rely on numerous third-party vendors for services like payment processing and shipping [16]. Each connection creates potential security risks, as demonstrated by the 2013 Target breach where attackers gained access through a heating and cooling vendor's compromised credentials [17]. Recent research has shown that blockchain technology could help secure these supply chain connections [18,19].

This paper examines how e-commerce platforms can protect themselves in this challenging environment. Through analysis of real-world security incidents [20] and current best practices [21], we provide a practical roadmap for implementing effective security measures. Our recommendations focus on balancing robust security with business efficiency [22], helping organizations protect sensitive data while maintaining smooth operations.

The need for comprehensive security awareness and training has become increasingly critical [22,23]. As cyber threats evolve, organizations must adopt both technological solutions and human-centric approaches to security [24]. The integration of advanced security measures with employee training creates a more robust defense against modern cyber threats [25].

The remainder of this paper is organized as follows: Sect. 2 analyzes the current state of e-commerce security, examining present threats and vulnerabilities. Section 3 explores the legal framework governing e-commerce security, including GDPR, CCPA, and other key regulations. Section 4 presents a comprehensive policy framework for implementing security measures. Section 5 discusses the challenges and risks facing e-commerce platforms, while Sect. 6 details the current threat landscape. Sections 7 and 8 focus on emerging threats and implementation strategies, respectively. Sections 9 and 10 cover key regulations and compliance monitoring. The paper concludes with case studies, recommendations, future directions, and concluding thoughts in Sects. 11, 12, 13 and 14.

2 Current State and Analysis

The landscape of e-commerce security in 2024 presents a complex web of challenges that extend beyond simple cyber defense [1,7]. Our analysis reveals three critical dimensions of the current threat environment: evolving attack patterns, technological shifts in criminal tactics, and systemic vulnerabilities in the e-commerce ecosystem [2].

First, traditional cyber threats have grown both in sophistication and frequency [20]. While phishing attacks remain a primary concern - accounting for 36% of global breaches in 2022 [26] - their nature has evolved significantly. Modern phishing attempts no longer rely on obvious spelling errors or suspicious links. Instead, they utilize social engineering tactics combined with stolen corporate branding and legitimate-looking domains [3]. Ransomware attacks have also shifted from simple encryption-based extortion to sophisticated double-extortion tactics, where criminals both encrypt data and threaten to publish sensitive information unless paid [9].

The technological battlefield has transformed dramatically with the emergence of AI-powered cyber threats [5,8]. Criminal organizations are now deploying machine learning algorithms to automate attack patterns and bypass security systems. These AI systems can analyze patterns in security responses and adapt in real-time, making them particularly challenging to defend against [6]. For instance, recent cases have shown AI-generated phishing emails achieving success rates up to 30% higher than traditional methods [4]. Even more concerning is the rise of deepfake technology in fraud attempts, where artificial intelligence creates convincing audio or video impersonations of executives to authorize fraudulent transactions or access [6,24].

The third and perhaps most challenging dimension lies in the interconnected nature of e-commerce operations [15]. Modern online retail platforms typically integrate with 20–30 third-party services, from payment processors to inventory management systems [16]. Each integration represents a potential security vulnerability. The infamous 2013 Target breach exemplifies this risk perfectly - attackers gained access not through Target's own security systems but through an HVAC vendor's compromised credentials [17]. This incident sparked a fundamental shift in how businesses approach supply chain security [18], yet similar vulnerabilities persist across the industry.

In response to these challenges, leading e-commerce platforms have begun implementing multi-layered security strategies [21]. Advanced encryption protocols protect data both at rest and in transit, while blockchain technology is being explored for securing transaction records and supply chain integrity [19]. Multi-factor authentication has become standard practice, though implementation quality varies significantly across platforms [16].

However, technological solutions alone have proven insufficient [22]. Our analysis reveals a critical gap in human-centric security measures. While 67% of surveyed e-commerce platforms reported substantial investments in security technology, only 23% maintained comprehensive security training programs for their staff [23]. This disparity helps explain why human error continues to play a role in over 85% of successful cyber breaches [26].

On top of these challenges, keeping up with regulations has become quite a juggling act [10,11]. Online businesses now have to deal with a whole range of international rules - from Europe's GDPR [12] to California's CCPA [13], plus various local privacy laws [14]. To handle this complexity, many companies are turning to systems that combine security measures with compliance tracking [9].

The explosion of mobile shopping has brought its own security headaches [25]. These days, more than 60% of people buy things on their phones, but keeping these purchases safe isn't simple. We're seeing new threats pop up specifically targeting mobile users - think counterfeit apps, risky public WiFi connections, and tricky text message scams (smishing) that need special security solutions [2].

To tackle all these challenges effectively, businesses need a security plan that covers everything - from technical safeguards to human behavior and supply chain vulnerabilities [15,20]. Let's dive into some specific strategies and frameworks that can help build this kind of comprehensive protection.

3 Legal Framework and Regulatory Compliance

Operating in the global e-commerce space requires navigating an increasingly complex legal environment, where regulations vary by region but have global consequences. This section outlines the foundational legal frameworks that govern e-commerce platforms and explores the practical challenges businesses face when trying to remain compliant across jurisdictions.

3.1 Global Regulatory Landscape

Online businesses today must comply with a patchwork of overlapping and evolving regulations [10,11]. Many of these laws apply not based on where the business is located, but where the customer resides. This extraterritorial application means a company based in one country may be held accountable under another country's laws. Regulatory enforcement is becoming stricter, with higher financial penalties and a greater emphasis on protecting consumer rights, particularly concerning transparency and control over personal data [1,10,11,16,23].

As a result, companies often find themselves juggling compliance with multiple regulations simultaneously, each with its own set of requirements. The

GDPR, for instance, mandates explicit user consent and timely breach notification, while the CCPA focuses on consumer opt-out rights and grants individuals the power to sue for data breaches [9,14]. These frameworks evolve rapidly, forcing organizations to continuously monitor changes and update systems accordingly [10,13]. Implementing these requirements often involves building out extensive technical infrastructure, which can be particularly burdensome for small to mid-sized enterprises [2,20].

General Data Protection Regulation (GDPR). The GDPR is considered the benchmark for data privacy laws globally [10,12]. It requires businesses to integrate privacy into the design of systems, offer data portability, and report data breaches within 72 h. Organizations must obtain clear consent before collecting personal data and provide mechanisms for users to control how their data is used.

Enforcement examples include Amazon's $812 million (€746 million) fine for improper use of consumer data, WhatsApp's $245 million (€225 million) penalty for transparency failures, and Google's $54 million (€50 million) fine related to consent management [23]. These cases underscore the financial and reputational risks associated with non-compliance.

California Consumer Privacy Act (CCPA). As the most comprehensive privacy law in the United States, the CCPA grants California residents rights similar to those under the GDPR, such as the ability to access, delete, and opt out of the sale of their personal data. The law also introduces specific disclosure and annual reporting requirements. Unlike the GDPR, the CCPA provides a private right of action, enabling consumers to sue companies for certain types of data breaches.

Practical Challenges

Despite their intentions, regulations like the GDPR and CCPA pose several practical challenges. The most significant include: Compliance Across Borders: E-commerce platforms operating internationally must interpret and implement regulations for each market they serve. This often results in duplicated efforts, redundant systems, and increased operational costs.

Interpreting Ambiguities: Regulatory language can be vague or open to interpretation. For example, defining what constitutes "reasonable" security measures varies between regions and industries.

Resource Allocation: Achieving compliance is resource-intensive, particularly for companies that lack dedicated legal or compliance teams. Hiring experts, implementing new technology, and training staff represent significant investments.

Keeping Pace with Change: Laws are not static. They are amended and updated regularly, creating an ongoing compliance burden. Companies must stay updated to avoid falling behind.

3.2 Industry-Specific Standards

In addition to general data protection laws, e-commerce platforms must adhere to industry-specific standards. These frameworks often provide more detailed guidance on topics such as payment processing and data security. For example, the Payment Card Industry Data Security Standard (PCI DSS) imposes rigorous controls for handling credit card information, including encryption, authentication, and continuous monitoring.

India's proposed Personal Data Protection Bill (PDPB) adds another layer of complexity, requiring data localization and introducing new consent models. As countries develop their own privacy laws, global businesses must adapt quickly or risk penalties.

3.3 Case Study Alignment

Case studies such as the 2013 Target breach and the 2023 Alibaba Cloud incident provide real-world illustrations of the regulatory landscape in action. Although not perfect examples, they are close representations of how gaps in compliance and vendor oversight can lead to catastrophic breaches. Target's breach exposed weaknesses in vendor management and internal segmentation, while Alibaba's highlighted the risks tied to third-party software and cloud security. Both cases spurred regulatory responses and industry-wide reforms, reinforcing the need for stringent legal and security compliance.

The 2024 AT&T data breach adds another instructive example. Occurring in April 2024, it compromised call and text metadata of nearly all AT&T wireless customers, due to unauthorized access via Snowflake cloud accounts. Though no message content was breached, the incident revealed severe lapses in cloud credential security and emphasized the need for enforcing multi-factor authentication. AT&T's timely disclosure and subsequent collaboration with law enforcement showcased good crisis response, but the breach itself exposed new weaknesses in third-party cloud reliance. This case illustrates the ongoing challenges of balancing operational efficiency, customer trust, and regulatory obligations in large-scale data ecosystems.

These examples demonstrate how businesses that fail to adopt a comprehensive and proactive compliance strategy are vulnerable to both attacks and the significant financial consequences that follow. By understanding the legal frameworks and anticipating challenges, e-commerce platforms can better position themselves for secure and compliant operations in the digital economy.

3.4 Compliance Strategy Framework

To navigate these complex requirements effectively, organizations should adopt a structured approach [9,16]. Table 1 summarizes the regulatory compliance strategy framework [23].

Table 1. Major Legal Frameworks Impacting E-commerce Platforms

Framework	Region	Focus Area	Penalties for Non-Compliance
GDPR [12]	European Union	Transparency, Data minimization	Up to €20M or 4% annual turnover
CCPA [13]	California (USA)	Consumer rights, Opt-out mechanisms	Up to $7,500 per intentional violation
PCI DSS [16]	Global	Payment security, Fraud prevention	Service restrictions, Financial penalties
PDPB [14]	India	Data localization, Consent management	Up to $2M or 4% of turnover
LGPD	Brazil	Data protection, Consumer rights	Up to 2% revenue or $10M per violation

3.5 Emerging Trends and Future Considerations

Several developments are shaping the future of e-commerce regulation [7,8]:

- **Cross-border Data Flows**: Growing complexity in international data transfer requirements [10]
- **AI Regulation**: Emerging frameworks for artificial intelligence and automated decision-making [6,8]
- **Privacy-Enhancing Technologies**: Requirements for privacy-preserving techniques in data processing [25]
- **IoT Security**: New standards for connected devices and smart retail systems [15,19]

3.6 Practical Implementation Guidelines

Successful implementation of security measures in e-commerce platforms requires a comprehensive and systematic approach to compliance and risk management [9,22]. Organizations must begin by establishing a unified compliance program that effectively addresses overlapping requirements across various regulations [14]. This involves creating standardized processes that satisfy multiple regulatory frameworks simultaneously while maintaining detailed documentation of all security measures and procedures [16,20].

Clear accountability structures form the backbone of effective implementation [21]. Organizations should designate dedicated compliance officers who oversee the entire security framework and coordinate with different departments [22]. These officers need well-defined roles and responsibilities, supported by robust reporting mechanisms that ensure swift communication of potential security issues [2]. The reporting structure should facilitate both routine updates and urgent security notifications, enabling quick response to emerging threats [1].

4 Policy Framework

To succeed in today's online marketplace, businesses need security that's both thorough and forward-thinking [9,20]. Let's look at the key policies that can help protect your business while keeping customers' trust and staying on the right side of regulations [22].

4.1 Zero Trust Architecture (ZTA)

Google's BeyondCorp framework exemplifies Zero Trust architecture in practice, removing traditional perimeter-based security and enforcing contextual access control [27]. The old way of handling security - trusting people once they're inside the system - just doesn't cut it anymore [21]. That's where Zero Trust comes in. Think of it as a "trust no one" approach: every user and device has to prove themselves every time, no matter where they are or what access they've had before [2].

This might sound strict, but it works. Take the case of a major online store that switched to this approach - they saw security incidents drop by 60% after implementing things like two-factor authentication, constant monitoring, and smart access controls [16]. By checking credentials at every step, from first login to accessing specific data, they've made it much harder for both outside attackers and potential internal threats to cause harm [3].

4.2 AI-Driven Security and Ethics

Mastercard, for example, has successfully used AI to improve fraud detection and reduce false positives, demonstrating AI's dual impact on security and user experience [28]. While AI has become a game-changer in catching fraud [7], we can't just throw technology at the problem without thinking about the ethical implications [8]. Security systems need to be effective, but they also need to be fair and transparent [6].

Here's a real example: an online platform started using AI to spot fraud and managed to cut fraudulent transactions by 40% [5]. Great news, right? But then they noticed the system was showing some bias in its decisions. Instead of ignoring the problem, they added fairness checks and gave users more control over how the AI analyzed their activities [4]. It's a perfect example of why we need to balance powerful technology with ethical considerations.

4.3 Supply Chain Security

E-commerce platforms rely heavily on third-party vendors, making supply chain security crucial [15]. After the Target data breach, which originated from compromised vendor credentials [17], the industry has shifted toward more stringent vendor management [18]. Effective policies now include detailed vendor security assessments, regular audits, and clear security requirements in all contracts [16]. Many platforms have implemented real-time monitoring systems to track vendor access and detect unusual patterns [1], helping prevent similar breaches before they occur.

4.4 Data Management and Privacy

Privacy regulations like GDPR [12] and CCPA [13] have made data management policies essential for e-commerce success [10,11]. Modern platforms must clearly explain how they collect and use customer data, while giving users control over their information [14]. This approach goes beyond mere compliance it builds trust [23]. For example, when a leading global e-commerce company updated its privacy policies to exceed GDPR requirements, it saw increased customer engagement and fewer privacy-related complaints [19].

4.5 Implementation Strategy

Successful policy implementation requires coordination across all business areas [9]. Organizations should:

- Start with clear documentation of all security requirements [16]
- Train employees regularly on security procedures [22]
- Monitor compliance continuously [2]
- Update policies based on new threats and regulations [25]
- Maintain open communication with customers about security measures [10]

When all these pieces work together, online stores get better protection against cyber threats while building trust with their customers [20]. But it's not a set-it-and-forget-it solution - regular check-ups and updates are crucial as both technology and threats keep changing [7]. The trick is finding that sweet spot between tight security and user-friendly experience [9] - after all, even the best security isn't helpful if it makes your platform too frustrating to use.

By weaving in cutting-edge technologies like blockchain [18] and AI [8], online businesses can build an even stronger security foundation. This forward-thinking approach doesn't just handle today's threats - it helps businesses stay ahead of whatever challenges tomorrow might bring [21].

Table 2. Common Cybersecurity Threats and Their Impact on E-commerce

Threat Type	Primary Impact	Real-World Example	Estimated Cost
Phishing	Customer data theft, Account takeover	PayPal impersonation scams (2023)	$2.7B annually
DDoS	Service disruption, Lost sales	AWS DDoS Attack (2023)	$100K per hour
Ransomware	Data encryption, System lockout	WannaCry Attack	$4B globally
SQL Injection	Database breach, Data theft	LinkedIn Data Breach	$3M per incident

5 Challenges and Risks

Running an online business today is like playing a never-ending game of digital chess [1,7]. The threats we're facing aren't just simple technical problems anymore - cybercriminals have seriously upped their game, moving from basic hacking to sophisticated attacks using phishing and ransomware [4]. Even big companies with deep pockets for security aren't immune [9]. Here's a sobering fact: a 2023 study found that about 40% of data breaches still happen because of simple human mistakes, even in well-protected systems [26].

Selling globally adds another layer of complexity [14]. Imagine trying to follow a different set of rules for every region you sell in - from Europe's GDPR [12] to California's CCPA [13]. Each has its own specific requirements for handling customer data [10,11], and getting it wrong can cost you big time - both in fines and lost customer trust [23].

Then there's the supply chain puzzle [15]. Most online stores work with dozens of other companies for everything from payments to shipping [16], and each connection is like a potential door for attackers [18]. Let's look at Table 2 for a breakdown of the most common cybersecurity threats and what they mean for online businesses. These challenges create the foundation for understanding today's evolving threat landscape—a topic explored in the following section.

6 Threat Landscape

Today's cyber threats target e-commerce platforms in increasingly sophisticated ways [2,6]. The threat landscape reveals multiple dimensions of risk that businesses must address to protect their operations and customers [20]. The most common attack remains phishing, where criminals create convincing fake emails and websites to steal login details and payment information [3]. A recent study showed that 46% of e-commerce attacks start with phishing attempts [26]. DDoS attacks pose another significant threat, flooding websites with traffic to shut them down [9], costing businesses an average of $100,000 per hour of downtime. Ransomware has grown more aggressive, with attackers not only locking companies out of their data but also threatening to release customer information unless paid [1]. As shown in Table 2, common cybersecurity threats can be categorized by their impact and estimated costs.

The financial impact of cyberattacks continues to rise [23], with the average cost of a data breach in e-commerce reaching $4.2 million in 2023 [26]. Larger breaches can exceed $50 million when accounting for all costs, including customer compensation, legal fees, and long-term reputation damage [16]. Small businesses face particular vulnerability [22], with 60% closing within six months of a major cyber attack.

Criminals exploit multiple weaknesses to breach systems [25], with compromised employee credentials accounting for 32% of breaches [21]. Unpatched software vulnerabilities create openings in 24% of cases [9], while third-party vendor access leads to 21% of breaches [15]. Misconfigured cloud services round out the major entry points [2], causing 14% of security incidents.

Beyond technical attacks, criminals increasingly rely on psychological manipulation through social engineering [5]. Business Email Compromise (BEC) scams have evolved to become highly targeted operations [4], with attackers thoroughly researching companies before attempting fraud. These sophisticated attacks often target employees who can authorize payments or access sensitive data [22], making human judgment a critical factor in security.

AI and machine learning have introduced new dimensions to these threats [7,8]. Attackers now use these technologies to automate and enhance their operations [6], while defenders must similarly adopt advanced technologies to protect their systems [24]. The emergence of blockchain technology offers new security possibilities [19], but also introduces unique challenges in implementation and integration [18].

This complex threat landscape requires organizations to maintain vigilance across multiple fronts [9], combining technical security measures with employee training and awareness [22]. Understanding these threats helps businesses develop more effective security strategies and allocate resources appropriately to protect against evolving cyber risks [21]. A breakdown of the prevalence of these threats is shown in Fig. 1. The variety and cost of threats underscore the need for proactive, layered defense strategies that evolve with the landscape.

7 Emerging Threats

Technology advancement brings new security challenges to e-commerce [7]. Attackers now use artificial intelligence to create more effective attacks [6]. AI helps criminals automate their attacks, adapt to security measures, and create more convincing fake emails and websites [5,8]. This requires companies to develop equally sophisticated AI-based defenses [24].

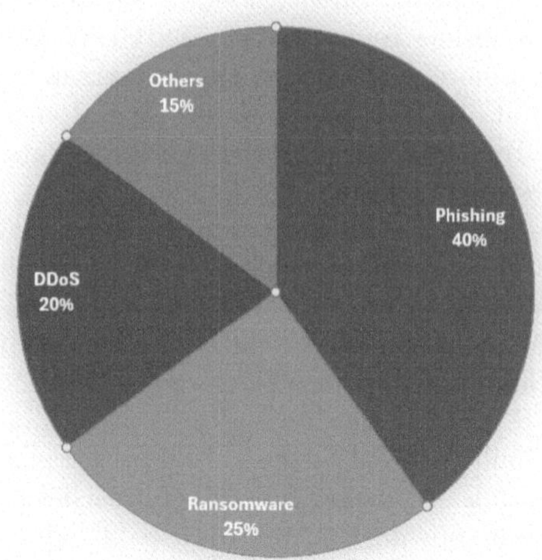

Fig. 1. Global Threat Landscape for E-Commerce Platforms.

The growth of connected devices in retail and shipping has created new weaknesses [15]. Many IoT devices use simple passwords and outdated software, making them easy targets for attackers [20]. Once compromised, these devices can provide access to larger networks and sensitive data [2].

Deepfake technology poses a growing threat to e-commerce security [6]. Criminals can create realistic fake videos and voice recordings to trick employees into providing access or transferring money [4]. Some attackers have successfully impersonated executives in video calls, highlighting the need for new verification methods [9].

8 Implementation Strategy

Organizations can strengthen their implementation plans by referencing frameworks such as NIST Cybersecurity Framework (CSF) and NIST SP 800-53, which offer structured, scalable guidance. [29] Implementing effective security requires a systematic approach that builds security from the ground up [16]. This strategy can be implemented in six phases, with specific timelines and key activities for each phase:

Phase 1 - Risk Assessment: involves identifying security gaps within the platform [9], focusing on vulnerabilities in payment systems, customer data storage, and third-party service connections [25]. This phase also includes reviewing regulatory requirements such as GDPR, CCPA, and PCI DSS to assess compliance shortcomings [10,11]. Additionally, any past security incidents are analyzed to identify risks and potential areas for improvement [1].

Phase 2 - Policy Development: focuses on creating clear and transparent security policies [21]. These policies address critical areas such as payment processing, customer data protection, and supply chain security management [15]. Rules are established for data encryption, password management, and secure vendor management [16].

Phase 3 - Technology Development: ensures the implementation of advanced security measures [20]. This includes deploying secure payment gateways that comply with PCI DSS standards [16]. Tools like encryption, firewalls, and multi-factor authentication are introduced to safeguard information [2].

Phase 4 - Staff Training: emphasizes building cybersecurity awareness across teams [22]. Training sessions cover best practices, phishing detection, and responses to fraudulent activity [3]. Employees are taught how to recognize and report suspicious activities [9].

Phase 5 - Testing and Monitoring: involves conducting regular penetration tests [2] to evaluate the effectiveness of security measures and uncover potential weaknesses. Real-time monitoring tools are employed to oversee platform activities [1].

Phase 6 - Compliance Verification: ensures the platform adheres to regulatory requirements [12,13] through regular audits. These audits confirm compliance with standards like PCI DSS, GDPR, and CCPA [14]. External experts are engaged periodically to evaluate security measures [23].

Table 3 summarizes the implementation timeline with specific activities and expected outcomes for each phase. A secure implementation is only effective if continuously supported by regulatory awareness and ongoing compliance monitoring.

Table 3. Implementation Timeline for Security Measures

Phase	Key Activities	Implementation Period	Expected Outcomes
Risk Assessment	Vulnerability scanning, Compliance review	1–2 Months	Security gaps identified, Risk report
Policy Development	Security policy creation, Documentation	2–3 Months	Policy framework, Guidelines
Technology Development	Security tools deployment, Integration	3–4 Months	Enhanced security infrastructure
Staff Training	Security awareness, Role-specific training	Ongoing, Quarterly	Improved security awareness
Testing and Monitoring	Penetration tests, System monitoring	Ongoing, Monthly	Vulnerability insights
Compliance Verification	External audits, Gap analysis	Bi-annual	Compliance confirmation

9 Key Regulations

The regulatory landscape shapes how e-commerce platforms protect customer data [10]. GDPR [12] sets strict rules for companies handling European customer data, requiring clear permission for data collection and use [14]. Companies must tell customers what data they collect and why, with significant fines for breaking these rules [23]. CCPA [13] gives California residents similar rights [11], letting them know what data companies collect and request its deletion.

PCI DSS focuses specifically on protecting payment information [16]. These standards require businesses to use encryption, control who can access payment data, and regularly check their security measures [20]. Following these rules helps prevent credit card fraud and unauthorized access to financial information [9]. Figure 2, llustrates the comparative analysis between GDPR and CCPA requirements, highlighting their key differences and similarities in data protection approaches. While both frameworks emphasize consumer privacy rights, they implement distinct mechanisms for enforcement and compliance [10,11].

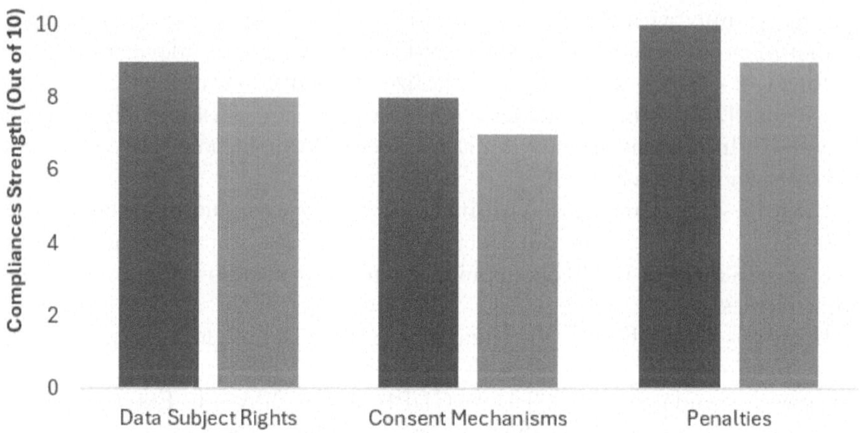

Fig. 2. Comparative Analysis of GDPR and CCPA Requirements

10 Compliance and Monitoring

Maintaining security in e-commerce requires constant vigilance through compliance checks and monitoring [2,9]. Regular security audits help ensure platforms follow key regulations like GDPR [12], CCPA [13], and PCI DSS [16]. These checks, combined with security testing, help find and fix weaknesses before criminals can exploit them [1]. Modern monitoring tools watch network activity around the clock, quickly spotting unusual behavior that might signal an attack [20,21].

Companies need clear ways to measure their security performance [23]. This includes tracking how well data is encrypted, how many security rules are broken, and how quickly teams respond to problems [2]. Automated tools help by constantly checking systems against security requirements, making it easier to find and fix issues quickly [3,9].

Watching supply chain partners has become equally important [15]. Companies regularly review their vendors' security practices and use special tools to monitor how these partners connect to their systems [17]. Strong contracts help ensure vendors maintain good security practices [16]. Employee training plays a vital role too [22], with regular sessions teaching workers how to spot fake emails [4], handle sensitive data safely [10], and report security problems.

Good security also means keeping detailed records of any problems or suspected breaches [14]. By carefully studying what went wrong after each incident [1], companies can make improvements to prevent similar problems in the future [25]. The integration of AI-powered monitoring tools [7,8] has further enhanced the ability to detect and respond to security incidents in real-time [5,6]. Effective compliance is not a one-time task but an ongoing discipline supported by monitoring, training, and governance.

11 Case Studies in E-Commerce Security

These case studies reflect three primary types of vulnerabilities: third-party/vendor risk, internal process failures, and cloud misconfigurations.

The 2013 Target breach stands as a classic example of supply chain vulnerability. Attackers stole credentials from a heating and cooling vendor, using this access to infiltrate Target's network. They installed malware on payment systems, compromising 40 million credit cards and 70 million customers' personal information during the holiday shopping season. The breach cost Target over $292 million in settlements and damages. Key security failures included weak vendor security practices, insufficient network segmentation, and delayed response to security alerts. Target's response involved a complete security overhaul, including stricter vendor controls and enhanced monitoring systems.

The evolution of e-commerce security breaches can be understood through two significant incidents: the 2013 Target breach and the 2023 Alibaba supply chain attack. These cases demonstrate how security threats have evolved while highlighting persistent vulnerabilities in e-commerce systems.

More recently, the 2023 Alibaba Cloud breach demonstrates how supply chain attacks have evolved. Attackers exploited a vulnerability in Alibaba's software supply chain, potentially affecting thousands of businesses using their cloud services. The breach occurred through a compromised third-party software package, allowing attackers to access customer data and cloud resources. Initial estimates suggest damages exceeding $500 million, affecting over 100,000 businesses worldwide. The incident revealed new challenges in cloud security and software supply chain integrity.

These cases highlight several enduring lessons. First, vendor security remains critical - both traditional vendors like Target's HVAC supplier and modern cloud service providers can introduce significant risks. Second, rapid detection and response are essential; both cases showed how delayed reactions can amplify damage. Third, comprehensive security requires multiple layers of protection, from vendor screening to network segmentation and continuous monitoring. Finally, the evolution from Target to Alibaba shows how attacks have grown more sophisticated, targeting not just individual companies but entire digital supply chains.

Both incidents led to significant changes in e-commerce security practices. Target's breach spurred the adoption of chip-based payment cards in the U.S., while the Alibaba incident has accelerated the implementation of software supply chain security measures across the industry. These cases continue to influence how companies approach security, compliance, and vendor management in the digital marketplace.

12 Recommendations and Best Practices

Zero-Trust Architecture (ZTA) represents a fundamental shift in security paradigms, requiring continuous verification of all users and devices to effectively reduce both internal and external threats [21]. Implementation through multi-factor authentication and real-time monitoring [9] establishes comprehensive access control while minimizing unauthorized activity risks [2]. The evolution of global regulatory frameworks, notably GDPR [12] and CCPA [13], has established more uniform international standards [10,14], facilitating consistent data protection across jurisdictions [11].

Data protection strategies have evolved to incorporate robust encryption and tokenization methodologies [16,20], substantially reducing breach risks [1]. Tokenization enhances security by substituting sensitive data with non-sensitive equivalents, ensuring alignment with PCI DSS requirements [16]. Advanced AI-driven threat detection systems [7,8] now provide real-time analysis of user behavior patterns, identifying anomalies such as credential abuse and unusual access patterns [6], significantly reducing potential attack impacts [5].

Secure API management has become critical for third-party service integration [15,18], necessitating robust encryption, rate limiting, and authentication protocols [3]. However, technical solutions must be complemented by comprehensive employee training programs [22]. Regular security awareness training, particularly focusing on phishing prevention [4] and data handling protocols, enables staff to actively participate in security maintenance [9].

Incident response planning remains fundamental to effective security management [23]. Well-defined protocols for containment, recovery, and post-incident analysis [2] minimize operational disruption and protect organizational reputation [17]. Modern security frameworks increasingly incorporate blockchain technology [18] and advanced monitoring systems [25], creating more resilient security architectures.

13 Future Directions

The evolution of e-commerce security necessitates both technological advancement and enhanced inter-organizational collaboration [7,8]. Blockchain technology provides immutable transaction records [18,19], strengthening fraud prevention and fostering trust between transacting parties. Additionally, industry initiatives toward unified global security standards [9] facilitate more effective cross-border data protection mechanisms [10].

Artificial intelligence and machine learning capabilities continue to advance threat detection [5]. These systems demonstrate superior efficiency in identifying fraudulent activities through pattern analysis of user behaviors and transaction characteristics [6,24]. Enhanced threat intelligence sharing among organizations [2] enables more rapid and coordinated responses to emerging attack vectors [1].

The proliferation of mobile commerce has accelerated edge computing adoption [15]. This architectural shift enables localized data processing, improving both performance and security, particularly in supply chain operations [20]. User education has emerged as a critical security component [22], addressing the persistent vulnerability of social engineering attacks [4].

14 Conclusion

E-commerce security has become fundamental to organizational sustainability in the digital marketplace [9,16]. Effective protection of customer data and payment systems, coupled with robust supply chain security [15], directly impacts business continuity and stakeholder trust [23].

Security strategy development requires comprehensive risk assessment and policy formulation [21]. Implementation encompasses appropriate technology deployment [25], personnel training [22], and continuous control validation [2]. Adherence to security standards like PCI DSS [16] and privacy regulations including GDPR [12] and CCPA [13] enhances both compliance and customer confidence [14].

While emerging technologies such as blockchain [18] and AI-enhanced security solutions [7] provide advanced protection mechanisms [8], human factors remain crucial. Comprehensive security awareness programs for employees and customers [3], combined with robust technical controls and clear procedures [9], establish a resilient security framework.

The sustainability of e-commerce operations depends on balancing security requirements with operational efficiency [1]. As threat landscapes evolve [6], organizations must maintain security agility [24] while preserving stakeholder trust [23]. This balanced approach contributes to a more secure digital commerce ecosystem [21].

References

1. Neto, N.N., Madnick, S., De Paula, A.M.G., Borges, N.M.: Developing a global data breach database and the challenges encountered. J. Data Inform. Qual. **13**(1), 1–33 (2021). https://doi.org/10.1145/3439873
2. González-Granadillo, G., González-Zarzosa, S., Diaz, R.: Security information and event management (SIEM): Analysis, trends, and usage in critical infrastructures. Sensors (2021). https://doi.org/10.3390/s21144759
3. Rahimi, N.: Security consideration in peer-to-peer networks with a case study application. Int. J. Netw. Secur. Appl. **12** (2020)
4. Rahimi, N., Maynor, J., Gupta, B.: Adversarial machine learning: difficulties in applying machine learning to existing cybersecurity systems. In: CATA, pp. 40–47 (2020)
5. Juneja, A., Juneja, S., Bali, V., Jain, V., Upadhyay, H.: Artificial intelligence and cybersecurity: Current trends and future prospects. In: The Smart Cyber Ecosystem for Sustainable Development, pp. 431–441 (2021). https://doi.org/10.5281/zenodo.13333202
6. Sharma, M., Kaur, M.: A review of deepfake technology: an emerging AI threat. In: Soft Computing for Security Applications: Proceedings of ICSCS 2021, pp. 605–619 (2022). https://doi.org/10.1007/978-981-16-5301-8_44
7. Camacho, N.G.: The role of AI in cybersecurity: addressing threats in the digital age. J. Artif. Intell. General Sci. **3**(1), 143–154 (2024). https://doi.org/10.60087/jaigs.v3i1.75
8. Radanliev, P., Santos, O., Brandon-Jones, A., Joinson, A.: Ethics and responsible AI deployment. Front. Artif. Intell. **7** (2024). https://doi.org/10.3389/frai.2024.1377011
9. Hans, J., Brandtweiner, R.: Best practices for vulnerability management in large enterprises: a critical view on the common vulnerability scoring system. In: Risk Analysis, Hazard Mitigation, and Safety and Security Engineering XIII, vol. 214 (2022). Retrieved January 2025. https://www.witpress.com/Secure/elibrary/papers/SSR22/SSR22010FU1.pdf
10. Hoofnagle, C.J., Van Der Sloot, B., Borgesius, F.Z.: The European Union general data protection regulation: What it is and what it means. Inform. Commun. Technol. Law **28**(1), 65–98 (2019). https://doi.org/10.1080/13600834.2019.1573501
11. Goldman, E.: An introduction to the California Consumer Privacy Act (CCPA). Research paper, Santa Clara Univ. Legal Studies (2020). https://doi.org/10.2139/ssrn.3211013
12. GDPR: General Data Protection Regulation. Regulation (EU) 679. Retrieved January 2025 (2016). https://ec.europa.eu/info/law/law-topic/data-protection_en
13. California Office of the Attorney General: California Consumer Privacy Act (CCPA). Retrieved January 2025 (2023). https://oag.ca.gov/privacy/ccpa
14. Mulligan, S.P., Freeman, W.C., Linebaugh, C.D.: Data protection law: an overview. Technical Report 45631, Congressional Research Service (2019). Retrieved January 2025. https://www.everycrsreport.com/files/20190325_R45631_6d0338c074bf38e6cddac62f94b5d143ed3a18d0.pdf

15. Murad, S.A., Rahimi, N., Roy, I., Gupta, B.: Security challenges in the industrial Internet of Things (IIoT). In: Quality Assessment and Security in Industrial Internet of Things, pp. 20–31. CRC Press (2021)
16. Williams, B., Adamson, J.: PCI Compliance: Understand and Implement Effective PCI Data Security Standard Compliance. CRC Press (2022). https://doi.org/10.1201/9781003100300
17. Plachkinova, M., Maurer, C.: Security breach at target. J. Inf. Syst. Educ. **29**(1), 11–20 (2018)
18. Rahimi, N., Roy, I., Gupta, B., Bhandari, P., Debnath, N.C.: Blockchain technology and its emerging applications. In: Blockchain Technology for Data Privacy Management, pp. 133–157. CRC Press (2021)
19. Rahimi, N., Gudapati, S.S.V.: Emergence of blockchain technology in the healthcare and insurance industries. In: Blockchain Technology Solutions for the Security of IoT-Based Healthcare Systems, pp. 167–182. Academic Press (2023)
20. Rahimi, N.: A study of the landscape of security issues, vulnerabilities, and defense mechanisms in Web based applications. In: 2021 International Conference on Computational Science and Computational Intelligence (CSCI), pp. 806–811. IEEE (2021)
21. Seaman, J.: Zero trust security strategies and guideline. In: Digital Transformation in Policing: The Promise, Perils and Solutions, pp. 149–168. Springer, Cham (2023). https://doi.org/10.1007/978-3-031-09691-4_9
22. Tolossa, D.: Importance of cybersecurity awareness training for employees in business. Vidya-A J. Gujarat Univ. **2**, 104–107 (2023). ISSN: 2321-1520 E
23. Charoen, D.: Revisiting the (disappearing) cost of data breach disclosures. Digital Policy, Regulation and Governance (2024). https://doi.org/10.1108/DPRG-02-2024-0033 . ahead-of-print
24. Rahimi, N., Murad, S.A., Lee, S.B.: Entrepreneurship opportunities in cybersecurity. In: Generating Entrepreneurial Ideas With AI, pp. 172–191. IGI Global (2024)
25. Rahimi, N., Gupta, B.: Security issues, vulnerabilities, and defense mechanisms in wireless sensor networks: State of the art and recommendation. Integration of WSNs into Internet of Things, pp. 1–15 (2021)
26. Data breach investigations report 2023. Technical report, Verizon (2023). Retrieved January 2025. https://www.verizon.com/business/resources/reports/dbir/
27. Google LLC: BeyondCorp — A New Approach to Enterprise Security. Retrieved January 2025 (2023). https://cloud.google.com/beyondcorp
28. Mastercard: Mastercard uses AI to reduce fraud and false declines. Retrieved January 2025 (2023). https://newsroom.mastercard.com/press-releases/mastercard-ai
29. National Institute of Standards and Technology (NIST): Framework for improving critical infrastructure cybersecurity, version 1.1. Technical report, U.S. Department of Commerce, National Institute of Standards and Technology (NIST) (2018). Retrieved January 2025. https://www.nist.gov/cyberframework

AI Horizons — Models, Methods and Applications

Prediction of Adverse Side Effects of Antidepressants Using an Expert System

Paulina Becerra-Vivanco[1], Lucila I. Castro-Pastrana[2], and Roberto Rosas-Romero[1(✉)]

[1] Electrical and Computer Engineering Department, Universidad de las Américas-Puebla, San Andrés Cholula, 72810 Puebla, Mexico
roberto.rosas@udlap.mx

[2] Chemical and Biological Sciences Department, Universidad de las Américas-Puebla, San Andrés Cholula, 72810 Puebla, Mexico

Abstract. An expert system is proposed to predict the risk of adverse side effects in patients taking antidepressants. The system predicts the risk of seven adverse drug reactions (*serotonin syndrome*, *nausea*, *suicidal ideation*, *somnolence*, *gastrointestinal symptoms*, *sexual dysfunctions* and *hepatobiliary disorders*) based on physiological information, including *sex*, *age* and pharmacological factors such as *type of antidepressant* and *concurrent medication*. The risk of side effects is categorized into four degrees: *none*, *low*, *medium*, and *high*. The effectiveness of the proposed expert system was validated using 54 patient profiles compiled from databases with high classification rates, especially for hepatobiliary disorders and gastrointestinal symptoms. The system utilizes fuzzy logic to handle the inherent imprecision to identify a patient's age group. Furthermore, fuzzy logic does not require extensive databases and provides clinicians with an understanding of the rationale for the recommendation.

Keywords: Pharmacovigilance · antidepressive agents · adverse effects · fuzzy system · membership function · defuzzification

1 Introduction

Depression is the most prevalent mental health disorder, with an estimated global prevalence of 3.8%, affecting over 280 million people worldwide [1]. Despite the availability of effective treatments, two-thirds of patients exhibit suboptimal responses to antidepressant treatments, often due to the high incidence of *adverse drug reactions* (ADRs) [2,3]. Post-marketing surveillance and *pharmacovigilance* throughout a drug's lifecycle are crucial for identifying and evaluating ADRs, thus expanding the knowledge needed for their prevention [5].

Pharmacovigilance is a clinical discipline focused on the detection, evaluation, understanding, and prevention of adverse drug effects [4]. It involves continuous monitoring of medications, vaccines, and biological products to ensure

I. Roy et al. (Eds.): CAINE 2025, CCIS 2740, pp. 265–276, 2026.
https://doi.org/10.1007/978-3-032-14893-3_18

the safe and effective use of antidepressant medications [6]. Pharmacovigilance aims to be preventive and predictive by detecting and analyzing patterns from databases composed of case reports from healthcare providers, patients, and the pharmaceutical industry. Efforts to develop predictive models for ADRs specifically associated with antidepressant treatments have been made, yet work and research remain to be done since traditional approaches in pharmacovigilance can be costly and time-consuming [7–9].

Real-world evidence databases containing patient-specific ADR data have enabled the development of second-generation artificial intelligence tools, useful for analyzing large amounts of data and making predictions. Pharmacovigilance databases, such as the World Health Organization's VigiBase or the European Medicines Agency's EudraVigilance, are useful for developing AI-based tools capable of analyzing ADR reports [10].

Artificial intelligence methods have been applied to analyze and understand user-generated text for pharmacovigilance, where 14 published papers reported positive findings related to the identification of ADRs [11]. These studies reported positive results on the effectiveness and accuracy of using *Natural Language Processing* that is applied to user-generated digital content to identify ADRs for various diseases such as HIV [12], cancer [13,14], asthma, cystic fibrosis, arthritis, diabetes [15]. One method reported results for the identification of ADRs for five antidepressants: citalopram, chlorpromazine, cyclobenzaprine, dulexetine, and promethazine [16]. The data sources in these studies were social media (tweets), forums and drug reviews. The most widely used classifier has been the support vector machine [12–15]; and some methods have used ADABoost [12], decision trees [12], artificial neural networks [12], and conditional random fields [17].

First-generation AI methods, such as expert systems and fuzzy logic, are useful AI systems that adapt to the complexities of the real world without the need for large datasets. Fuzzy logic can adapt to these scenarios without requiring a large amount of data. Knowledge about adverse drug effects is constantly evolving, and first-generation AI tools adapt more flexibly to these changes than second- and third-generation AI tools. Furthermore, these models provide physicians and clinical staff with an interpretation of the results. On the other hand, machine learning models require large datasets for effective training and lack interpretability, making their decision-making process difficult to understand. Without further information, the binary outputs of a black-box algorithm will have limited clinical utility, as clinicians need to understand the basis for the recommendation.

This work presents an expert system designed to predict the risk of seven types of adverse side effects commonly induced by antidepressants by considering four patient variables: sex, age, type of antidepressant, and concurrent medication usage. By incorporating these variables into the predictive system, potential adverse side effects, such as serotonin syndrome, suicidal behavior, nausea, gastrointestinal issues, and sleep disorders, are predicted.

2 Methodology

In this work, a system was designed and implemented to emulate the decision-making process conducted by a human expert to predict *adverse drug reactions* (ADRs) induced by antidepressants. This system is based on *Fuzzy Logic*, which is a mathematical framework for rule-based reasoning under uncertainty, allowing degrees of truth. This makes fuzzy logic ideal for identifying an age group for a patient: childhood, adolescence, old age.

The proposed system was designed in three stages: (1) Definition of input variables that potentially impact the ADRs a patient may experience, (2) Definition of the output variables that correspond to ADRs as a result of taking antidepressant medications, (3) Definition of a rule base, which infers the values of the output variables and makes decisions based on the values of the input variables. Figure 1 depicts how these three elements interact.

Fig. 1. Three main stages to design an expert system that predicts ADRs induced by antidepressants.

2.1 Dataset

To design and test the system, 54 patient profiles were raised from VigiBase [18], AdisInsight [19], and FDA's Adverse Event Reporting System (FAERS) [20]. VigiBase is the World Health Organization (WHO) global dataset that serves as a resource for detecting and preventing ADRs related to medicines and vaccines with 35 millions of anonymized reports from various countries. VigiBase is managed by the Uppsala Monitoring Centre [21] on behalf of WHO, which is an independent, self-funded, non-profit foundation, dedicated to the safer use of medicines and vaccines. AdisInsight is a drug development intelligence database developed by Springer Nature for the pharmaceutical drug development market with over half a million users annually.The FDA's Adverse Event Reporting System (FAERS) is an interactive web-based tool that allows queries of the general public to search for information related to human ADRs reported to the FDA by the pharmaceutical industry, healthcare providers and consumers.

Fuzzy logic uses "if-then" rules for decision-making, when information is imprecise or uncertain. Within these rules, a proposition is a statement involving variables without clearly defined boundaries, such as the patient's age group. A genetic algorithm was used to determine the boundaries of each group, and 24 patient profiles were used to run the genetic algorithm, while 30 profiles were used to assess the performance of the expert system.

2.2 Input Variables

Patient information in the databases is presented as a narrative, for example: *"A 49-year-old man developed fatigue, sleep disturbances, as well as anxiety during treatment with xanax for depression. Additionally, he received an off-label treatment with remdesivir for COVID-19 infection ..."*. The input variables were extracted from these narratives, and the variables considered to predict adverse reactions to antidepressants are [22,23]: (1) *type of antidepressant*, (2) *other drugs* that are simultaneously taken by the patient, (3) *sex*, and (4) *age group*. Table 1 shows the values (column 2) assigned to each variable (column 1).

Table 1. Input variables to be considered by the expert system.

Input Variable	Variations
Type of Antidepressant	Selective serotonin reuptake inhibitors (SSRIs)
	Serotonin-noradrenaline reuptake inhibitors (SNRIs)
	Tricyclic antidepressants (TCAs)
	Monoamine oxidase inhibitors (MAOIs)
	Serotonin Modulators (SMs)
	Atypical Antidepressants (ADs)
Other Drugs	YES
	NO
Sex	Women
	Men
Age group	Child (0–12)
	Adolescent (9–18)
	Adult (15–65)
	Elderly (60+)

Three input variables are categorical variables: *other drugs*, *sex* and *antidepressant type*. Each of these variables is grouped into mutually exclusive categories, rather than measured numerically. Thus, crisp sets were used for these variables since each input value was clearly either in or out of their respective sets. For the case of *other drugs*, there are two mutually exclusive elements, YES or NO. For the variable of sex, there were two mutually exclusive sets, where $x \in \{MEN, WOMEN\}$. The classification of antidepressant types was considered according to Sheffler and colleagues [23].

Biologically speaking, the age group, such as childhood, adolescence, adulthood, and old age, is not precisely delineated due to the continuous nature of human development. While certain biological markers, such as puberty or menopause, signify transitions between stages, the boundaries between these sets are inherently fuzzy [24]. In *crisp sets*, the transition for an element between full-membership and non-membership is abrupt and well-defined. In fuzzy sets, such as age groups, this transition can be gradual, since the boundaries of fuzzy sets are not abrupt and could overlap. Individuals in an age group can also belong to another group to different degrees. Therefore, an individual's age-group membership is measured by a continuous function that maps a real number in $[0, 1]$ to an age x: $\mu(x) \in [0, 1]$; where $\mu(x) = 1$ represents full membership for x. Based on these conditions, the best fit for an age-group membership function was determined to be of trapezoidal shape as shown in Fig. 2. A genetic algorithm was used to obtain the size and position of the trapezoidal membership functions for each age group.

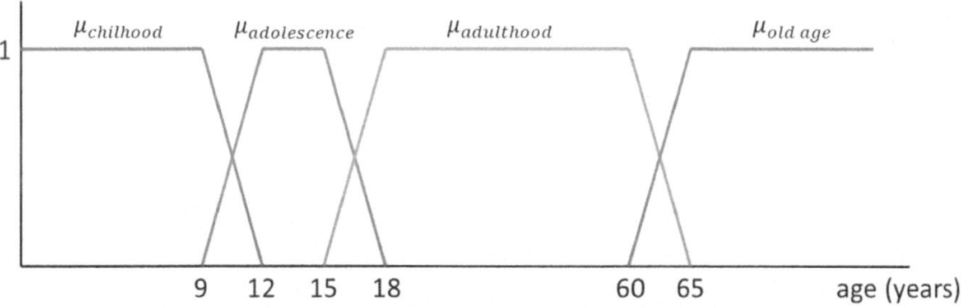

Fig. 2. Membership functions for four age groups: *childhood*, *adolescence*, *adulthood*, and *old age*.

2.3 Output Variables

A list of the seven main ADRs that are most frequently reported for each of the types of antidepressants was compiled after reviewing drug information and pharmacovigilance databases [18,25–33]. ADRs were categorized according to the Medical Dictionary for Regulatory Activities Terminology (MedDRA), which is used specifically for coding in pharmacovigilance [34]: (1) *serotonin syndrome* (RSS), (2) *nausea* (RN), (3) *suicidal and self-injurious behavior* (RST), (4) *sleep disorders and disturbances* (RDS), (5) *gastrointestinal disorders* (RSI), (6) *sexual function and fertility disorders* (RSD), (7) *hepatobiliary disorders* (RHB).

The output variables (ADRs) and the input variables (sex, age, type of antidepressant, and other medications) were correlated so that the risk of each effect was categorized based on the system input as NONE, LOW, MEDIUM, and HIGH. Since the goal is to determine a patient's risk of experiencing ADRs, three crisp sets were defined for three risk levels that were assigned values within the system (LOW = 1, MEDIUM = 2, HIGH = 3).

2.4 Rules and Defuzzification

In the field of expert systems, knowledge is most commonly represented with natural language expressions of the type *IF premise THEN conclusion.* This deductive form expresses an *inference* such that if a fact or premise (consequent) is known, then another fact called a conclusion (consequent) can be derived.

A series of IF-THEN rules was developed to predict each ADR based on the combinations and variations of input variables, thus obtaining premises that combine *sex*, *age group*, *type of antidepressant* and considering whether *any other medication is taken.* The premise P of a rule (IF P THEN Q) is a proposition that can be either true or false, and the consequent Q is executed if and only if the antecedent is true. The premise is a compound proposition, where atomic propositions are combined using two logical connectives, AND ($A \wedge B$) and OR ($A \vee B$), and the connective *THEN* was used to establish the relation with the ADR risk levels (conclusion, consequent).

A premise P is numerically evaluated using a *membership function* μ. This membership function takes values on the interval 0 to 1 ($\mu(x) \in [0, 1]$) for the case of the fuzzy variable *age group*. For the case of the crisp variables (*antidepressant*, *other drugs*), the membership function is Boolean with two numerical value types interpreted as logical values, 0 = FALSE and 1 = TRUE. The evaluation of the compound propositions $A \wedge B$ and $A \vee B$ was numerically computed as $min(\mu_A, \mu_B)$ and $max(\mu_A, \mu_B)$, respectively. As an example, the computational evaluation of the compound proposition *"an adolescent AND takes SNRIs OR SSRIs"* was implemented as $min(\mu_{adolescent}, max(\mu_{SNRI}, \mu_{SSRI}))$.

The risk levels were assigned values, according to LOW = 1, MEDIUM = 2, HIGH = 3. The final risk level value, assigned by the rule base, is determined by calculating a weighted average of these risk level values, according to

$$risk = \frac{1 \times \alpha_{LOW} + 2 \times \alpha_{MEDIUM} + 3 \times \alpha_{HIGH}}{\alpha_{LOW} + \alpha_{MEDIUM} + \alpha_{HIGH}}; \tag{1}$$

where the process to compute this weighted average is called *defuzzification.* The expert system was supplied with input data or facts to which the rules were applied. The facts were numerically evaluated to obtain a risk value and deduce a conclusion from this result. There is no risk if $risk \in [0.0.5]$, there is LOW risk if $risk \in (0.5, 1.5]$, there is MEDIUM risk if $risk \in (1.5, 2.5]$, there is HIGH risk if $risk \in (2.5, 3]$. The risk calculation in Eq. 1 was defined for each population (female patients and male patients) to infer the risk level of each ADR by calculating α_{LOW}, α_{MEDIUM} and α_{HIGH}.

3 Results and Discussion

The development of the expert system was conducted in MATLAB R2020a on an Intel Xeon(R) processor at 2.71 GHz and with a 32-GB RAM and the code is found in a GitHub site [35]. To establish a strict set of rules to imitate the decision-making processes of experts in the field of Pharmacovigilance, this project was supported by a Ph.D. in Pharmaceutical Chemistry-Biology with a specialty in pharmacovigilance, who helped identify: (1) the variables with a significant impact on the presence of side effects, (2) the different drug reaction mechanisms of medications, and (3) the outline of the project approach. To implement and test the system, 54 patient profiles were collected, of which the output results were already known. Three of these patient profiles are shown in Table 2. For these examples, the input data was defined and specified so that the patients met different characteristics, belonging to different age groups, to highlight the differences between the conditions that different organisms might present. The respective expected and predicted ADR risk levels for each patient profile are shown in Table 3.

Table 2. Three patient profiles.

Patient Variables	Patient 1	Patient 2	Patient 3
Sex	Female	Male	Female
Age	26	35	6
Type of antidepressant	SSRI	SNRI	SSRI
Other Drugs	No	No	No

Table 3. Expected and predicted ADR risk levels for each patient profile.

Patient Variables	Patient 1	Patient 2	Patient 3
Serotonin Syndrome	Medium	Medium	Low
Nausea	Medium	High	Medium
Suicidal Ideation	Medium	Medium	Low
Drowsiness/Somnolence	Low	Medium	Medium
Gastrointestinal Symptoms	Low	None	High
Sexual disorder and infertility	High	High	None
Hepatobiliary Disorders	Low	Medium	Low

The trapezoidal membership function for an age group is characterized by three parameters: (1) the *support* is the region characterized by *nonzero membership* $\mu(x) > 0$, (2) the *core* is the region characterized by *full membership* $\mu(x) = 1$, and (3) the *boundaries* are the regions containing ages with non-zero membership, but not full membership, $0 < \mu(x) < 1$. The membership function for childhood was limited to have only one boundary to the right of the support for the fuzzy variable, while the function for old age was restricted to having one boundary on the left side of the range. To determine the shape and position of the age-group membership functions, the core and boundary had to be calculated. The left and right boundaries were considered to be the same for both adolescence and adulthood. The unknowns were the support and boundary of the four membership functions, which were encoded as bit strings. Four-bit binary strings were used to define each unknown in each membership function. These strings were concatenated to give a 32-bit $(4 \times 2 \times 4)$ string. The genetic algorithm started with an initial population of 24 strings. The binary values were mapped to decimal values for the fuzzy variables and these decimal numbers were rounded as integers. To determine the fitness of the combination of membership functions in each string, the square of the errors in the outputs was measured, given the inputs. The fitness values were used to select string as parents for the next generation. After crossover, the new strings were generated. The process of selecting and generating strings was repeated until convergence. To run the genetic algorithm, 24 patient profiles were used.

Figure 3 shows seven confusion matrices for seven ADRs to antidepressants, obtained from 30 collected patient profiles. Each matrix column corresponds to the number of predictions for each risk level (low, medium and high risk); each row presents the instances in the actual risk level; and the diagonal elements represent the *correct classification rate* (CCR) for each risk level. It is observed that the highest classification rates were reached for the case of hepatobiliary disorders and gastrointestinal symptoms.

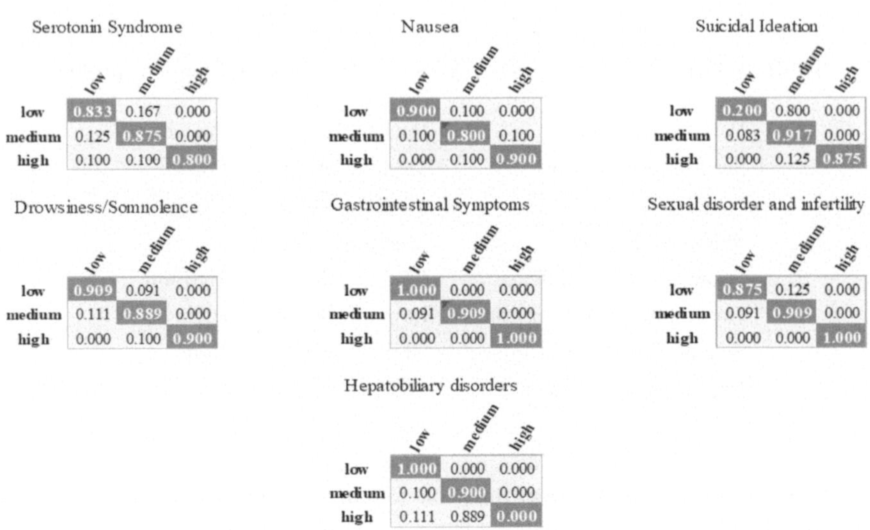

Fig. 3. Confusion matrices obtained for seven adverse effects to antidepressants for 30 patient cases.

Table 4 shows different methods to predict ADRs in patients on antidepressant medication, including the proposed one. The purpose of this Table is to show how AI-based frameworks have been used to predict adverse effects to drugs. It is observed that the proposed method is the only one that is based on the framework of crisp set logic and fuzzy set logic.

Table 4. A comparison with different methods.

Author and year	Causes of variations in patient response	Output Variables	Framework
Preskorn, 2014 [36]	Genetics, age, disease, environment	Clinical response (drug effect)	Algebraic equation
Zhao et al., 2023 [37]	Molecule graph, semantic features of ADRs, latent representations, SMILES sequence	Seven classes of serious clinical outcomes	Deep Learning
Campos et al., 2023 [38]	Polygenic risk scores for BMI, insomnia and chronic headaches	Weight gain, insomnia and headaches	Structural equation modelling, polygenic risk scoring and regressions
Li et al.; 2024 [39]	259 predictors spanning genetic, clinical, illness, demographic, and antidepressant information	30 ADRs, the most prevalent: low sexual desire, dry mouth, sleepiness/drowsiness	Elastic net, random forest and XGBoost
Dragioti et al., 2019 [40]	Age, adverse health outcomes, exposure and nonexposure, illnesses examined	13 Adverse health outcome domains	Statistical Analysis
Proposed	Sex, age, type of antidepressant, other medications	Four risks levels for seven adverse drug reactions	Expert System & Fuzzy Logic

4 Conclusion

The objective of the current work is to propose an expert system that predicts the risk level (none, low, medium, high) in seven adverse drug reactions (ADRs) to antidepressants by feeding the system with relevant facts about the patient, specifically type of antidepressant (selective serotonin reuptake inhibitors, serotonin-noradrenaline reuptake inhibitors, tricyclic antidepressants, monoamine oxidase inhibitors, serotonin modulators, atypical antidepressants), sex, age and whether another medication is being taken. By incorporating this data into the system, it provides predictions of possible ADRs such as serotonin syndrome, suicidal behavior, nausea, gastrointestinal symptoms, drowsiness, sexual disorders and infertility, and hepatobiliary disorders. This expert system imitates the logical process of a Pharmacovigilance expert to give an advice on the use of antidepressants by using facts and rules and following a series of inferences that lead to a decision. Fuzzy logic was used to handle the uncertainties generated when assigning an age group (child, adolescent, adult, elderly) to a patient. Thus, the system consists of three modules, (1) a fuzzifier that assigns a numerical age value to fuzzy sets (children, adolescent, adult, elderly) with some degree of membership, (2) a fuzzy inference machine that uses if-then rules to produce a fuzzy output, (3) a defuzzifier that converts the fuzzy output, generated by the fuzzy inference engine, into a crisp output.

Through this approach, the objective is to enhance the understanding and management of antidepressant therapy by introducing a tool for predicting and mitigating potential risks associated with medication usage. These tools also have the benefit of being easily updatable as clinical evidence evolves. Over time, pharmacovigilance experts and medical literature may report new adverse reactions with higher prevalence in patients taking antidepressants or an increase in the incidence of adverse reactions in specific patient groups. Finally, if access to databases of reported adverse reactions over time is available, whether from a hospital, a healthcare subsystem, or a national pharmacovigilance center, the risk rating made by expert systems could be personalized by analyzing these datasets with patient information at the individual level, and even scaled to a machine learning system.

References

1. World Health Organization (2023). Depressive disorder. https://www.who.int/news-room/fact-sheets/detail/depression
2. Zhu, L., Ferries, E., Suthoff, E., Namjoshi, M., Bera, R.: Economic burden and antidepressant treatment patterns among patients with major depressive disorder in the United States. J. Manag. Care Spec. Pharm. **28**, 2–13 (2022)
3. Cipriani, A., et al.: Comparative efficacy and acceptability of 21 antidepressant drugs for the acute treatment of adults with major depressive disorder: a systematic review and network meta-analysis. Lancet **391**, 1357–1366 (2018)
4. World Health Organization: The importance of pharmacovigilance - Safety monitoring of medicinal products (2002). https://www.who.int/publications/i/item/10665-42493
5. Lavertu, A., Vora, B., Giacomini, K.M., Altman, R., Rensi, S.: A new era in pharmacovigilance: toward real-world data and digital monitoring. Clin. Pharmacol. Ther. **109**, 1197–1202 (2021)
6. Sánchez-de-Paz, M.P., et al.: Pharmacovigilance in neuroscience: Farmacovigilancia en neurociencia. J. Appl. Cogn. Neurosci. **2**, 00163800 (2021)
7. Strawn, J.R., Mills, J.A., Poweleit, E.A., Ramsey, L.B., Croarkin, P.E.: Adverse effects of antidepressant medications and their management in children and adolescents. Pharmacotherapy **43**(7), 675–690 (2023)
8. Braund, T.A., Tillman, G., Palmer, D.M., Gordon, E., Rush, A.J., Harris, A.W.F.: Antidepressant side effects and their impact on treatment outcome in people with major depressive disorder: an iSPOT-D report. Transl. Psychiatry **11**, 417 (2021)
9. Güneç, S.S., et al.: Primum non nocere: in silico prediction of adverse drug reactions of antidepressant drugs. Comput. Toxicol. **18**(100165) (2021). https://doi.org/10.1016/j.comtox.2021.100165
10. Aronson, J.K.: Artificial intelligence in pharmacovigilance: an introduction to terms, concepts, applications, and limitations. Drug Saf. **45**(5), 407–418 (2022)
11. Pilipiec, P., Liwicki, M., Bota, A.: Using machine learning for pharmacovigilance: a systematic review. Pharmaceutics **14**(2) (2022). https://doi.org/10.3390/pharmaceutics14020266
12. Adrover, C., Bodnar, T., Huang, Z., Telenti, A., Salathé, M.: Identifying adverse effects of HIV drug treatment and associated sentiments using twitter. JMIR Public Health Surveill. **1**(10) (2015). https://doi.org/10.2196/publichealth.4488
13. Bian, J., Topaloglu, U., Yu, F.: Towards large-scale twitter mining for drug-related adverse events. In: Proceedings of the 2012 International Workshop on Smart Health and Wellbeing, SHB 2012; Maui, HI, USA, 29 October 2012, pp. 25–32 (2012). https://doi.org/10.1145/2389707.2389713

14. Mishra, A., Malviya, A., Aggarwal, S.: Towards automatic pharmacovigilance: analysing patient reviews and sentiment on oncological drugs. In: Proceedings of the 2015 IEEE 15th International Conference on Data Mining Workshops; Atlantic City, NJ, USA, 14–17 November 2015, pp. 1402–1409 (2015). https://doi.org/10.1109/ICDMW.2015.2
15. Ru, B., Harris, K., Yao, L.: A content analysis of patient-reported medication outcomes on social media. In: Proceedings of the 2015 IEEE International Conference on Data Mining Workshop; Atlantic City, NJ, USA, 14–17 November 2015, pp. 472–479 (2015). https://doi.org/10.1109/ICDMW.2015.150
16. Akay, A., Dragomir, A., Erlandsson, B.E., Akay, M.: Assessing anti-depressants using intelligent data monitoring and mining of online fora. IEEE J. Biomed. Health Inform. **20**, 977–986 (2016). https://doi.org/10.1109/JBHI.2016.2539972
17. Nikfarjam, A., Sarker, A., O'Connor, K., Ginn, R., Gonzalez, G.: Pharmacovigilance from social media: mining adverse drug reaction mentions using sequence labeling with word embedding cluster features. J. Am. Med. Inform. Assoc. **22**(3), 671–681 (2015)
18. World Health Organization. WHO collaborating center for international drug monitoring (2024). https://www.vigiaccess.org/
19. Springer. AdisInsight (2025). https://adisinsight.springer.com/
20. U.S. Food and Drug Administration. FDA Adverse Event Reporting System (FAERS) Public Dashboard (2023). https://www.fda.gov/drugs/fdas-adverse-event-reporting-system-faers/fda-adverse-event-reporting-system-faers-public-dashboard
21. Uppsale Monitoring Centre. About Uppsala Monitoring Centre (2024). https://who-umc.org/about-uppsala-monitoring-centre/
22. Ran, C., Zhou, H., Tan, C., Tan, J., Zhang, Z., Zhao, W.: Detection and evaluation of adverse drug reaction signals of antidepressants based on FDA adverse event reporting system database. Open J. Depression 17–25 (2020). https://doi.org/10.4236/ojd.2020.92002
23. Sheffler, Z.M., Patel, P., Abdijadid, S.: Antidepressants. StatPearls Publishing, Treasure Island (FL) (2024)
24. Balasundaram, P., Avulakunta, I.: Human growth and development. StatPearls (2024). https://www.ncbi.nlm.nih.gov/books/NBK567767
25. Miller, C., Taskiran, S.: Medication for kids with depression. Care Education Science (2024). https://childmind.org/article/medication-for-kids-with-depression/
26. Nishizawa, S., et al.: Differences between males and females in rates of serotonin synthesis in human brain. Proc. Natl. Acad. Sci. **94**, 5308–5313 (1997)
27. Hashimoto, K.: Impact of age on optimal dose of antidepressants. EClinicalMedicine **18**, 1–2 (2020)
28. Ellis, R.: What are the side effects of antidepressants? WebMD (2023). https://www.webmd.com/depression/side-effects-antidepressants
29. Wichniak, A., Wierzbicka, A., Walecka, M., Jernajczyk, W.: Effects of antidepressants on sleep. Curr. Psychiatry Rep. **19**(9) (2017). https://doi.org/10.1007/s11920-017-0816-4
30. US Food and Drug Administration. Finding and learning about side effects (2022). https://www.fda.gov/drugs/information-consumers-and-patients-drugs/finding-and-learning-about-side-effects-adverse-reactions
31. Johnson, D.: Side effects of antidepressants in females – what women should know? Lindner Center of Hope (2024). https://lindnercenterofhope.org/blog/side-effects-antidepressants-women/

32. Find Drugs & Conditions. Drugs.com (2024). https://www.drugs.com
33. Böhm, R., et al.: OpenVigil FDA - inspection of U.S. American adverse drug events pharmacovigilance data and novel clinical applications. PloS One **11**(6), 0157753 (2016). https://doi.org/10.1371/journal.pone.0157753
34. NCBO. Medical dictionary for regulatory activities terminology (MedDRA). BioPortal (2024). https://bioportal.bioontology.org/ontologies/MEDDRA
35. GitHub. Prediction of adverse side effects of antidepressants using an expert system (2024). https://github.com/rrosasr/Prediction-of-adverse-side-effects-of-antidepressants-using-an-expert-system
36. Preskorn, S.H.: Prediction of individual response to antidepressants and antipsychotics: an integrated concept. Dialogues Clin. Neurosci. **11**(417), 545–554 (2014)
37. Zhao, H., et al.: Identifying the serious clinical outcomes of adverse reactions to drugs by a multi-task deep learning framework. Commun. Biol. (2023)
38. Campos, A.I., et al.: Understanding genetic risk factors for common side effects of antidepressant medications. Commun Med. **1**(45) (2023). https://doi.org/10.1038/s43856-021-00046-8
39. Li, D., et al.: Prediction of antidepressant side effects in the genetic link to anxiety and depression study. MedRXiv (2024)
40. Dragioti, E., et al.: Association of antidepressant use with adverse health outcomes: a systematic umbrella review. JAMA Psychiat. **76**(12), 1241–1255 (2019)

Elitist Country-Segmented Memetic Algorithm: Integrating Multithreading and SIMD for Efficient Solution Search

Piero Pastor, Manuel Tupia(✉), and Rony Cueva

Pontificia Universidad Católica del Perú, Av. Universitaria 1801, San Miguel, Peru
{a20210836,tupia.mf,cueva.r}@pucp.edu.pe

Abstract. This paper introduces a novel memetic optimization approach that segments the population into multiple semi-independent subpopulations, referred to as countries, each evolving in parallel through concurrent local search cycles. The proposed Country-Segmented Memetic Algorithm (CSMA) leverages multithreading and SIMD capabilities to enhance computational efficiency without compromising solution quality. By decentralizing elitism and enabling controlled migration between subpopulations, the method achieves a better balance between exploration and exploitation. Experimental evaluations on standard benchmark functions demonstrate superior convergence speed and robustness compared to classical memetic algorithms. Additionally, theoretical scalability assessments using Amdahl's Law, Gustafson's Law, Karp-Flatt's metric, and isoefficiency models confirm the algorithm's effectiveness in parallel environments.

Keywords: Memetic Algorithms · Multithreading · SIMD Optimization · Parallel Metaheuristics · Population Segmentation

1 Introduction

The increasing demand for solving large-scale and computationally intensive optimization problems has led to the widespread adoption of metaheuristic algorithms. Among them, Memetic Algorithms (MAs) have gained considerable attention due to their hybrid nature—combining the global search capabilities of evolutionary algorithms with the fine-tuned exploitation provided by local search procedures. However, traditional MAs are often implemented in a sequential fashion, limiting their ability to fully exploit the computational power offered by modern parallel and multicore architectures.

As algorithmic complexity grows and datasets increase in dimensionality, the need for concurrent and efficient optimization frameworks becomes more pressing. In this context, parallelism can be harnessed at multiple levels—from low-level instruction sets such as SIMD (Single Instruction, Multiple Data) to high-level concurrency models using multithreading. While several efforts have been made to parallelize evolutionary algorithms, few approaches address population structure, elitism bias, and local refinement in a cohesive and scalable manner [16].

I. Roy et al. (Eds.): CAINE 2025, CCIS 2740, pp. 277–298, 2026.
https://doi.org/10.1007/978-3-032-14893-3_19

This work introduces a new perspective on memetic optimization by segmenting the population into countries, each functioning as a semi-autonomous subpopulation with its own evolutionary cycle. This country-segmented architecture enables the use of parallel threads, each independently managing a portion of the search space. A controlled migration mechanism promotes information exchange between countries while maintaining diversity and preventing premature convergence. Furthermore, the proposed model supports multi-objective optimization by assigning distinct fitness functions to each cluster running in parallel.

To validate the effectiveness of the proposed Country-Segmented Memetic Algorithm (CSMA), we perform a comprehensive set of experiments, including both single- and multi-objective optimization tasks. Benchmark results are analyzed not only in terms of solution quality but also through rigorous parallel performance metrics including Amdahl's Law, Gustafson's Law, Karp-Flatt's metric, and isoefficiency analysis [17]. The remainder of this paper is structured as follows:

Section 2 provides a literature review of memetic algorithms, SIMD parallelism, and multithreading techniques.

Section 3 presents the proposed methodology, including the design of the CSMA, algorithmic structure, and formal definitions.

Section 4 details the experimental setup and reports results, including performance benchmarks and multithreading evaluations.

Section 5 concludes the paper with final remarks and directions for future research.

2 Literature Review

Metaheuristic algorithms have become indispensable tools for tackling complex optimization problems across various domains. Among these, Memetic Algorithms (MAs) stand out due to their hybrid nature, combining global search strategies with local refinement techniques to balance exploration and exploitation effectively.

Recent advancements have seen the development of adaptive and hybrid MAs tailored for specific applications. For instance, Truong Dao et al. (2024) introduced a hybrid metaheuristic combining Particle Swarm Optimization and Generalized Normal Distribution Optimization, demonstrating improved performance in predicting antimicrobial peptide toxicity. Similarly, Kovács et al. (2025) proposed the Colonial Bacterial Memetic Algorithm, integrating cultural algorithms and co-evolutionary dynamics for robotic applications, showcasing enhanced adaptability and solution quality.

Parallel computing has further augmented the capabilities of MAs. Sun et al. (2022) provided a comprehensive overview of parallel computing strategies in metaheuristic algorithms, highlighting the benefits of both absolute and virtual parallelism in enhancing computational efficiency. Moreover, the integration of SIMD (Single Instruction, Multiple Data) architectures has been explored to accelerate numerical computations within metaheuristics, as discussed by various researchers.

Originally proposed by Pablo Moscato in 1989 [15], memetic algorithms are conceptually rooted in Richard Dawkins' theory of memes, which are defined as cultural units transmitted and refined through imitation and learning. In algorithmic terms, this metaphor is embodied by embedding local search or problem-specific heuristics into population-based global optimization methods, thus improving convergence behaviour and solution accuracy.

Over the years, memetic algorithms have developed into a versatile class of hybrid methods applied across a wide spectrum of domains, including operations research, artificial intelligence, bioinformatics, and engineering optimization. The following review presents a synthesis of foundational concepts, seminal contributions, and emerging directions in memetic computation, offering context for their growing importance in contemporary optimization research.

2.1 Foundations of Memetic Algorithm

As previously noted, the foundational concept of memetic algorithms was introduced by Pablo Moscato in 1989. Building upon the principles of genetic algorithms (GAs), MAs were conceptualized to combine the global search capabilities of evolutionary strategies with local refinement techniques. This hybridization is inspired by the idea of memes from Richard Dawkins' "The Selfish Gene" (1976), where cultural information (memes) is transmitted and evolves through imitation. In this context, the algorithm mirrors this by using local search procedures as "memes" to improve solutions iteratively within the global search process.

2.2 The Evolution of Memetic Algorithms

Generally, MAs are defined as hybrid stochastic methods that embed one or more intensive local search algorithms within an evolutionary framework [1]. The rationale behind this hybridization is that, after new individuals are generated via mutation, they can be further refined through local search before the selection process. This strategy allows the retention of individuals with low initial fitness but high potential for improvement, who would otherwise be discarded prematurely in a classical evolutionary approach.

Over time, memetic algorithms have demonstrated superior performance in complex problem domains, especially those characterized by "big valley" structures—featuring multiple local optima and a gradient guiding towards the global optimum. In such contexts, the performance of simple evolutionary algorithms, such as the $(1 + 1)$ EA, deteriorates rapidly with increasing search landscape difficulty. Conversely, the $(1 + 1)$ MA maintains expected optimization times on the order of polynomial time due to its ability to efficiently escape local optima and navigate between different basins of attraction [2].

It is important to note that, although the term "memetic algorithm" was not explicitly used in earlier studies, foundational concepts were already present. For instance, Costa and Oliveira (2003) proposed an elitist approach within multi-objective genetic algorithms by employing a secondary population to store non-dominated solutions. This mechanism—aimed at preventing the loss of high-quality solutions and maintaining adequate diversity along the Pareto front—clearly embodies intensification and information retention principles typical of MAs [3].

Overall, the evolution of memetic algorithms has been marked by increasing sophistication, incorporating elements such as iterated local search, crossover operators, adaptive strategies, and diversity control mechanisms. These developments have established MAs as one of the most robust strategies in combinatorial and metaheuristic optimization.

2.3 Elitist and Multimodal Algorithms

Elitist memetic algorithms constitute an advanced class of metaheuristic methods that combine global evolutionary search with local refinement strategies. Their primary strength lies in integrating elitism, a mechanism that guarantees the preservation of high-quality solutions across generations, thus preventing the loss of optimal candidates due to stochastic effects [5].

In the domain of multi-objective and multimodal optimization, elitism has been shown to significantly enhance algorithmic performance. Groşan et al. [4] demonstrated that elitist frameworks—such as SPEA and NSGA-II—achieve superior convergence to the true Pareto front compared to non-elitist counterparts. Their experiments revealed that non-elitist approaches, such as VEGA, are prone to premature convergence and stagnation at suboptimal solutions, particularly in complex search landscapes [4].

Shahin introduced the Memetic Elitist Pareto Evolutionary Algorithm for Virtual Network Embedding (MEPE-VNE), which incorporates NSGA-II, elitist selection, and local search. The algorithm consistently outperformed baseline methods by maintaining solution quality, increasing acceptance ratios, and reducing operational costs. Elitism in MEPE-VNE ensured that the best solutions were preserved intact, enhancing both stability and effectiveness [5].

Moreover, the integration of the local search procedure from $(1 + 1)$-PAES within the M-PAES framework enables the individual improvement of each candidate solution prior to participating in the recombination phase. This process involves systematically evaluating neighboring solutions through mutations and selecting those that are non-dominated, based on a reference archive of previously discovered solutions. In this way, it ensures that each member of the population evolves toward promising regions of the objective space, thereby enhancing both the local exploitation and the global diversity of the obtained solution set [6].

2.4 Island Segmentation in Evolutionary Algorithms

Island models are a prominent approach in evolutionary computation, designed to enhance diversity and parallelism. In these models, the global population is partitioned into subpopulations (islands) that evolve independently and periodically exchange individuals. This setup prevents premature convergence and supports exploration across the solution space.

The Multikulti algorithm exemplifies a diversity-driven migration strategy by selecting individuals for migration based on genotypic dissimilarity to the target subpopulation, rather than fitness alone. This method has been shown to outperform conventional migration approaches in various optimization tasks [7].

Expanding on this, the Dynamic Island Model based on Spectral Clustering (DIM-SP) dynamically adjusts the number of islands by clustering individuals according to their similarity. This allows for adaptive control of exploration and exploitation, improving performance and avoiding redundant convergence across islands [8].

In memetic algorithms, Salcedo-Sanz et al. [9] proposed an island-based parallelization scheme that combines global evolutionary search with local refinement, demonstrating robust performance in fuzzy clustering problems.

Migration policy design—including frequency, selection criteria, replacement strategy, and network topology—has a critical influence on performance. Araujo et al. [7] and others emphasize that asynchronous and diversity-promoting migrations lead to more effective search dynamics.

Finally, Díaz et al. [10] introduced a lightweight parallelization strategy using a "collector" instance that aggregates the best solutions from other nodes. Though not exclusive to island models, this method complements them well, enabling efficient coordination with minimal communication overhead.

These developments underscore the value of island segmentation for building scalable, diverse, and high-performing evolutionary systems.

2.5 SIMD (Single Instruction Multiple Data)

In modern high-performance computing, short vector Single Instruction, Multiple Data (SIMD) extensions have become essential for accelerating compute-intensive applications. Initially introduced for multimedia workloads, SIMD extensions such as Intel's SSE family, AMD's 3DNow!, Motorola's AltiVec, and IBM's BlueGene/L vector instructions expanded to support both single and double-precision floating-point operations, enabling their adoption in scientific computing [11]. The main idea behind these extensions is to allow multiple floating-point operations to be performed in parallel by packing several data elements into vector registers. However, despite the theoretical speedups offered by SIMD, effective utilization is often hindered by constrained memory access patterns, data alignment requirements, and significant vectorization overhead in general-purpose compilers.

2.6 CPU Multithreading and Multicore Architecture

Multithreading and multicore architecture are fundamental in the design of modern concurrent systems. Efficient utilization of multiple physical and logical CPU cores enables the parallel execution of threads, leading to substantial performance improvements.

Yi (2024) explains that multithreading divides tasks into smaller subtasks that can be executed concurrently, either through context switching on a single core or in true parallelism across multiple cores. Technologies such as hyper-threading enable modern CPUs to run multiple logical threads per physical core, effectively increasing parallel workload capacity [12].

From a formal verification perspective, Hobor and Gherghina (2012) introduce a concurrent separation logic tailored to barrier synchronization—a mechanism widely used in shared-memory multicore programs. Their logic supports reasoning about resource ownership redistribution across threads, a crucial aspect for ensuring data race freedom in pipeline-style parallel computations [13].

On the algorithmic side, Mascarell (2019) proposes two mutual exclusion algorithms for shared-memory systems, one using a coordinator process and another fully symmetric. These algorithms ensure fair and exclusive access to critical sections, addressing typical concurrency issues in multicore environments [14].

Together, these works illustrate the synergy between hardware-level multithreading capabilities and the software mechanisms—both formal and algorithmic—necessary to exploit them safely and efficiently.

3 Methodology

3.1 Motivation

Memetic algorithms combine the global exploration capabilities of metaheuristics with the local exploitation of deterministic methods, achieving an effective balance between both phases. However, their traditional sequential execution limits the utilization of modern computational resources and can lead to high execution times. Moreover, classical approaches tend to rely solely on the globally best individuals for local refinement, which may introduce bias and result in incomplete exploration of the search space.

To improve computational efficiency and exploratory diversity, we propose a segmented and parallelizable version of the memetic algorithm inspired by a geopolitical structure: countries. Each country operates as a semi-independent subpopulation, and an independent objective function and improvement, enabling the use of multithreading techniques to reduce execution time without compromising algorithmic performance.

3.2 Objective

The objective is to propose an innovative approach to classical memetic algorithms, where the population is not treated as a single entity, but rather divided into clusters, or more specifically, countries with its own distinct cultural identity. The goal is to optimize exploration by having each segment work individually, thus taking advantage of concurrency techniques such as multithreading to improve the overall temporal efficiency of the algorithm.

This new approach is enriched by segmented exploration, allowing for more comprehensive exploration. Rather than always relying on the top n individuals for local improvement, this depends on each cluster and the objective function assigned to it. Furthermore, by performing a migration or shuffle of individuals between clusters, the algorithm continuously explores countries that do not receive elite individuals while exploiting countries that hold a higher percentage of elite individuals; and furthermore, since there's the possibility of assigning a different objective function per cluster (cultural diversity by country), it's possible to reach a more optimal result for different functions and a much better overall fitness, as long as they share the same range of results and domain. In summary, the approach aims to enhance both computational efficiency by reducing execution times and algorithmic performance through improved optimization.

3.3 Country-Segmented Memetic Algorithm (CSMA)

The proposed algorithm, named the Country-Segmented Memetic Algorithm (CSMA), divides the total population into k subpopulations, referred to as countries. Each country runs its own local memetic cycle (selection, variation, and local search) independently, allowing for:

- Segmented exploration: each country explores a distinct region of the search space.
- Contextualized local refinement: individuals selected for local improvements are not globally best, but the best within their own cluster (country). Similarly, each cluster is capable of undergoing a distinct refinement process. This approach not only facilitates population-level exploration through elite variation but also enhances diversity by applying tailored improvements to each cluster. Consequently, this contributes to a more diverse exploration of the search space. This strategy aligns with the concept of cultural variation within the algorithm, reflecting the idea that 'each country operates in its own way.'
- Controlled migration: a periodic migration phase is implemented, to control the algorithm's exploration, in which individuals (both elite and random) are exchanged between countries. This allows continuous exploration of underperforming regions while intensifying search in high-quality areas. For this migration to occur, all threads must synchronize at the start and wait until the migration is fully completed by one of them (randomly selected) before proceeding.
- Support for Multi-Objective: In clustered environments, each processing unit can be trained to optimize a distinct objective function, allowing for concurrent multi-objective optimization. This enables the system to approximate optimal solutions across several functions simultaneously. For consistent results and convergence, it is advisable that the functions have similar ranges and share a common global minimum.

Due to the relative independence between countries, each one can be executed on a separate thread, leveraging parallel architectures for improved efficiency.

3.4 Algorithm Description

See Fig. 1.

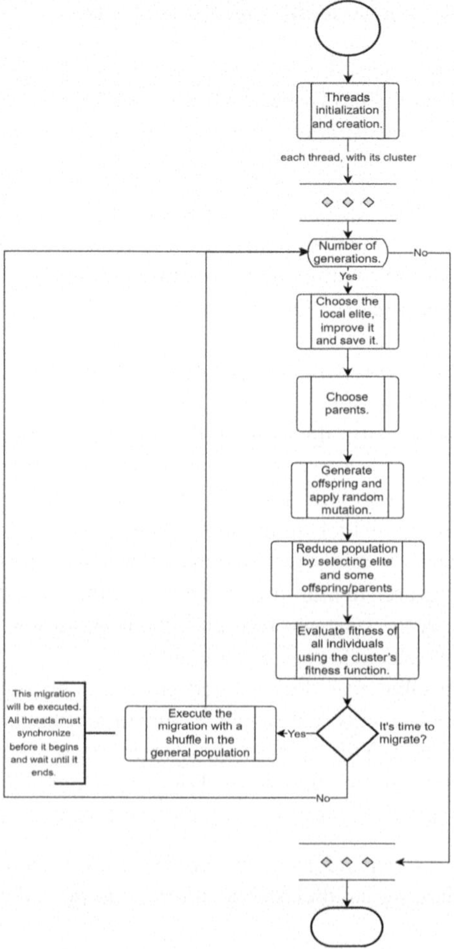

Fig. 1. Algorithm Description

3.5 Pseudocode

CSMA Pseudocode

See Fig. 2.

Algorithm 1 CSMA

```
1: G: (∀i, ∀j : 0 ≤ i < PopulationSize y 0 ≤ j < Dimensions : LowBounds ≤ Population[i][j] ≤
   HighBounds)
2: Require:
3:   Pre: Dimensions ≥ 1 y NClusters ≥ 1 y PopulationSize ≥ NClusters y Generations ≥ 1
4: proc CSMA(value Functions : array of procedures, value Improvements : array of
   Class, value PopulationSize : integer, value Dimensions : integer, value NClusters :
   integer, value Generations : integer, value ClusterSize : array of integer)
5: Population := InitializePopulation(PopulationSize)
6: j := 0
7: do j < PopulationSize →
8:   Fitness[j] := Functions[i](Population[j], Dimensions)
9: od
10: i, EndCluster, BegCluster := 0
11: do i < NClusters →
12:   EndCluster := EndCluster + ClusterSize[i]
13:   Throw Thread → CSMAThread(Population, Fitness, Functions, Improvements[i],
   Dimensions, PopulationSize, Generations, i, BegCluster, EndCluster)
14:   BegCluster := BegCluster + ClusterSize[i]
15: od
16: Join Threads
17: ret Best(Population, Fitness)
18:
19: Ensure:
20:   Post: G ∧ ¬∃i : 0 ≤ i < PopulationSize : BestFitness > Fitness[i]
```

Fig. 2. CSMA Pseudocode.

CSMA Thread Pseudocode

See Fig. 3.

Algorithm 2 CSMAThread

```
1: G: (∀i, ∀j : 0 ≤ i < PopulationSize y 0 ≤ j < Dimensions : LowBounds ≤ Population[i][j] ≤
   HighBounds)
2: Require:
3:   Pre: Dimensions ≥ 1 y Generations ≥ 1 y BegCluster < EndCluster
4: proc CSMAThread(value Population : array of float, value Fitness : array of float,
   value Functions : array of procedures, value Improvement : ImprovementDefinitions,
   value Dimensions : integer, value PopulationSize : integer, value Generations :
   integer, value Thread : integer, value BegCluster : integer, value EndCluster :
   integer)
5: iteration := 0
6: Parents := []
7: Sons := []
8: {invariant : G}
9: {bound : Generations}
10: do iteration < Generations →
11:   Sort(Population, Fitness, BegCluster, EndCluster)
12:   {postcondition : ∀i : BegCluster ≤ i < EndCluster : Fitness[i] ≤ Fitness[i + 1]}
13:   counter := BegCluster
14:   {invariant : G}
15:   {bound : EliteSize ∧ EndCluster}
16:   do counter < EliteSize + BegCluster ∧ counter < EndCluster →
17:     Improvement(Population[counter], Fitness[counter], Dimensions, Functions[Thread])
18:   od
19:   {postcondition : ∀i : BegCluster ≤ i < EliteSize ∧ BegCluster ≤ i < EndCluster :
   Fitness[i] ≤ LastFitness[i]}
20:   GetParents(Parents, ParentSize)
21:   {postcondition : ∀i, ∃j : 0 ≤ i < ParentSize ∧ BegCluster ≤ j < EndCluster : Parents[i] =
   Population[j]}
22:   GetSons(Sons, SonsSize)
23:   {postcondition : ∀i, ∃j ∧ ∃k : 0 ≤ i < SonsSize ∧ BegCluster ≤ j, k < EndCluster : Sons[i] =
   Combine(Parents[j], Parents[k], ParentPercent)}
24:   Mutation(Sons, SonsSize)
25:   {postcondition : ∀i ∧ ∀j : 0 ≤ i < SonsSize ∧ 0 ≤ j < Dimensions : Pr(Sons[i][j] ≠
   LastSons[i][j]) = MutationRate}
26:   CutPopulation(Sons, Parents, Population, EliteSize)
27:   {postcondition : ∀i, ∀j : BegCluster ≤ i < EndCluster ∧ BegCluster ≤ j < EliteSize :
   Pr(Population[i] ∈ Sons) = ReproductionRate ∧ Pr(Population[i] ∈ Parents) = 1 −
   ReproductionRate ∧ LastPopulation[j] ∈ Population[i]}
28:   counter := BegCluster
29:   do counter < EndCluster →
30:     Fitness[counter] = Functions[Thread](Population, Dimensions)
31:   od
32:   if iteration % MigrationRatio = MigrationRatio-1 → PreparingMigration(Population,
   Fitness) fi
33:   {postcondition : Permutation(Population, Fitness, PopulationSize)}
34: od
35:   {postcondition : ∀i, ∃j : BegCluster ≤ i, j < EndCluster : Fitness[j] ≤ Fitness[i]}
36:
37: Ensure:
38:   Post: G
```

Fig. 3. CSMA Thread Pseudocode.

3.6 Formal Definition

Let the objective be:

$$\min_{x \in \mathbb{R}^n} f_i(x) \tag{1}$$

where $f_i : \mathbb{R}^n \to \mathbb{R}$ is the cost function to minimize, and i is the cluster number to which the function is assigned.

The population P of size N is divided into k disjoint subsets:

$$P = C_1 \cup \ldots \cup C_k, C_i \cap C_j = \emptyset \, for \, i \neq j \tag{2}$$

Each C_i represents a country, which evolves independently through a memetic cycle.

Local Memetic Cycle per Country C_i. For each generation:

1. Local improvement: For a proportion $\alpha \cdot |C_i|$ of individuals from the current population. Let $E_i \subset C_i$ be this elite subset. Each $x \in C_i$ is refined using a local search operator $\mathcal{L}_i$, where i is the cluster number to which improvement function is assigned:

$$\forall x \in E_i \subset C_i, x' = \mathcal{L}_i(x) \tag{3}$$

2. Parents Selection: Select parents $x, y \in C_i$ via selection method (e.g., tournament). The project used a random-elitist selection strategy.
3. Reproduction: Via crossover and mutation to generate an offspring z':

$$z' = Mutate(Crossover(x, y)) \tag{4}$$

The resulting offspring replace some individuals in the population, usually the worst, maintaining a constant population size.

Migration Strategy. To balance exploration and exploitation among countries, a migration mechanism is introduced every T_m generation. Migration occurs by randomly shuffling the entire global population using a uniform permutation. The shuffled individuals are then reassigned to the available subpopulations $\{C_1, \ldots, C_k\}$ in a round-robin or sequential manner.

Let $P = \{x_1, ..., x_N\}$ be the global population. A random permutation $\wp$ is applying to get a new ordering.

$$P' = \wp(P) = \left\{x_{\wp(1)}, \ldots, x_{\wp(N)}\right\} \tag{5}$$

Then, the individuals are distributed evenly across the k countries:

$$C_j' = \left\{x_{\wp(j)} | i \equiv j(mod \; k)\right\} for \, j = 0, \ldots, k-1 \tag{6}$$

This means some individuals may remain in their original cluster, while others are randomly reassigned. The result is a lightweight, unbiased migration mechanism that maintains diversity without explicit selection or tracking of performance.

Computational Complexity. Assuming a fitness evaluation cost of C, and independent execution of countries on p threads, the algorithm's time complexity per iteration is:

$$O\left(\frac{N \cdot C}{p}\right) \tag{7}$$

This model exploits parallel architectures by distributing the workload across multiple threads, reducing overall runtime while maintaining or improving optimization performance.

Overhead Function. To evaluate the performance cost associated with parallelization, we define an overhead function that quantifies the additional time incurred due to factors such as thread management, synchronization, and communication, and the migration process (mutex and barriers, and the shuffle). This function is used to assess how the system scales with respect to the number of threads and the size of the population.

$$T_o(P, N) = \frac{1}{k}(\alpha N + \beta P) + \delta \log(P) + \gamma \frac{1}{N} \tag{8}$$

where:

- P is the number of threads.
- N is the total population size.
- k is the number of iterations between each migration.
- $\log(P)$ represents the intrinsic thread management cost, which grows with the number of threads due to the increased complexity of managing multiple execution units.
- α, β, γ and δ are constants used to adjust the influence of each variable in the model, ensuring proper scaling based on empirical results.

The analysis of the partial derivatives of $T_o(P, N)$ with respect to P and N offers valuable information for resource allocation strategies and performance optimization. The derivatives are computed as follows:

$$\left(\frac{\vartheta T_o}{\vartheta P}, \frac{\vartheta T_o}{\delta N}\right) = \left(\frac{\beta}{k} + \frac{\delta}{P}, \frac{\alpha}{k} - \frac{\gamma}{N^2}\right) \tag{9}$$

These expressions provide a formal basis for understanding the sensitivity of the overhead to variations in system configuration:

- The partial derivative with respect to P is strictly positive under the assumption that $\beta, \delta > 0$ indicating that increasing the number of processing units generally leads to an increase in overhead due to additive linear and logarithmic coordination costs.
- The partial derivative with respect to N exhibits a non-monotonic behavior. Setting it to zero yields the condition:

$$\frac{\alpha}{k} = \frac{\gamma}{N^2} \Rightarrow N = \sqrt{\frac{k\gamma}{\alpha}} \tag{10}$$

which identifies a problem size that locally minimizes the overhead by balancing the linear scaling term against the inverse penalty due to granularity. This critical point characterizes an efficient operating region where computational resources are most effectively utilized.

4 Experiment and Results

4.1 Experimental Setup

All experiments were conducted on a machine equipped with an Intel Pentium Silver N5030 processor (4 cores, 4 threads) and 8 GB of RAM. The algorithms were implemented in C++ and compiled using the native compiler of the Linux Mint operating system. To ensure statistical robustness, each experiment was independently repeated 100 times for performance evaluation. In the case of multithreaded benchmarking, the number of repetitions was reduced to 20 due to the increased computational overhead.

Four standard benchmark functions were used to assess the behavior of the algorithms under diverse optimization landscapes (Table 1):

Table 1. Functions for benchmark in $d = 5$.

Function	Range	Global minima
Ackley	$[-32.768, 32.768]$	0
Michalewicz	$[0, \pi]$	– 4.68765
Rosenbrock	$[-5, 10]$	0
Griewank	$[-600, 600]$	0

All functions were evaluated in a fixed dimensional setting with $d = 5$.

In addition to single-objective scenarios, the experimental protocol included evaluations under multi-objective conditions (Table 2):

Table 2. Functions for benchmark multi-objective.

Function	Range	Global minima
Sphere	$[-5.12, 5.12]$	0
Rastrigin	$[-5.12, 5.12]$	0

Specifically, performance was assessed after training on the Rastrigin function, and subsequently on a combined objective of Sphere and Rastrigin.

To analyze the performance of the proposed algorithm under parallel execution, additional experiments were conducted using only the Ackley function. This allowed for a controlled assessment of multithreading efficiency and scalability, particularly in relation to the theoretical bounds imposed by Amdahl's Law, Gustafson's Law, Karp-Flatt's Law and Isoefficiency.

Also, the hyperparameters selected for the CSMA where (Table 3):

Table 3. Algorithm's hyperparameters.

Hyperparameter	Value	Description
Population size	1500	Size of the population
Dimensions	5	Dimension of every individual
N Threads	4	Quantity of threads and clusters
Migration Ratio	100	How many generations pass before a migration
Generations	1000	The number of generations for each cluster
Mutation Rate	0.1	Probability to mutate
Marriage Rate	0.5	Percentage of parents in the global population
Parent Percent	0.5	Percentage of the chromosome that offspring will inherit from one parent
Elite Size	0.65	Global size of the elite that will be improved
Reproduction Rate	1	Percentage of the new population that will be sons (except the elite)
Mutation Bounds	0.1	Mutation for each chromosome

For the experiments, two improvement algorithms were used: Hill Climbing and Simulated Annealing. Each algorithm was employed with its respective hyperparameters (Tables 4 and 5):

Table 4. Hill Climbing's hyperparameters.

Hyperparameter	Value
Attempts	0.1
Step Size	10

Table 5. Simulated Annealing's hyperparameters.

Hyperparameter	Value
Temperature	100
Min Temperature	0.001
Alpha	0.98
Attempts	10
Step Size	0.1

4.2 Algorithm Performance

The performance of the proposed algorithm is assessed through comparative evaluations against established baseline methods across the four benchmark functions presented previously.

Table 6. Performance for each function $d = 5$.

Functions	CSMA	MA
Ackley	0.0076	0.7222
Michalewicz	– 4.6829	– 4.6728
Rosenbrock	0.1611	0.7004
Griewank	0.0070	0.1023

Table 7. t-student for every function.

Functions	t	p
Ackley	– 10.9028	0.0000
Michalewicz	– 3.9473	0.0001
Rosenbrock	– 13.3330	0.0000
Griewank	– 16.4905	0.0000

The results demonstrate that CSMA consistently achieves competitive or superior outcomes compared to the baselines. It exhibits strong performance on functions characterized by high multimodality and deceptive local optima, such as Ackley and Michalewicz.

At the same time, the performance of the CSMA was evaluated for the same functions but varying the migration ratio. The ratio of 100, which was used in the previous evaluation, was compared against a ratio of 1 for this new evaluation.

Table 8. Performance for each function $d = 5$.

Function	CSMA100	CSMA1
Ackley	0.0076	0.0273
Michalewicz	– 4.6829	– 4.5830
Rosenbrock	0.1611	0.4024
Griewank	0.0070	0.1023

Table 9. t-student for every function.

Functions	t	p
Ackley	– 19.5518	0.0000
Michalewicz	– 13.1767	0.0000
Rosenbrock	– 2.9712	0.0037
Griewank	– 4.3030	0.0000

According to the results, we can see that there are instances, where decreasing the migration ratio significantly increases exploration by altering the elite more constantly. As a result, some functions benefit from this, while others experience a decline in performance. However, when multiple improvement functions are available, the results can benefit more if one function is executed for a longer duration than another. This explains why using a ratio of 100 yields better results than a ratio of 1.

Simultaneously, the results of the evaluation of the algorithm trained with a single objective function (CSMAUni) are presented, along with its evaluation using two objective functions distributed across four clusters, two for each function (CSMAMul). The first algorithm was trained using the Rastrigin function, while the second was trained with both the Rastrigin function and the Sphere function. A boxplot is then presented, illustrating the performance of each algorithm when evaluating its best individual with each of the functions. The first two evaluations aim to determine whether, despite being trained with a single objective function, the algorithm performs effectively when evaluated with a different function. The remaining three evaluations assess its performance with the functions it was trained on and provide insight into the overall fitness achieved during the training process.

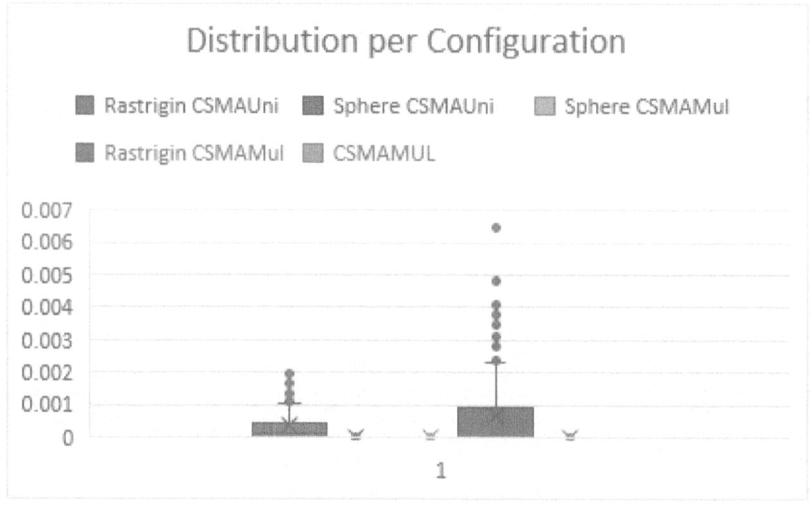

Fig. 4. Boxplots of the performance.

It can be observed that the results obtained under a single-objective fitness evaluation —represented by the first boxplot— correspond essentially to the overall fitness derived from a unique objective function. In contrast, the fifth result exhibits a substantially improved overall fitness compared to the former.

4.3 Execution Times

The execution time of the proposed algorithm is assessed through comparative evaluations against established baseline methods across the four benchmark functions presented previously and evaluated to get the performance.

Table 10. Execution time for each function $d = 5$.

Functions	CSMA (s)	MA (s)	Speedup
Ackley	9.3385	28.1489	3.0143
Michalewicz	14.8774	30.6095	2.0574
Rosenbrock	13.9271	27.3748	1.9656
Griewank	10.3032	25.3347	2.4589

Table 11. t-student for every function.

Functions	t	p
Ackley	– 22.7698	0.0000
Michalewicz	– 46.0563	0.0000
Rosenbrock	– 489.6820	0.0000
Griewank	– 141.8730	0.0000

Based on the results, a substantial difference is observed, with a clear indication that the CSMA implemented is significantly faster than the classical MA. In the subsequent section, its efficiency will be assessed under a multithreading paradigm to gain a deeper understanding of the underlying reasons for its superior speed.

4.4 Multithreading Benchmarking

To evaluate the impact of parallelization on computational efficiency, a series of multithreading experiments was conducted using only the Ackley function, as it represents a complex and computationally intensive fitness landscape. The proposed algorithm was executed under varying numbers of threads (from 1 to 4, aligned with the physical hardware limits of the experimental machine), and each configuration was repeated 20 times to obtain statistically meaningful results and to reduce the influence of outliers in execution time, the harmonic mean was used as a more robust central tendency measure.

Table 12. Speedup per number of threads.

Thread	Speedup
2	1.5758
3	2.0318
4	2.3399

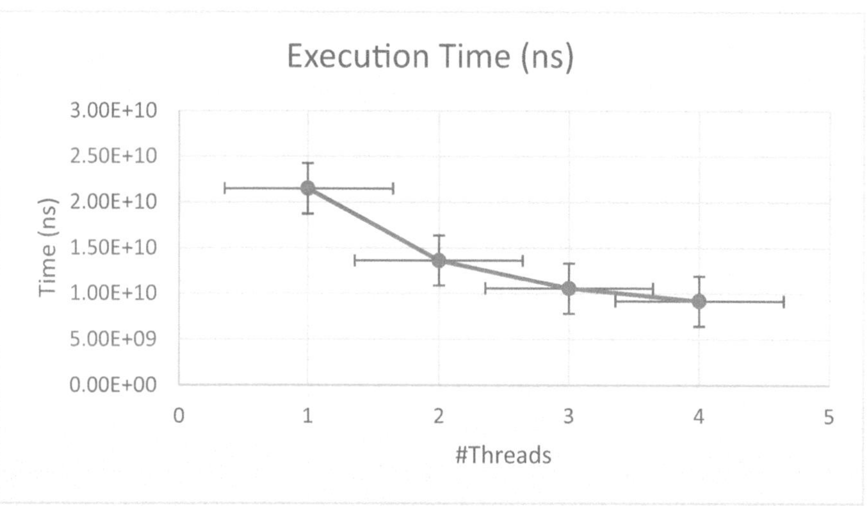

Fig. 5. Execution time per number of threads.

To complement the empirical speedup measurements, several theoretical models were employed to further characterize the scalability of the proposed algorithm under parallel execution. These models provide insight into the algorithm's potential for improvement, limitations due to serial components, and the relationship between problem size and parallel efficiency.

Karp-Flatt Metric. The Karp–Flatt metric was used to estimate the serial fraction $\in(n)$ of the algorithm as a function of the number of processors n and the observed speedup $S(n)$. It is defined as:

$$\in(n) = \frac{\frac{1}{S(n)} - \frac{1}{n}}{1 - \frac{1}{n}} \tag{11}$$

This metric offers a practical way to infer the algorithm's scalability bottlenecks from experimental data. The results:

Table 13. Sequential percentage.

Thread	$\in (n)$
2	0.2692
3	0.2382
4	0.2365

And this indicates that the serial fraction remains relatively low for small core counts, suggesting a well-parallelized implementation with limited sequential overhead.

Isoefficiency. To evaluate the scalability with increasing problem size, the isoefficiency was estimated. This function describes the growth rate of the problem size W to maintain an efficiency E as the number of processors increases.

$$K = \frac{1 - E}{E} \tag{12}$$

And the general isoefficiency equation is:

$$W = K \cdot T_o(n) \tag{13}$$

And $T_o(n)$ was defined before, but also has another definition from:

$$E = \frac{W}{W + T_o(n)} \Rightarrow T_o(n) = \frac{W}{E} - W \tag{14}$$

In this case, the problem size remains fixed at $W = 21.4778$, and the efficiency was measured empirically for 2, 3, and 4 threads. Using these values, we compute the associated overheads:

$$E(n) = \frac{1}{1 + \frac{T_o(n)}{n \cdot T_n - T_o(n)}} \tag{15}$$

where T_n is the execution time for n threads. Executing these functions with their respective number of threads yields different efficiency results, given that the problem size was kept constant.

Table 14. Isoefficiency analysis.

Threads	E	T_O	W
2	0.7879	5.7811	21.4778
3	0.6773	10.2339	21.4778
4	0.5849	15.2375	21.4778

As the number of threads increases, the overhead becomes more significant. To maintain a parallel efficiency of 0.7879, the problem size would need to scale accordingly. This analysis illustrates the growth rate of W required for the algorithm to remain efficient as concurrency increases.

Amdhal's Law. For theoretical upper-bound estimation, Amdahl's Law was applied using the classical formulation:

$$S(n) = \frac{1}{\in(n) + \frac{1 - \in(n)}{n}} \tag{16}$$

where $\in$ represents the sequential fraction of the algorithm and we will use the previously obtained values. Using values inferred from empirical speedup data, the resulting model confirms the presence of diminishing returns as the number of threads increases:

Table 15. Amdhal's Law.

Thread	$S(n)$
2	1.5758
3	2.0318
4	2.3399

Gustafson's Law. In contrast, Gustafson's Law was employed to model scalability in scenarios where the problem size increases proportionally with the number of processors. It is defined as:

$$sS(n) = n + (1 - n) \cdot \in(n) \tag{17}$$

where $\in$ represents the sequential fraction of the algorithm and we will use the previously obtained values. This formulation assumes that the parallel workload scales with n and thus offers a more optimistic projection for real-world applications involving large data or adaptive workloads. The results suggest that the proposed method retains high scalability potential under such conditions.

Table 16. Gustafson's Law.

Thread	$sS(n)$
2	1.7529
3	2.5058
4	3.287

Here, we can observe that the theoretical maximum speed up, according to both Amdahl's and Gustafson's laws, is high. This is due to the small percentage of sequential operations, which primarily depends on the objective function or the tuners, in our case, only hill climbing for this test.

5 Conclusions

This paper introduced the Country-Segmented Memetic Algorithm (CSMA), a novel approach to memetic optimization that segments the population into independent clusters—referred to as countries—executed in parallel using multithreading and SIMD-based acceleration. The proposed architecture addresses two key limitations of traditional memetic algorithms: the inefficient use of computational resources due to sequential execution, and the bias introduced by relying exclusively on globally elite individuals for local refinement.

Experimental results demonstrated that the CSMA outperforms classical memetic approaches across several benchmark functions in both single- and multi-objective optimization contexts. The segmented structure enabled broader exploration of the solution space, while the controlled migration mechanism-maintained diversity and promoted effective information sharing between subpopulations.

5.1 Performance Evaluation

The performance of the proposed Country-Segmented Memetic Algorithm (CSMA) was evaluated using four standard benchmark functions: Ackley, Michalewicz, Rosenbrock, and Griewank. Each function was tested in a 5-dimensional setting over 100 independent runs. The results, summarized in Table 6, indicate that CSMA consistently outperforms the classical Memetic Algorithm (MA) across all functions in terms of solution quality.

Statistical validation using the t-test (Table 7) confirms the significance of these differences, with p-values well below the 0.05 threshold. Notably, CSMA demonstrates strong performance on highly multimodal functions such as Ackley and Michalewicz, where traditional approaches often struggle due to local optima entrapment.

5.2 Effect of Migration Frequency

To evaluate the impact of the migration strategy, additional experiments were conducted comparing two migration ratios: one migration every 100 generations (CSMA100) and one every generation (CSMA1). As shown in Table 8, CSMA100 consistently delivers better performance, particularly on functions requiring exploitation stability. The statistical analysis (Table 9) further supports this, indicating that more frequent migrations (CSMA1) may introduce excessive randomness, undermining convergence in some cases.

These findings suggest that a moderate migration interval enhances both the exploration and exploitation capabilities of the algorithm, particularly when different improvement functions operate simultaneously.

5.3 Multi-objective Performance

The performance of CSMA was also tested in a multi-objective setup using the Sphere and Rastrigin functions. Two configurations were evaluated: CSMAUni (trained solely on Rastrigin) and CSMAMul (trained concurrently on Sphere and Rastrigin across different clusters). The results, visualized in the boxplots (Fig. 4), show that CSMAMul achieves improved overall fitness, highlighting the benefit of segmented multi-objective training.

This outcome confirms that country-based segmentation enables the algorithm to simultaneously optimize diverse objectives, yielding higher-quality solutions even when the functions differ in structure and landscape.

5.4 Execution Time and Parallel Efficiency

Execution times were measured for both CSMA and the baseline MA across all benchmark functions (Table 10). CSMA exhibited substantial speedups ranging from 2x to 3x, attributed to its parallel structure. Statistical tests (Table 11) affirm the significance of these time differences.

Further benchmarking focused on multithreading performance using the Ackley function. Speedups were computed for 2, 3, and 4 threads (Table 12), with results indicating a near-linear improvement up to the hardware limit. Figure 5 visualizes execution time reductions as thread count increases.

5.5 Scalability Analysis

To complement empirical results, several scalability models were applied:

- Karp-Flatt Metric: Revealed a low sequential fraction (approx. 23–27%) across configurations (Table 13), suggesting effective parallelization.
- Isoefficiency Analysis: Demonstrated that maintaining high parallel efficiency requires increasing problem size proportionally with the number of threads (Table 14).
- Amdahl's Law: Confirmed the diminishing returns at higher core counts (Table 15), as expected.
- Gustafson's Law: Indicated better scalability under workload expansion scenarios (Table 16), reinforcing the algorithm's potential in large-scale optimization.

References

1. Nguyen, P.T.H., Sudholt, D.: Memetic algorithms beat evolutionary algorithms on the class of hurdle problems, pp. 1071–1078 (2018). https://doi.org/10.1145/3205455.3205456
2. Nguyen, P.T.H., Sudholt, D.: Memetic algorithms outperform evolutionary algorithms in multimodal optimization. Artif. Intell. **287**, 103345 (2020)
3. Costa, L., Oliveira, P.: An elitist genetic algorithm for multiobjective optimization. In: Metaheuristics: Computer Decision-Making, vol. 86 217–236, Springer, US, Boston, MA, (2003)
4. Groşan, C., Oltean, M., Oltean, M.: The role of elitism in multiobjective optimization with evolutionary algorithms. Acta Universitatis Apulensis (2025)
5. Shahin, A.A.: Memetic elitist Pareto evolutionary algorithm for virtual network embedding. CIS **8**, 73 (2015)
6. Knowles, J.D., Corne, D.W.: M-PAES: a memetic algorithm for multiobjective optimization. In: Proceedings of the 2000 Congress on Evolutionary Computation. CEC00 (Cat. No.00TH8512), vol. 1 325–332. IEEE, La Jolla, CA, USA (2000)

7. Araujo, L., Guervos, J.J.M., Cotta, C., Vega, F.F.D.: MultiKulti algorithm: migrating the most different genotypes in an island model (2008). https://doi.org/10.48550/arXiv.0806.2843
8. Meng, Q., Wu, J., Ellisy, J., Kennedy, P.J.: Dynamic island model based on spectral clustering in genetic algorithm, pp. 1724–1731 (2017). https://doi.org/10.1109/IJCNN.2017.7966059
9. Salcedo-Sanz, S., Del Ser, J., Geem, Z.W.: An island grouping genetic algorithm for fuzzy partitioning problems. Sci. World J. **2014**, 1–15 (2014)
10. Díaz, D., Valledor, P., Ena, B., Iglesias, M., Menéndez, C.: Improved method for parallelization of evolutionary metaheuristics. Mathematics **8**, 1476 (2020)
11. Franchetti, F., Kral, S., Lorenz, J., Ueberhuber, C.W.: Efficient utilization of SIMD extensions. Proc. IEEE **93**, 409–425 (2005)
12. Yi, X.: A study of performance programming of CPU, GPU accelerated computers and SIMD architecture (2024). https://doi.org/10.48550/arXiv.2409.10661
13. Hobor, A., Gherghina, C.: Barriers in concurrent separation logic: now with tool support! Logical Methods Comput. Sci. **8**(2) (2012)
14. Mascarell, J.B.: Two mutual exclusion algorithms for shared memory (2019). https://doi.org/10.48550/arXiv.1903.10402
15. Moscato, P.: On evolution, search, optimization, genetic algorithms and martial arts: towards memetic algorithms, Caltech concurrent computation program, California Institute of Technology (2000). https://www.researchgate.net/publication/2354457
16. Subburaj, B., Amali, S.M.J.: A fuzzy system based self-adaptive memetic algorithm using population diversity control for evolutionary multi-objective optimization. Sci. Rep. **15**, (2025). Article number: 5735. https://doi.org/10.1038/s41598-025-89289-2
17. Palubeckis, G.: A fast local search based memetic algorithm for the parallel row ordering problem. Appl. Math. Comput. **486**, 129040 (2025). https://doi.org/10.1016/j.amc.2024.129040

Bert_DPRCNN: A Novel Approach to Text Classification Using Hybrid Deep Learning Models

Yogeshvar Reddy Kallam[1](✉), Lovely Yeswanth Panchumarthi[2](✉), Lavanya Parchuri[2], Sriya Padmanabhuni[2], Satya Krishna Nunna[2], and Krishna Kambhampaty[1]

[1] Department of Computer Science, The Pennsylvania State University, University Park, PA 16801, USA
{yvk5381,kvk6012}@psu.edu

[2] Department of Computer Science, SRM University AP, Amaravati 522240, Andhra Pradesh, India
{yeswanthpanchumarthi,lavanya_parchuri,sriya_padmanabhuni, satyakrishna.n}@srmap.edu.in

Abstract. Text classification, a fundamental technique in information retrieval, categorizes textual data to streamline search operations, significantly enhancing efficiency. This research introduces CNN, a novel hybrid deep learning model that advances text classification by synergistically combining the strengths of transformer models, CNNs, RNNs, and LSTMs. The model stands out with its innovative design, inputting 'input ids' and 'attention mask' into a pretrained transformer, followed by a series of operations leading to robust feature extraction. CNN outperforms baseline models, demonstrating superior accuracy in text classification. Extensive experimentation on both Chinese and English datasets substantiates this, with the model achieving the highest accuracies of 94.54% and 91.89%, respectively. This paper underscores the effectiveness of the proposed model in enhancing text classification, paving the way for more efficient information retrieval.

Keywords: Text Classification · CNN · Deep Learning Models · AG's News Dataset

1 Introduction

Text classification [1–5] is a fundamental technique in information retrieval, serving as a vital tool for organizing and categorizing textual data. By automatically assigning predefined categories or labels, it significantly enhances the efficiency of search operations and information access, playing a crucial role in tasks such as document indexing,

L. Parchuri, S. Padmanabhuni, S. K. Nunna, K. Kambhampaty—These authors contributed equally to this work.

I. Roy et al. (Eds.): CAINE 2025, CCIS 2740, pp. 299–308, 2026.
https://doi.org/10.1007/978-3-032-14893-3_20

information filtering, and content recommendation. Ongoing advancements in text classification methods, from traditional machine learning to deep learning, continuously improve the accuracy and scalability of information retrieval systems.

The field employs a diverse array of algorithms, including Decision Trees, Gradient Boosting Machines (GBM) [6], Random Forests, Convolutional Neural Networks (CNNs) [7], and Recurrent Neural Networks (RNNs) [8]. Pretrained models like BERT [9] and GPT [10], and their variants, have revolutionized text classification by leveraging largescale pretraining for improved performance. While non pretrained models such as Text RCNN [3], FastText DPCNN [11], and Text RNN [12] have demonstrated efficacy in capturing textual patterns, a persistent challenge lies in effectively synergizing the complementary strengths of these diverse architectural paradigms. Optimal performance often necessitates a comprehensive approach that can capture various textual patterns—ranging from global contextual understanding to intricate local features and sequential dependencies.

This research addresses this challenge by introducing CNN, a novel hybrid deep learning model designed to advance the state of the art in text classification. The primary objective is to achieve robust feature extraction and improved classification accuracy across diverse datasets by synergistically combining the inherent strengths of transformer models, Convolutional Neural Networks (CNNs), Recurrent Neural Networks (RNNs), and Long Short-Term Memory (LSTM) networks. The CNN model distinguishes itself through its innovative hybrid architecture, which integrates these components in a meticulously designed sequence of operations to capture a more robust and comprehensive representation of text. The efficacy of the CNN model is substantiated by its superior performance in text classification, consistently outperforming various baseline and state of the art models in extensive experimentation on both Chinese and English datasets.

2 Dataset

The AG's news topic classification dataset is a collection of more than 1 million news articles gathered from over 2000 news sources by the academic news search engine ComeToMyHead, which has been active since July 2004. This dataset is made available to the academic community for research purposes in various fields, including data mining, information retrieval, and data compression. The dataset used for text classification benchmarking contains 4 largest classes selected from the original corpus, with each class consisting of 30,000 training samples and 1,900 testing samples. In total, there are 120,000 training samples and 7,600 testing samples. The file classes.txt contains a list of classes corresponding to each label. This dataset is utilized as a benchmark in the paper" Character-level Convolutional Networks for Text Classification" by Xiang Zhang, Junbo Zhao, and Yann LeCun, presented at the Advances in Neural Information Processing Systems 28 (NIPS 2015) conference.

3 Related Works

The TextRCNN architecture combines recurrent and convolutional structures [13]. It begins with word embeddings as input, which are then processed by a bidirectional recurrent layer to capture contextual information from both directions. The outputs of this layer are concatenated with the original word embeddings and passed through a convolutional layer, where multiple filters extract key features. Finally, a max pooling layer captures the most important features for text classification. This architecture efficiently captures diverse contextual information without the need for human designed features.

The DPCNN is a deep CNN with a pyramid structure, designed for efficient representation of longrange text associations [11]. It maintains a constant feature map count, reducing computational cost while increasing depth. Using pooling with stride 2 after each convolution block, it halves data size and doubles effective coverage. With 15 weight layers, DPCNN outperforms previous models on six benchmark datasets for sentiment classification and topic categorization. It also enhances text region embedding with unsupervised embeddings, improving accuracy by leveraging diverse information from the text.

The fastText architecture [14] uses a word representation model that considers the internal structure of words, making it ideal for morphologically rich languages like Turkish. Words are represented as bags of character n-grams, enabling effective handling of out of vocabulary words. The model averages word vectors to create a document vector, which is then used in a hidden layer for classification. This approach allows for efficient text classification without requiring language specific preprocessing steps.

The text RNN architecture [15] in the paper uses LSTM cells in a bidirectional setup to effectively process sequential data, ensuring accuracy by considering both past and future context. For text generation, n-gram tokens are employed, and input sequences are padded for uniform length. During training, words are embedded, dropout is applied for regularization, and optimization is done using the Adam optimizer with categorical crossentropy as the loss function.

The RCNN_ATT model [16], utilized in this study, is an advanced architecture for text classification. It integrates a bidirectional LSTM network with an attention mechanism to improve feature extraction from intricate texts. The bidirectional LSTM captures contextual information, while the attention mechanism calculates category contribution scores. This fusion enables the model to efficiently classify technical requirement texts, enhancing supply and demand matching efficiency. The architecture's efficacy is underscored by its superior performance compared to other neural network models.

The BERT architecture discussed in this paper [17] is a large pretrained Transformer network designed for natural language processing tasks. It consists of multiple layers, each with multiple attention heads capturing different language aspects. The attention mechanism allows focusing on relevant parts of input, learning complex language patterns. BERT's pretraining includes masked language modeling and next sentence prediction, aiding context and relationship understanding. This architecture enables BERT to achieve high accuracy when fine tuned on various tasks.

The BERT-CNN architecture combines BERT's deep bidirectional transformer knowledge with CNN's text classification capabilities [18]. It utilizes contextualized embeddings from BERT's last four hidden layers, which are fed into several CNN convolution layers. These layers apply convolution operations using filters of varying sizes to classify text. Following convolution, a Global Max-Pooling operation is applied, and the results are passed through a dense layer with a Sigmoid function to obtain the final binary label. This model structure effectively identifies offensive speech on social media.

The BERT-RNN architecture combines BERT's contextual understanding of sentences with an RNN's sequential processing [19]. BERT processes entire sentences simultaneously, considering both preceding and following inputs for each word. The RNN further processes this rich contextual information sequentially, enhancing the model's ability to handle sequences and predict outcomes. This integration improves performance in natural language processing tasks, particularly in sentiment analysis, achieving high accuracy and precision in classifying sentiments as positive, negative, or neutral.

The BERT-DPCNN model [20] synergizes BERT's language representation abilities with DPCNN's efficient long-range association representation for text categorization. BERT pretrains deep bidirectional representations by conditioning on both left and right context in all layers. In contrast, DPCNN is a low-complexity, word-level convolutional neural network that efficiently represents long-range associations in text. The DPCNN architecture alternates a convolution block and a downsampling layer, creating a deep network where internal data size shrinks in a pyramid shape.

Human trust in machines and artificial intelligence may not differ significantly from trust in other people [21].

In the context of AI, trust has gained importance for several reasons—for example, the use of robots in high-risk military operations or autonomous vehicles has helped save lives. Trust is essential for the adoption and success of new technologies, influencing both consumer behavior and business outcomes, as seen with digital assistants like Amazon Alexa and Google Assistant. This paper explores the concept of trust in relation to technology, proposes strategies to strengthen it, and concludes that trust in machines can be developed in much the same way as trust in humans.

Detecting malicious users can help prevent or reduce the impact of attacks by enabling timely intervention before unauthorized or illegal actions occur. This approach is effective in countering insider and masquerade attacks and can be applied across mobile, cloud, and pervasive computing environments [22].

4 Methodology

The proposed model Fig. 1 is a sophisticated deep learning architecture that leverages the power of transformer models, convolutional neural networks (CNNs), recurrent neural networks (RNNs), and long short-term memory (LSTM) networks. The model begins by inputting 'input ids' and 'attention mask' into a pretrained transformer model, which predicts masked tokens in the input sequence. The output from the transformer is then processed through a series of operations including softmax application, quantization, and embedding to yield dense vector representations. These embeddings are further processed through multiple convolutional layers for feature extraction. Each convolutional layer is equipped with distinct filter sizes and applies a convolution operation and ReLU activation to the embeddings. The outputs of these layers are then max-pooled and concatenated to form the final feature representation.

This representation is subsequently fed into both an RNN and an LSTM layer for further processing. The final encoded output is formed by concatenating the last outputs of the RNN and LSTM layers. This resulting tensor represents the encoded sequences and can be further utilized for subsequent processing, such as feeding into a linear layer to obtain final predictions.

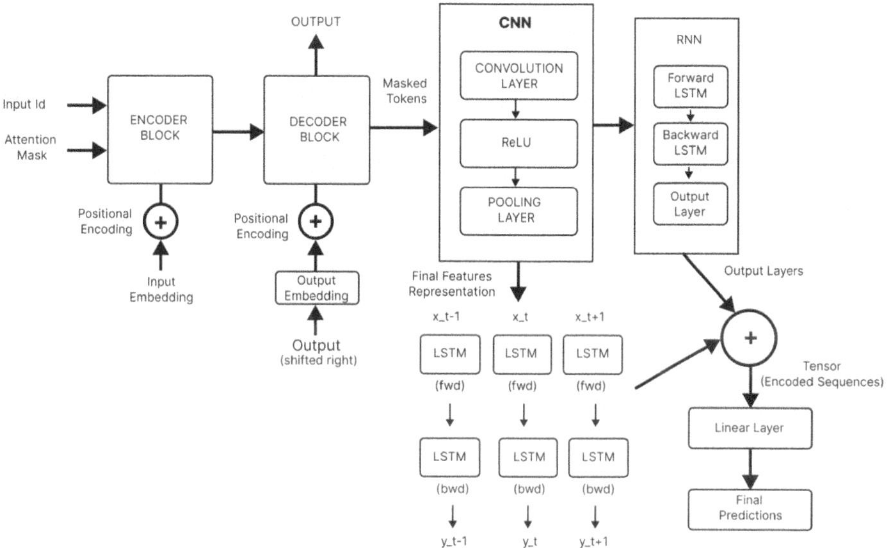

Fig. 1. Proposed CNN Model Architecture

This model is designed to handle complex tasks and extract meaningful features from the input data. It combines the strengths of various neural network architectures to deliver robust and accurate predictions. For a detailed understanding of the model and its operations, please refer to the mathematical equations provided above. They offer a comprehensive view of the model's inner workings and the transformations applied to the data at each stage. This detailed view can help in understanding the model's functionality and its potential applications.

The process begins by inputting 'input ids' and 'attention mask' into a transformer model. This transformer is a pretrained masked language model, predicting masked tokens in the input sequence. The resulting output from the transformer is represented as a tensor $L \in \mathrm{R}^{n \times v}$, where n is the sequence length and v is the vocabulary size.

4.1 Softmax Normalization and Quantization

To prepare transformer outputs for subsequent processing stages, we apply a two-step transformation involving normalization and discretization. First, the logits are masked to eliminate contributions from padded positions, ensuring that only meaningful token representations participate in further processing:

$$\mathbf{L}'_{\mathrm{i,j}} = \begin{pmatrix} \mathbf{L}_{\mathrm{i,j}} & \text{if } \mathbf{M}_i = 1 \\ -\infty & \text{if } \mathbf{M}_i = 0 \end{pmatrix} \tag{1}$$

The masked logits undergo softmax normalization to convert them into probability distributions, enabling stable numerical processing:

$$p_{i,j} = \frac{\exp(\mathbf{L}'_{i,j})}{\mathbf{L}_{k=1}^{v} \exp(\mathbf{L}'_{i,k})} \text{ where } \mathbf{p} \in \mathrm{R}^{\mathrm{n} \times \mathrm{v}} \tag{2}$$

The resulting probabilities are then quantized into discrete integer representations suitable for embedding layer processing:

$$q_{i,j} = \text{round}(p_{i,j} \times (v - 1)) \text{ where } \mathbf{q} \in \mathrm{Z}^{n \times v} \tag{3}$$

Finally, the quantized values are converted to long integer format for computational compatibility:

$$\mathbf{q}_{\text{long}} = \text{long}(\mathbf{q}) \text{ where } \mathbf{q}_{\text{long}} \in \mathrm{Z}^{n \times v} \tag{4}$$

This quantization process serves as a crucial bridge between continuous transformer representations and discrete embedding processing, enabling effective integration of transformer outputs with convolutional architectures.

4.2 Quantized Probability Embeddings

The discretized probability representations are transformed into dense vector embeddings through a learnable embedding layer. This transformation converts sparse quantized probabilities into rich, dense representations suitable for convolutional processing.

Given an embedding weight matrix $\mathbf{W}_{\text{emb}} \in \mathrm{R}^{v \times d}$ where d is the embedding dimension, the embedding transformation is computed as:

$$\mathbf{X} = \mathbf{q}_{\text{long}} \times \mathbf{W}_{\text{emb}} \tag{5}$$

where $\mathbf{X} \in \mathrm{R}^{n \times d}$ represents the embedded sequence representations.

To prepare these embeddings for convolutional processing, a channel dimension is added and ReLU activation is applied:

$$\mathbf{X}' = \mathbf{X}.\text{unsqueeze}(1) \in \mathrm{R}^{n \times 1 \times d} \tag{6}$$

$$\mathbf{Y} = \text{ReLU}(\mathbf{X}') = \max(0, \mathbf{X}') \in \mathrm{R}^{n \times 1 \times d} \tag{7}$$

The ReLU activation introduces non-linearity while maintaining the spatial structure necessary for effective convolutional feature extraction.

4.3 Multi-scale Convolutional Feature Extraction

The embedded representations undergo hierarchical feature extraction through multiple convolutional layers with varying filter sizes. This multi-scale approach captures textual patterns at different granularities, from local n-gram features to broader contextual patterns.

For a set of filter sizes $\{f_1, f_2, ..., f_m\}$, each filter f_i applies convolution followed by ReLU activation:

$$\mathbf{H}_i = \text{ReLU}(\text{Conv1D}(\mathbf{Y}, \mathbf{W}_{fi})) \tag{8}$$

where $\mathbf{W}_{fi} \in \mathrm{R}^{fi \times 1 \times d}$ represents the convolutional weights for filter size f_i, and $\mathbf{H}_i \in \mathrm{R}^{(n-fi+1) \times 1 \times d}$ contains the resulting feature maps.

Global max-pooling is applied to each feature map to extract the most salient features:

$$\mathbf{M}_i = \text{MaxPool}(\mathbf{H}_i, \text{kernel\_ size} = n - f_i + 1) \tag{9}$$

resulting in $\mathbf{M}_i \in \mathrm{R}^{1 \times 1 \times d}$ for each filter.

The feature representations from all filters are concatenated to form a comprehensive multi-scale feature vector:

$$\mathbf{Z} = \text{Concat}(\mathbf{M}_1, \mathbf{M}_2, ..., \mathbf{M}_m) \in \mathrm{R}^{m \times 1 \times d} \tag{10}$$

The concatenated features are reshaped for input to recurrent layers:

$$\mathbf{O} = \text{Reshape}(\mathbf{Z}) \in \mathrm{R}^{md \times n} \tag{11}$$

This multi-scale convolutional processing enables the model to capture both fine-grained local patterns and broader textual structures essential for accurate classification.

4.4 Dual-Path Recurrent Processing

The convolutional features undergo parallel processing through both RNN and LSTM networks to capture complementary sequential dependencies. This dual-path approach leverages the efficiency of RNNs and the long-term memory capabilities of LSTMs.

The convolutional output is transposed to match the expected input format for recurrent layers:

$$\mathbf{R}_{\text{in}} = \mathbf{O}^T \in \mathrm{R}^{n \times md} \tag{12}$$

RNN Processing Path: The RNN path processes the sequence with initialized hidden states:

$$\mathbf{h}_0^{\text{RNN}} = \mathbf{0} \in \mathrm{R}^{1 \times n \times p} \tag{13}$$

$$\mathbf{R}_{\text{out}}, \mathbf{h}_t^{\text{RNN}} = \text{RNN}(\mathbf{R}_{\text{in}}, \mathbf{h}_0^{\text{RNN}}) \tag{14}$$

where p represents the number of RNN hidden units and $\mathbf{R}_{\text{out}} \in \mathrm{R}^{n \times p}$.

LSTM Processing Path: The LSTM path processes the same input with both hidden and cell state initialization:

$$\mathbf{h}_0^{\text{LSTM}} = \mathbf{0} \in \mathrm{R}^{1 \times n \times q}, \mathbf{c}_0^{\text{LSTM}} = \mathbf{0} \in \mathrm{R}^{1 \times n \times q} \tag{15}$$

$$\mathbf{L}_{\text{out}}, (\mathbf{h}_t^{\text{LSTM}}, \mathbf{c}_t^{\text{LSTM}}) = \text{LSTM}(\mathbf{R}_{\text{in}}, (\mathbf{h}_0^{\text{LSTM}}, \mathbf{c}_0^{\text{LSTM}})) \tag{16}$$

where q represents the number of LSTM hidden units and $\mathbf{L}_{\text{out}} \in \mathrm{R}^{n \times q}$.

Feature Integration: The outputs from both recurrent paths are concatenated to form the final encoded representation:

$$\mathbf{F}_{\text{encoded}} = \text{Concat}(\mathbf{R}_{\text{out}}, \mathbf{L}_{\text{out}}) \in \mathrm{R}^{n \times (p+q)} \tag{17}$$

This dual-path recurrent processing captures both short-term sequential patterns through the RNN and long-term dependencies through the LSTM, providing comprehensive sequential modeling capabilities.

4.5 Classification Layer

The final encoded features are processed through a linear classification layer to produce class predictions:

$$\mathbf{Y}_{\text{logits}} = \mathbf{F}_{\text{encoded}} \times \mathbf{W}_{\text{class}} + \mathbf{b}_{\text{class}} \tag{18}$$

where $\mathbf{W}_{\text{class}} \in \mathrm{R}^{(p+q) \times c}$ and $\mathbf{b}_{\text{class}} \in \mathrm{R}^{c}$ represent the classification weights and biases respectively, with c being the number of target classes. Final predictions are obtained through softmax normalization:

$$\mathbf{Y}_{\text{pred}} = \text{Softmax}(\mathbf{Y}_{\text{logits}}) \tag{19}$$

This comprehensive mathematical framework establishes the complete Bert DPRCNN processing pipeline, systematically integrating transformer contextualization, convolutional feature extraction, and recurrent sequence modeling to achieve superior text classification performance.

5 Experimentation Results

The Table 1 presents the performance of various models on Chinese and English datasets. The performance is measured in terms of accuracy, with higher percentages indicating better performance.

The model's novel hybrid architecture, which synergistically combines the strengths of transformer models, convolutional neural networks, recurrent neural networks, and long short-term memory networks, represents a significant advancement in feature extraction and sequential processing for textual data. The extensive experimentation results unequivocally demonstrate the superior performance of the Bert_DPRCNN model. It consistently achieved the highest accuracies on both Chinese (94.54%) and English (91.89%) datasets, outperforming various baseline and state of the art models by substantial margins. This consistent and significant outperformance across different languages underscores the robustness and generalizability of the proposed hybrid architecture. The model's capacity to extract language-agnostic, high-level features by integrating diverse types of information—contextual, local, and sequential—validates its design and positions it as a highly promising solution for complex, cross-lingual text classification challenges. These refinements collectively enhance the academic rigor and practical impact of the research paper.

Table 1. Model Performance

Model	Chinese	English
Transformer	89.91%	83.69%
TextRCLSFM	90.54%	79.13%
TextRNN_att	90.90%	79.91%
Fastext DPCNN	91.04%	76.46%
TextRNN	91.12%	77.64%
BERT_MR	91.12%	88.47%
TextCNN	91.22%	64.03%
BERT	91.22%	89.64%
DPCNN	91.25%	76.46%
TextRCNN	91.54%	78.59%
TextCRNN_Att	91.73%	80.46%
FastText	92.23%	73.29%
robert DPCNN	94.44%	88.47%
bert_DPRCNN	**94.54%**	**91.89%**

6 Conclusion

In conclusion, the bert_DPRCNN model demonstrated the best performance on both the Chinese and English datasets among the models evaluated. However, the performance of the models varied, and the choice of model may depend on the specific requirements of the task at hand. For instance, if computational resources are a concern, simpler models like TextRNN or FastText might be preferred despite their slightly lower performance. On the other hand, if the highest possible accuracy is the priority, more complex models like bert DPRCNN would be the best choice.

References

1. Sun, A., Lim, E.-P.: Hierarchical text classification and evaluation. In: Proceedings 2001 IEEE International Conference on Data Mining, pp. 521–528. IEEE (2001)
2. Uysal, A.K., Gunal, S.: The impact of preprocessing on text classification. Inf. Process. Manage. **50**(1), 104–112 (2014)
3. Lai, S., Xu, L., Liu, K., Zhao, J.: Recurrent convolutional neural networks for text classification. In: Proceedings of the AAAI Conference on Artificial Intelligence, vol. 29 (2015)
4. Kowsari, K., Jafari Meimandi, K., Heidarysafa, M., Mendu, S., Barnes, L., Brown, D.: Text classification algorithms: a survey. Information **10**(4), 150 (2019)
5. Minaee, S., Kalchbrenner, N., Cambria, E., Nikzad, N., Chenaghlu, M., Gao, J.: Deep learning–based text classification: a comprehensive review. ACM Comput. Surv. (CSUR) **54**(3), 1–40 (2021)

6. Friedman, J.H.: Greedy function approximation: a gradient boosting machine. Ann. Stat., 1189–1232 (2001)
7. Alzubaidi, L., et al.: Review of deep learning: concepts, cnn architectures, challenges, applications, future directions. J. Big Data **8**, 1–74 (2021)
8. Sherstinsky, A.: Fundamentals of recurrent neural network (RNN) and long short-term memory (LSTM) network. Physica D **404**, 132306 (2020)
9. Devlin, J., Chang, M.-W., Lee, K., Toutanova, K.: BERT: pre-training of deep bidirectional transformers for language understanding. arXiv preprint arXiv:1810.04805 (2018)
10. Floridi, L., Chiriatti, M.: GPT-3: its nature, scope, limits, and consequences. Mind. Mach. **30**, 681–694 (2020)
11. Johnson, R., Zhang, T.: Deep pyramid convolutional neural networks for text categorization. In: Proceedings of the 55th Annual Meeting of the Association for Computational Linguistics (Volume 1: Long Papers), pp. 562–570 (2017)
12. Abujar, S., Masum, A.K.M., Sanzidul Islam, M., Faisal, F., Hossain, S.A.: A bengali text generation approach in context of abstractive text summarization using RNN. In: Innovations in Computer Science and Engineering: Proceedings of 7th ICICSE, pp. 509–518 (2020)
13. Zhang, X., Zhao, J.J., LeCun, Y.: Character-level convolutional networks for text classification. CoRR **abs/1509.01626** (2015) 1509.01626
14. Joulin, A., Grave, E., Bojanowski, P., Douze, M., Jégou, H., Mikolov, T.: Fast-Text. zip: compressing text classification models. arXiv preprint arXiv:1612.03651 (2016)
15. Liu, P., Qiu, X., Huang, X.: Recurrent neural network for text classification with multi-task learning. arXiv preprint arXiv:1605.05101 (2016)
16. Ma, R., Teragawa, S., Fu, Z.: Text sentiment classification based on improved BiLSTM-CNN. In: 2020 Asia-Pacific Conference on Image Processing, Electronics and Computers (IPEC), pp. 1–4 (2020)
17. Liu, Y., et al.: RoBERTa: a robustly optimized BERT pretraining approach. CoRR **abs/1907.11692** (2019) 1907.11692
18. Zheng, S., Yang, M.: A new method of improving BERT for text classification. In: Intelligence Science and Big Data Engineering. Big Data and Machine Learning: 9th International Conference, IScIDE 2019, Nanjing, China, October 17–20, 2019, Proceedings, Part II 9, pp. 442–452. Springer (2019)
19. Sutskever, I., Vinyals, O., Le, Q.V.: Sequence to sequence learning with neural networks. CoRR **abs/1409.3215** (2014) 1409.3215
20. Conneau, A., Schwenk, H., Barrault, L., LeCun, Y.: Very deep convolutional networks for natural language processing. CoRR **abs/1606.01781** (2016) 1606.01781
21. Kambhampaty, K., Alruwaythi, M., Chowdhury, M.M., Nygard, K.E.: Trust and its influence on technology. In: The Midwest Instruction and Computing Symposium (2019)
22. Kambhampaty, K., Nygard, K.E.: Identifying insider and masquerade attackers in cloud computing and IoT devices. CSREA

Bug Seeding in Open-Source: An OpenStack Study

Praveen Kumar Sayani(✉)

Vienna, Austria
sayanipraveen@gmail.com

Abstract. This study investigates the phenomenon of bug seeding in open-source software development, with a focus on how and when bugs are introduced into the source code. Previous research often assumes that the bug is introduced in the commit preceding the bug fix. However, our large-scale observational study on the OpenStack project challenges this assumption. By analyzing 459 bug reports and tracing their origins, we reveal that only 50% of the previous commits caused the system failure, while other bugs resulted from unrelated changes such as API modifications. The study provides insights into patterns of bug introduction and lays the groundwork for future efforts to classify and prevent bugs that are not directly linked to recent commits. These findings aim to improve bug detection and prevention mechanisms in open-source development.

Keywords: Bug introduction · bug seeding · SZZ algorithm · previous commit

1 Introduction

The software engineering community has long been interested in understanding the mechanisms behind the introduction of bugs into source code. As software evolves, it undergoes numerous modifications, often as part of the maintenance process aimed at improving functionality or fixing defects. These modifications, however, can sometimes introduce new defects. Identifying the exact moment and origin of such defects is critical for improving software reliability and maintaining system stability.

A widely held notion in this context is that the most recent modification to a line of code—specifically, the *previous commit*—is responsible for seeding a bug. This idea underpins many techniques in software repository mining, particularly those related to bug localization and blame assignment. The concept of *bug seeding* revolves around the hypothesis that the last developer who modified a line of buggy code is the one who introduced the defect.

Although numerous studies adopt this assumption, accurately determining when and where a bug originated remains a non-trivial challenge. In practice, not all bugs are introduced in the most recent commit. Bugs may arise from indirect causes, such as changes in dependent APIs, configuration updates, or

I. Roy et al. (Eds.): CAINE 2025, CCIS 2740, pp. 309–320, 2026.
https://doi.org/10.1007/978-3-032-14893-3_21

behavioral shifts in external modules. These types of issues are rarely addressed in depth by the existing literature. Nevertheless, several foundational studies have operated under this presumption:

- In bug seeding analyses, for instance: *"This earlier change is the one that caused the later fixed"* [1], and *"The lines affected in the process of fixing a bug are the same one that originated or seeded that bug"* [2].
- Regarding bug fix patterns: *"The version before the bug fix revision is the bug version"* [3].
- In tools intended for preventing bugs: *"We assume that a change/commit is buggy if its modifications has been later altered by a bug-fix commit"* [4].

The only empirical support for this assumption is a small-scale manual inspection of 25 randomly selected bug-fix commits. This study, using an improved SZZ algorithm, concluded that the prior change typically corresponds to the bug introduction [1]. While this offers preliminary validation, the sample size is insufficient to draw generalizable conclusions. Given how frequently this assumption appears in the literature, broader empirical validation is warranted.

To that end, we examine this assumption within the scope of a large-scale software project—OpenStack. With its complex architecture and highly active development history, OpenStack provides an ideal environment to analyze bug seeding patterns. Our goal is to trace the origins of bugs and evaluate the validity of associating them solely with the preceding commit.

One concrete example extracted from OpenStack demonstrates how a bug was introduced not due to a faulty change in logic, but due to changes in external expectations—specifically an API enhancement. In this case, a feature became necessary due to community-driven design evolution. The following code snippets and the version history illustrated in Fig. 1 highlight this scenario.

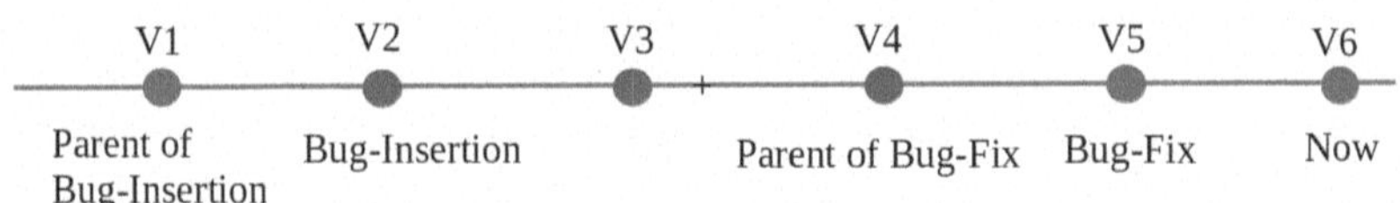

Fig. 1. Control version history with the bug-insertion commit and the bug-fix commit

State Before Bug Introduction (Version V1):

```
dbb854635 xioaxi 2013-07-11 def _check_backup_service(self, volume
    ):
afd69a95b victor 2013-08-28 """Verifies the availability of a
    backup service."""
dbb854635 xioaxi 2013-07-11 topic = CONF.backup_topic
                              ...
```

```
dbb854635 xioaxi 2013-07-11     srv['host'] == volume['host'] and
    not srv['disable']
```

Bug-Introduction Commit (Version V2):

```
-d6dd5cdfa keniche 2013-09-05 def _is_backup_service_enabled(self,
     volume):
 afd69a95b victor 2013-08-28 """Verifies the availability of a
     backup service."""
 dbb854635 xioaxi 2013-07-11 topic = CONF.backup_topic
                                  ...
-db b854635 xioaxi 2013-07-11     srv['host'] == volume['host']
    and not srv['disable']
```

Bug Resolution Commit (Version V5):

```
+bd5c3f5a0 Edward 2013-09-26 def _is_backup_service_enabled(self,
    volume, volume_host):
 afd69a95b victor 2013-08-28 """Verifies the availability of a
     backup service."""
 dbb854635 xioaxi 2013-07-11 topic = CONF.backup_topic
                                  ...
+bd5c3f5a0 Edward 2013-09-26     srv['host'] == volume_host and
    not srv['disable']
```

Using the standard SZZ algorithm, the commits *d6dd5cdfa* and *dbb854635* are marked as responsible for introducing the bug. However, the enhanced SZZ method proposed by [10] refines this process by filtering false positives—flagging *dbb854635* as the actual source of the bug, since *d6dd5cdfa* only renamed an API method. Yet, upon deeper analysis, this assumption fails. The function added in *V1* operated correctly at the time. The bug emerged due to a change in system requirements, not faulty logic in the earlier commit.

The primary objective of this study is to empirically determine the extent to which bugs can be reliably attributed to previous commits, and to uncover the underlying factors that lead to defect introduction. We thus conducted an extensive observational study based on issue tracking data from OpenStack's most active components. To perform this, it was first essential to distinguish genuine bug reports from the broader set of issue notifications.

We begin by addressing two foundational research questions:

- **RQ1:** How can genuine bug reports be reliably identified from issue tracking systems?
- **RQ2:** To what extent is the previous commit responsible for seeding a bug?

The insights gained from these questions lay the groundwork for a broader research agenda aimed at improving software quality by uncovering the systemic patterns that lead to defects. In future work, we aim to address additional questions such as:

- **RQ3:** Are there identifiable patterns in how bugs are seeded across projects?
- **RQ4:** Do these patterns manifest consistently across different programming languages?
- **RQ5:** What preventative mechanisms could be derived to avoid similar bugs in the future?

The rest of this paper is organized as follows. Section 2 outlines the related work in this area. In Sect. 3, we explain the two-stage methodological approach of our study. The findings are presented and analyzed in Sect. 4. Finally, Sect. 5 discusses implications, future work, and potential applications of our findings. This paper proceeds as follows: Sect. 2 reviews related work, Sect. 3 outlines our methodology, Sect. 4 presents empirical results, and Sect. 5 discusses implications and future directions.

2 Related Work

One of the most widely adopted techniques for automatically identifying bug-inducing changes in source code is the SZZ algorithm, first introduced by Sliwerski et al. [5]. This method integrates data from version control systems (VCS) and issue tracking systems (ITS) to establish links between bug-fix commits and their corresponding bug-introducing commits. The algorithm builds upon earlier efforts in the area of change management and mining software repositories [6–8] and has since become a foundational technique in empirical software engineering research.

Despite its widespread adoption, the SZZ algorithm is not without limitations. One primary shortcoming arises from its reliance on heuristic keyword matching within commit messages (e.g., terms such as "fix" or "bug") to identify bug-fix commits [9]. As a result, any bug-fix commit that omits such keywords may be overlooked by the algorithm, leading to incomplete or inaccurate tracing of bug-introducing changes.

To overcome these limitations, several enhancements have been proposed. Notably, Kim et al. [10] introduced an improved approach that leverages annotation graphs, replacing the original use of CVS annotations. Their method more accurately identifies changes across different versions by filtering out non-functional modifications such as formatting changes, comment edits, or white-space adjustments—thereby reducing the likelihood of false positives.

Building on the SZZ foundation, a number of studies have explored its effectiveness in various contexts. Williams et al. [1] revisited the algorithm to enhance the detection of bug-inducing commits and proposed classifications for different types of changes. Yang et al. [11] applied the SZZ approach to investigate which bug-inducing changes are more likely to result in severe issues when later labeled as bug-fix commits.

Bavota et al. [12] conducted an empirical investigation to understand the relationship between refactoring activities and fault introduction. Their findings indicated that only a small fraction of bugs could be attributed directly to refactoring operations. Similarly, Kim et al. [10] developed a technique for classifying changes in files as either buggy or clean by analyzing change metadata and token-level characteristics from source code.

Beyond fault localization, the SZZ algorithm has also been utilized for evaluating software quality assurance processes. Matsumoto et al. [13] studied the impact of developer activity on software reliability, concluding that modules modified by a larger number of developers tend to contain more defects. In a related line of research, Kamei et al. [14] incorporated SZZ into a validation framework for effort-aware defect prediction models. Their study examined process metrics such as lines of code (LOC) and evaluated predictions at both the package and file levels.

Further, Eyolfson et al. [15] used SZZ to investigate how factors like the time of day and developer experience influence the probability of a commit introducing a bug. Izquierdo et al. [2] applied the algorithm to analyze whether developers were more likely to fix bugs they had originally introduced themselves, offering insights into responsibility and code ownership dynamics.

Complementary to text-based methods like SZZ, other approaches have been proposed to improve the accuracy of bug origin detection. For example, Sinha et al. [16] introduced a dependency-based technique that tracks the semantic impact of bug fixes on program behavior. Their method does not rely on commit message keywords; instead, it examines the effects of code changes on program dependencies. By capturing semantic-level modifications, their approach can more precisely attribute the root cause of bugs. Both the SZZ and the dependency-based techniques follow a similar high-level strategy: (1) identify changes made during bug-fix commits by comparing the buggy and fixed versions of the code, and (2) trace back through version control history to find the last modification to the lines involved in the bug-fix.

Collectively, these studies underscore the importance of accurately identifying bug-introducing changes, not only to better understand defect lifecycles but also to improve tools for fault prediction, quality assurance, and automated debugging. Continued enhancements to these algorithms and alternative methods remain crucial for advancing the state-of-the-art in software maintenance and evolution.

3 Methodology

To investigate the origins of software bugs and the validity of associating them with the most recent commit, we conducted a systematic empirical study focusing on the OpenStack software ecosystem. Our study analyzed a total of 459 issue reports, commonly referred to as *tickets*, which were randomly sampled from four of the most active OpenStack repositories: `Nova`, `Cinder`, `Neutron`, and `Horizon`.

The selection of OpenStack as the subject of our investigation is driven by its extensive scale, vibrant developer ecosystem, and significant practical relevance. OpenStack is a widely adopted open-source platform for cloud infrastructure, utilized by some of the world's largest enterprises. With a contributor base exceeding 5,000 developers, the project embodies a highly dynamic and collaborative software development environment.

The development landscape of OpenStack is both broad and complex. It comprises more than 184,000 issue reports, out of which over 144,000 have been resolved through the efforts of more than 6,000 individual contributors. The codebase has evolved through over 233,000 commits and now includes more than two million lines of source code.

An important aspect of OpenStack's appeal as a research subject lies in the transparency of its development process. All essential development artifacts—including source code history, bug tracking data, and peer review feedback—are publicly available. This openness provides a valuable opportunity for conducting in-depth empirical analysis using real-world data drawn from an active and large-scale software ecosystem.

Our research methodology is structured in two primary stages, each with specific objectives and processes. The overall goal was to accurately classify issue reports and investigate the underlying cause of the identified bugs. In the first stage, a double-blind classification process was used to determine whether tickets were genuine bug reports. In the second stage, we conducted an in-depth analysis to understand the actual cause of the bug in the context of the codebase.

3.1 First Stage: Classification of Issue Reports

The primary objective of the first stage was to determine whether each issue ticket in the Launchpad repository qualified as a valid bug report. This task proved to be challenging due to the need to correlate and interpret data from multiple heterogeneous sources—namely, Launchpad (issue tracking), Gerrit (code review), and Git (version control). Because the context and intent of many tickets were not always explicitly documented, the analysis had to be performed manually to ensure the accuracy of classification.

To assist with this labor-intensive task, we developed a custom web-based application designed to streamline the classification process. This tool automatically aggregates relevant information from OpenStack repositories and presents it via a user-friendly interface. Key features of the tool include support for collaborative classification, traceability of reviewer decisions, and transparent documentation of the analysis workflow.

For each ticket, the following parameters were carefully examined to support accurate categorization:

- **Issue Title:** Often provides a concise summary of the problem or feature request.
- **Issue Description:** Contains detailed information regarding the nature of the issue.
- **Fix Commit Message:** Offers context on what change was made to resolve the issue.
- **Code Changes:** In some cases, the issue and commit messages lacked sufficient clarity, requiring a direct examination of the modified code to understand the ticket's purpose.

Based on the information retrieved and analyzed, each issue was assigned to one of the following three categories:

1. **Group 1 (Bug Report):** The issue clearly describes a malfunction or defect in the system behavior and has been resolved via a corrective code change.
2. **Group 2 (Not a Bug Report):** The issue pertains to feature enhancements, code optimizations, test updates, or other non-bug-related changes.
3. **Group 3 (Undecided):** The issue contains ambiguous or insufficient information and could not be confidently categorized. This may result from vague descriptions or the inherently complex nature of the problem.

The classification process was carried out independently by three researchers with programming and software engineering expertise, using a double-blind protocol to minimize bias. Disagreements between reviewers were resolved through discussion and consensus. This rigorous methodology ensures that the results of the subsequent analysis are based on a reliable and validated dataset of bug reports.

3.2 Second Stage: Analysis of Bug-Introducing Commits

In the second phase of our methodology, we focused exclusively on analyzing the commit history of the tickets categorized as valid *Bug Reports* during the initial stage. The objective of this phase was to determine whether the *previous commit*—as defined in standard SZZ-based assumptions—was indeed responsible for introducing the bug.

This entailed identifying the specific lines of code modified in the bug-fix commit and then tracing back their history to locate the point in the project's timeline where those lines were initially introduced or altered. Specifically, we examined two versions of each relevant file: (1) the version as it appears in the bug-fix commit, and (2) the version in the commit immediately preceding the fix, referred to as the *parent commit*. This parent is not necessarily the previous commit to the file itself, but rather the most recent commit to the lines that were changed in the bug-fix.

This process ensured that our analysis focused on the actual moment a bug was introduced, rather than assuming it appeared immediately before the fix. Often, lines added or altered in the bug-fix commit had already existed in earlier revisions, suggesting they were seeded earlier, and attributing them to the most recent commit would yield a false positive.

To support this analysis, we employed tools such as `git blame` to attribute line-level changes to specific commits and used `diff` to visually inspect differences between file versions. This manual inspection allowed us to discard irrelevant modifications such as formatting updates, comment changes, or version number adjustments—none of which are typically related to actual bugs.

Ultimately, we constructed a refined list of prior commits for each ticket and conducted a detailed review to assess whether the lines introduced or modified in those commits could be causally linked to the observed software fault.

4 Empirical Results

We selected a sample of 459 issue reports (tickets) from the four primary OpenStack components: `Nova`, `Cinder`, `Horizon`, and `Neutron`. Each of the first three contributed 125 tickets to the dataset, while Neutron had 84 tickets due to data availability constraints.

4.1 First Stage Results

The classification process during the first stage involved three independent researchers, each with sufficient domain expertise. Tickets were analyzed using a double-blind method to mitigate reviewer bias, with each issue examined by at least two different researchers using the custom-built web tool.

Table 1 presents the outcomes of the classification for each researcher. Researcher R1 reviewed 334 tickets, R2 reviewed 249, and R3 reviewed 334. Notably, inter-reviewer agreement was high: 292 tickets (approximately 72% of the sample) received matching classifications from both reviewers. Among these, 209 were identified as valid bug reports, 74 as non-bug reports, and 9 were labeled as undecided due to ambiguity or complexity.

Table 1. Classification statistics from the first stage for each researcher. The "Agreement" row shows tickets where at least two researchers reached the same conclusion.

	Bug Report	Not Bug Report	Undecided	Total
R1	(184) 55%	(115) 34%	(35) 11%	334 (100%)
R2	(188) 76%	(54) 22%	(7) 3%	249 (100%)
R3	(188) 56%	(116) 35%	(30) 9%	334 (100%)
Agreement	(209) 72%	(74) 25%	(9) 3%	292 (100%)

These results enabled us to confidently answer our first research question (RQ1): the methodology allowed accurate identification of bug-related tickets. Approximately 72% of the issues reviewed were consistently recognized as legitimate bug reports.

4.2 Second Stage Results

For the second stage, we considered the 189 tickets previously classified as *Bug Reports* by both reviewers. The aim was to investigate whether the immediately preceding commits actually introduced the faults fixed in those bug reports.

During this phase, we identified a total of 348 prior commits linked to the affected lines. After filtering out false positives—such as those due to formatting, whitespace, comments, or version number changes—308 commits remained for analysis. From this refined set:

- 152 commits (49%) were confirmed as having caused the bug,
- 114 commits (37%) were determined not to have introduced the bug,
- 42 commits (14%) could not be conclusively classified.

To better understand the relationship between the number of prior commits and the likelihood of fault introduction, we analyzed how many prior commits were involved per bug-fix. Table 2 shows that 131 bug reports had a single prior commit, while 58 had more than one.

Table 2. Impact of single vs. multiple prior commits on the likelihood of bug introduction.

	One prior commit	Multiple prior commits
Bug Caused	(65) 50%	(86) 48%
Not Caused	(30) 23%	(82) 46%
Undecided	(36) 27%	(11) 6%

This allows us to answer our second research question (RQ2): less than half (50%) of the prior commits could be confidently identified as the root cause of the bugs, indicating that SZZ-style assumptions do not universally apply.

5 Discussion and Future Work

Our findings challenge the core assumption underpinning many SZZ-based techniques—that the bug-introducing commit is always the most recent one prior to the fix. Empirical evidence from our study suggests that this assumption holds true in only about half of the cases analyzed. In the remaining instances, bugs were introduced by earlier changes or through indirect mechanisms such as changes in API behavior.

This insight has significant implications for fault localization and software quality assurance. Understanding that not all bugs originate in the immediately preceding commit calls for more sophisticated methods to trace bug origins. A key direction for future research is the identification and modeling of different bug seeding patterns—distinguishing between direct faults, cascading bugs from dependencies, or API-related incompatibilities.

We plan to explore the third research question (RQ3) by applying machine learning classifiers and support vector machines (SVMs) to identify patterns in these non-trivial bug cases, building on prior work [17,18]. Drawing from related studies [10,13,14,19,20], we aim to create a feature-based classification model using variables such as:

- Timestamp of the changes.
- Nature of modifications (e.g., control structures, function calls, variable names).

- Number of lines altered during the fix.
- Experience and activity level of the developer who applied the fix.

This modeling will enable us to predict fault-prone commits and help prioritize review and testing efforts. If recurring patterns can be observed in non-SZZ bugs, especially those linked to API evolution, this could offer a proactive approach to bug prevention.

We also aim to address RQ4 by extending our study to projects written in other programming languages like Java and JavaScript. These languages differ in compilation behavior and runtime error handling compared to Python (the main language of OpenStack). Using datasets from prominent projects like Eclipse and Mozilla, we intend to determine whether our findings generalize beyond the OpenStack ecosystem.

Moreover, we plan to enhance our classification tool with automated capabilities for identifying bug-inducing commits and distinguishing bug reports from feature requests. Our existing tool has already been accepted for presentation at OSS 2016 and serves as a valuable foundation for further innovation in automated bug tracking and prevention.

6 Conclusion

This paper shows the commonly held assumption that the commit immediately preceding a bug fix is the one responsible for introducing the bug. Using a comprehensive observational study on the OpenStack project—one of the most active and complex open-source ecosystems—we analyzed 459 issue reports, of which 189 were confirmed as valid bug reports. Our manual, line-level inspection of 348 prior commits related to these bug reports revealed that only about half (49%) could be definitively linked to the introduction of the bug, while a significant portion (37%) were unrelated, and 14% remained undecidable.

These findings challenge the reliability of the SZZ algorithm's foundational assumption and indicate that bug origins can often trace back beyond the most recent code modification. In many cases, bugs arise due to deeper structural or semantic issues, including external API changes or long-standing design flaws, which are not captured by traditional keyword-based or version-diff-based methods.

This research contributes empirical evidence that underlines the complexity of accurately locating bug-inducing commits and emphasizes the need for more nuanced tools and approaches. It also demonstrates the value of combining manual analysis with automated tooling for deeper software repository mining.

Future work will focus on extending this analysis across multiple languages and projects to generalize the findings and develop predictive models using machine learning techniques to identify and categorize bug seeding patterns. Such models may eventually be integrated into automated review systems, helping developers identify and prevent fault-prone changes earlier in the development lifecycle.

Ultimately, improving our understanding of when, how, and why bugs are seeded into code has the potential to enhance software reliability, reduce maintenance costs, and support higher-quality software engineering practices across a variety of development environments.

References

1. Williams, C., Spacco, J., Pugh, W.: Backtracking through code to find the source of bugs. In: Proceedings of the 6th International Working Conference on Mining Software Repositories, pp. 15–18 (2008)
2. Izquierdo, D., Cabot, J., Cosentino, V.: Are developers fixing their own bugs? In: Proceedings of the 8th International Working Conference on Mining Software Repositories, pp. 173–182 (2011)
3. Pan, K., Kim, S., Whitehead, E.J., Jr.: Toward an understanding of bug fix patterns. Empir. Softw. Eng. **14**(3), 286–315 (2009)
4. Fejzer, M., Kasauli, R.: Supporting developer awareness using bug-fix patterns. In: 2015 IEEE 22nd International Conference on Software Analysis, Evolution, and Reengineering, pp. 565–569 (2015)
5. Sliwerski, J., Zimmermann, T., Zeller, A.: When do changes induce fixes? In: Proceedings of the 2nd International Workshop on Mining Software Repositories, pp. 1–5 (2005)
6. Čubranić, D., Murphy, G. C.: Hipikat: recommending pertinent software development artifacts. In: Proceedings of the 25th International Conference on Software Engineering, pp. 408–418 (2003)
7. Fischer, M., Pinzger, M., Gall, H. C.: Analyzing and relating bug report data for feature tracking. In: 10th Working Conference on Reverse Engineering (WCRE), pp. 90–99 (2003)
8. Fischer, M., Pinzger, M., Gall, H. C.: Populating a release history database from version control and bug tracking systems. In: Proceedings International Conference on Software Maintenance, pp. 23–32 (2003)
9. Kamei, Y., et al.: Effects of over- and under-sampling on software defect prediction performance. In: 1st International Workshop on Artificial Intelligence for Requirements Engineering (AIRE 2007), pp. 30–36 (2007)
10. Kim, S., Whitehead, E.J., Jr., Zhang, Y.: Classifying software changes: clean or buggy? IEEE Trans. Software Eng. **34**(2), 181–196 (2008)
11. Yang, Y., et al.: Characterizing and detecting buggy if conditions. In: 2016 IEEE 23rd International Conference on Software Analysis, Evolution, and Reengineering, pp. 604–608 (2016)
12. Bavota, G., et al.: Does refactoring induce bugs? Empir. Softw. Eng. **18**(4), 791–816 (2013)
13. Matsumoto, S., et al.: An analysis of developer's work patterns for successful software projects. In: Proceedings of the 2010 ACM-IEEE International Symposium on Empirical Software Engineering and Measurement, pp. 1–10 (2010)
14. Kamei, Y., et al.: Revisiting common bug prediction findings using effort-aware models. In: 2010 IEEE International Conference on Software Maintenance, pp. 1–10 (2010)
15. Eyolfson, J., Tan, L., Lam, P.: Do time of day and developer experience affect commit bugginess? In: Proceedings of the 8th Working Conference on Mining Software Repositories, pp. 153–162 (2011)

16. Sinha, A., Harrold, M.J.: Analysis and testing of programs with exception-handling constructs. IEEE Trans. Software Eng. **26**(9), 849–871 (2010)
17. Shivaji, S., et al.: Reducing features to improve code change-based bug prediction. IEEE Trans. Software Eng. **39**(4), 552–569 (2013)
18. Shivaji, S., et al.: Improving the accuracy of fault prediction using feature selection. Empir. Softw. Eng. **18**(3), 462–498 (2013)
19. Hata, H., Mizuno, O., Kikuno, T.: Bug prediction based on fine-grained module histories. In: 2012 34th International Conference on Software Engineering (ICSE), pp. 200–210 (2012)
20. Fujiwara, K., et al.: Assessing the cost-effectiveness of fault prediction models. Empir. Softw. Eng. **18**(3), 429–461 (2013)

Analyzing Darknet Address Information Using Machine Learning

Trenton Ward, Cheryl Hinds(✉), and Jonathan Graham

Norfolk State University, Norfolk, VA 23504, USA
t.m.ward101292@spartans.nsu.edu, {chinds,jmgraham}@nsu.edu

Abstract. Cybercrimes are consistently advancing in complexity. Cyber threat intelligence (CTI) is useful for analyzing, evaluating, and drawing conclusions for areas such as indicators of compromise (IOCs) and tactics, techniques, and procedures (TTPs). However, valuable CTI is not always utilized effectively, particularly from the unregulated Dark Web. Manually processing this vast and complex data is unrealistic. This research uses artificial intelligence (AI) and machine learning (ML) to provide automation and process a sizable amount of information in a reasonable time. Specifically, it will use the DUTA-10K Darknet dataset to develop ML models based on the Random Forest, Support Vector Machine, and XGBoost algorithms. These models distinguish illicit and benign addresses, along with their classes and language categories. The resulting models showed reliable performance, with XGBoost consistently producing more effective models overall for predicting illicit status and language.

Keywords: Cyber Threat Intelligence (CTI) · Dark Web · Machine Learning (ML) · Data Science · Random Forest · Support Vector Machine · XGBoost

1 Introduction

As remote work becomes more common and cybercrimes more prevalent, the need to safeguard digital assets and networks has risen. This has fueled the growth of the cyber threat intelligence (CTI) market, which is predicted to exceed 54 billion by 2034 [1].

CTI involves collecting, analyzing, and applying data related to cyber threats. It provides insights into indicators of compromise (IOCs) and identifies tactics, techniques, and procedures (TTPs) to fortify security [2]. CTI is gathered from various sources, including the Dark Web. A significant challenge is the volume and quality of CTI data [4, 5].

The Dark Web is a hidden part of the internet accessed via special software like Tor. While it is often associated with illegal activities, it also provides privacy for various groups. CTI leverages the Dark Web by monitoring it for threats, stolen data, and communications, allowing organizations to enhance security [2]. The vast and complex nature of this data makes manual evaluation unrealistic. AI and ML offer a solution by providing automation for data sorting, structuring, and analysis.

I. Roy et al. (Eds.): CAINE 2025, CCIS 2740, pp. 321–335, 2026.
https://doi.org/10.1007/978-3-032-14893-3_22

This research develops ML models to analyze and extract actionable intelligence from Dark Web onion addresses using the DUTA-10K dataset. The models predict an address's criminality, specific class categories (main and sub), and language. Unlike previous studies that relied on direct site access and content features, this research concentrates on predictive signals inherent in the onion address metadata, its associated language, and existing categorization. This approach provides a unique perspective for extracting actionable intelligence under constrained data access.

2 Literature Review

Advanced approaches to CTI are necessary due to the intensifying presence of cyber threats. This literature review summarizes known existing research on CTI.

Studies have focused on handling CTI data. Sarker emphasized that ML model effectiveness relies on input data quality [6]. Sangher et al. proposed a proactive CTI system for Dark Web forums to identify cybercrimes, addressing the ineffectiveness of conventional tools against the unstructured and anonymous nature of the Dark Web [2].

Comprehensive surveys highlight advancements and hurdles in the CTI field. Saeed et al. [7] recognized issues such as data quality and information overload. Kolluri [8] noted that traditional cybersecurity is often insufficient, advocating for AI-driven methods. Basheer and Alkhatib [9] highlighted the role of Dark Web investigation for proactive threat detection.

Novel approaches to utilizing CTI have also been introduced. Zhou et al. [10] proposed "CTI View," an analysis system leveraging natural language processing (NLP) to obtain IOCs and TTPs from unstructured CTI. Koloveas et al. [11] introduced INTIME, an ML-based framework for gathering web data for CTI. Gao et al. [12] developed THREATRAPTOR to enable efficient cyber threat hunting. These methods require continuous effort to keep up with evolving data and behaviors.

Specific research has focused on the DUTA-10K dataset. Al-Nabki et al. [13] extended the DUTA dataset and introduced ToRank, an algorithm for recognizing dominant suspect domains. Shin et al. [14] utilized DUTA-10K for classification using TextCNN. Sennad et al. [15] further improved Dark Web behavior cataloguing using embedding methods with XGBoost. These studies, however, primarily relied on textual content features obtained from direct site access. Al Nabki et al. [16] explored a supervised ranking approach using a subset of DUTA-10K with features from text, HTML, and network topology.

This research builds upon these works by addressing the limitation of lacking direct access to Dark Web sites. While previous studies used on-page features, this research focuses on predictive signals in the onion address metadata, its language, and existing categorization. This approach offers a differing perspective on extracting actionable intelligence for threat intelligence professionals.

3 Methodology

This section outlines the approach utilized for the development and evaluation of ML models aimed at classifying Dark Web activities based on limited metadata, specifically onion addresses and language. The methodology details the selection and implementation of ML algorithms, dataset characteristics and preparation, iterative model development, and anticipated results. The entire implementation process was conducted using Python.

3.1 Machine Learning Algorithms

The research applies ML algorithms to discern patterns within Dark Web onion addresses and associated metadata without direct site access. Supervised learning algorithms were chosen for classification. Random Forest, Support Vector Machine (SVM), and Extreme Gradient Boosting (XGBoost) were selected due to their proven effectiveness in classification, ability to handle complex feature relationships, and interpretability [17–19]. These algorithms support binary, multi-class, and multi-label classification, necessary for predicting illicit status, main class, sub class, and language.

The implementation was conducted in Python utilizing Scikit-learn for traditional algorithms. The developed models serve three primary functions: classifying an onion address's criminality, predicting its specific main and sub class categories, and identifying its likely language. This multi-faceted classification aims to provide insights based solely on the address.

A hierarchical model was used for each algorithm, cascading predictions from one section to the next. This allowed the output of the illicit status model, which conducted simpler binary classification, to be used as an additional feature for subsequent models, improving performance by narrowing down potential classes and languages. This resulted in a total of four models per algorithm.

3.2 Datasets

The primary dataset is the DUTA-10K from the University of León [13, 20]. This dataset provides onion addresses with categorizations and language, which is suitable for analysis without direct site access. The dataset was cleaned, and categories were encoded. Advanced feature engineering was performed on the onion addresses to derive structural and statistical features. This process resulted in a total of 526 engineered features, which were used to train the models.

With this cleaned dataset, statistics were calculated. Figure 1 shows key statistics, illustrating a significant focus on 'BENIGN' and English sites, but also a reasonable spread of data for valuable results.

For computer application, categories were feature encoded using one-hot encoding, where multi-label categories gained binary columns for each unique type. To manage high cardinality and class sparsity, types appearing less than 20 times were consolidated into an 'Other' label for main class, sub class, and language categories.

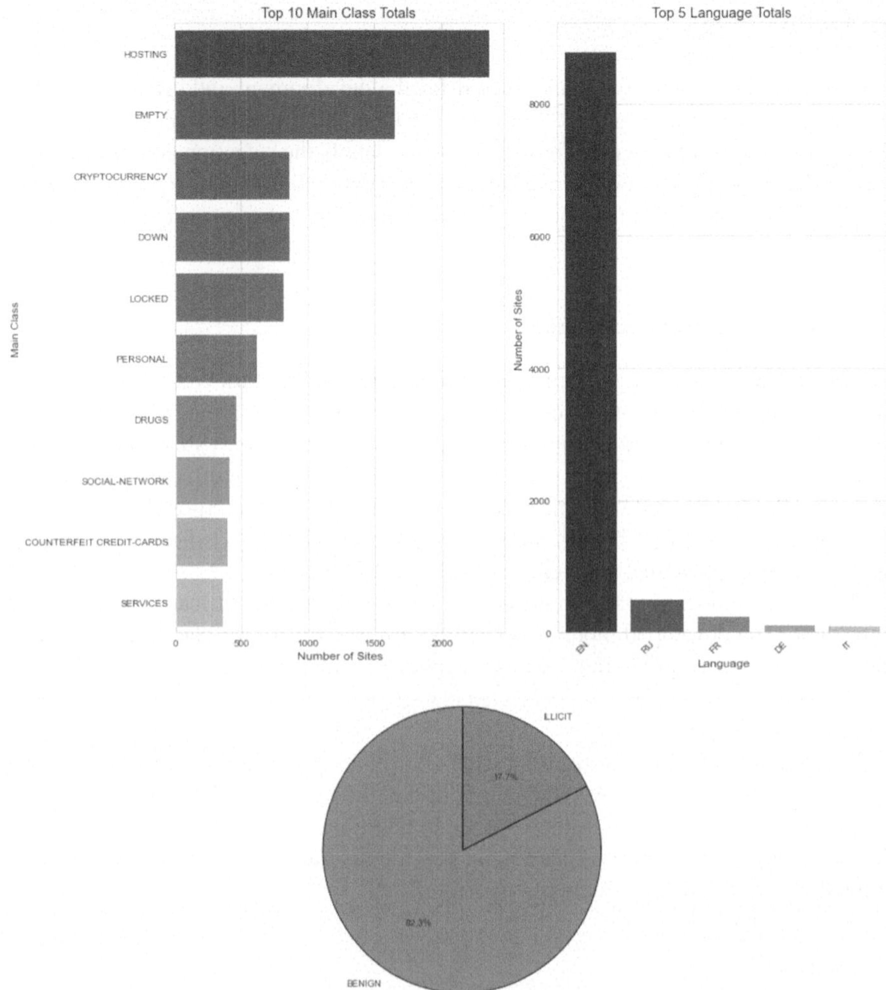

Fig. 1. DUTA-10K Dataset Statistics.

Advanced feature engineering was performed on the onion addresses. Figures 2 and 3 illustrate samples of the dataset's transformation from its original form to the engineered feature set.

	Onion_Address	Main_Class	Sub_Class	lang
1	darkw4u3xkeb5pzn.onion	Art	Music	en
2	tune4xs6mj2evcr6.onion	Art	Music	en
3	a4yedjgciupu7zzt.onion	Art	Music	en
4	76qugq4pb42lpgwx.onion	Art	Music	en
5	kurdox4tfzujxddq.onion	Art	Music	en

Fig. 2. DUTA_10K.xls Original Dataset Sample

Onion_Address	Main_Class	Sub_Class	lang	Illicit_Status	Digits_Ratio	Letters_Ratio	Address_Entropy	Unique_Ch
darkw4u3xkeb5pzn.onion	ART	MUSIC	EN	BENIGN	0.1875	0.8125	3.875	15
tune4xs6mj2evcr6.onion	ART	MUSIC	EN	BENIGN	0.25	0.75	3.75	14
a4yedjgciupu7zzt.onion	ART	MUSIC	EN	BENIGN	0.125	0.875	3.75	14
76qugq4pb42lpgwx.onion	ART	MUSIC	EN	BENIGN	0.3125	0.6875	3.5	12
kurdox4tfzujxddq.onion	ART	MUSIC	EN	BENIGN	0.0625	0.9375	3.452819531	12

Fig. 3. DUTA_10K_combined_features_targets.csv Final Dataset Sample Example

3.3 Developing Models

The ML model development followed a methodical approach:

Feature Selection and Engineering: Features included structural such as keyword flags, n-grams, longest common substrings, hex character values, and character position as well as statistical features comprising of digit/letter ratios, hex character frequency, entropy, and unique character count. All models applied the same base set of features.

Data Pre-Processing: Due to class imbalances, SMOTETomek was used in conjunction with inherent algorithm balancers to oversample minority classes and undersample majority classes [21]. This created a more balanced and cleaner dataset suitable for training. Any exceptions are reported in the results.

Model Development Order: Models for each algorithm were constructed sequentially with illicit status, then main class, then sub class, then language. The illicit status model's output was used as an additional feature for subsequent models.

Training: The preprocessed dataset was split 80/20 into training and validation sets. Cross-validation, consisting of RandomizedSearchCV with 5 randomly sampled hyperparameter combinations and 2-fold cross-validation totaling 10 fits per section per algorithm, ensured robustness and minimized overfitting.

Testing: Models were rigorously evaluated on a separate, unseen 20% test set for unbiased assessment of predictive capabilities.

Visualizing and Evaluating Results: Performance was assessed using accuracy, precision, recall, F1-score, and confusion matrices. Classification report plots and confusion matrix heatmaps provided comprehensive understanding. Feature importance was calculated and displayed with top features for interpretability.

Adjusting the Models: Iterative refinement involved fine-tuning parameters, exploring different feature combinations, and re-evaluating algorithm choices to optimize performance.

3.4 Implementation Details

The implementation was conducted in a Python 3.9.13 environment. Key libraries and their versions used include: pandas (version 1.5.3), numpy (version 1.23.5), scikit-learn (version 1.2.2), imbalanced-learn (version 0.10.1), xgboost (version 1.7.5), matplotlib (version 3.7.1), seaborn (version 0.12.2), and joblib (version 1.2.0).

4 Results and Discussion

In this section, the research results produced are presented and discussed, covering the findings of the training and testing of the various ML algorithms, including calculated performance and feature importance portrayed through figures and tables. The section then delves into the outcomes of the final evaluation phase to directly compare the algorithms and the models they produced.

4.1 Model Training and Testing

During model development, multiple outcomes were calculated and plotted to gain an understanding of their performance and interpretability. These results include accuracy, precision, recall, F1-score, and top features.

Table 1 summarizes the accuracy of all algorithms and their corresponding models during both training and testing. Generally, illicit status and language prediction accuracies were promising across all algorithms. However, testing accuracy for main class and sub class models was much lower than training accuracy, indicating potential overfitting. XGBoost demonstrated the highest results overall, with RF typically trailing slightly, and SVM consistently performing lower.

Table 1. ML Model Training and Testing Accuracy Results.

Algorithm	Illicit_Status	Main_Class	Sub_Class	Lang
Random Forest	Training: 0.9169 Testing: 0.8211	Training: 0.9530 Testing: 0.3014	Training: 0.9781 Testing: 0.4966	Training: 0.9965 Testing: 0.8404
Support Vector Machine	Training: 0.8373 Testing: 0.7401	Training: 0.9419 Testing: 0.1596	Training: 0.9625 Testing: 0.2777	Training: 0.9961 Testing: 0.8457
XGBoost	Training: 0.9643 Testing: 0.8221	Training: 0.9902 Testing: 0.2903	Training: 0.9946 Testing: 0.5497	Training: 0.9984 Testing: 0.8462

The first algorithm utilized was Random Forest (RF). The training report showed strong performance across both "BENIGN" and "ILLICIT" classes, with high precision, recall, and F1-scores. However, the testing report revealed a significant drop in performance for the "ILLICIT" class, particularly in recall at 0.07 and F1-score at 0.12, while the "BENIGN" class maintained high scores. This imbalance in performance was due to the model's tendency to predict addresses as "Benign" as discovered through examining the confusion matrix. The matrix showed that RF correctly classified 1678 "BENIGN" addresses but struggled significantly with "ILLICIT" addresses, misclassifying 342 illicit sites as benign while only correctly identifying 25 illicit sites. This strong bias towards predicting "BENIGN" largely explains the low recall and F1-score for the "ILLICIT" class despite the respectable overall accuracy. Table 2 presents the top 10 features of the RF models.

Table 2. ML Model Feature Importance for Random Forest.

Top Features	Illicit_Status	Main_Class	Sub_Class	Lang
1	Digits_Ratio - 0.045122	Predicted_Illicit_Status - 0.045929	Predicted_Illicit_Status - 0.046669	First_Char_Val - 0.029001
2	Max_Illicit_Substring_Length - 0.044817	First_Char_Val - 0.027907	First_Char_Val - 0.023018	Max_Illicit_Substring_Length - 0.016798
3	Hex_Char_Count_2 - 0.042914	ix - 0.018532	Last_Char_Val - 0.017199	Last_Char_Val - 0.014539
4	Hex_Char_Count_3 - 0.039933	se - 0.017109	Hex_Char_Count_3 - 0.013247	Hex_Char_Count_7 - 0.014245
5	Letters_Ratio - 0.039833	en - 0.017075	Has_Illi_Keyword - 0.012302	ug - 0.013785
6	Hex_Char_Count_5 - 0.038879	et - 0.015610	Max_Illicit_Substring_Length - 0.012215	Hex_Char_Count_2 - 0.012568
7	Hex_Char_Count_6 - 0.037943	li - 0.014299	Unique_Char_Count - 0.011319	ro - 0.012355
8	Hex_Char_Count_4 - 0.033625	Last_Char_Val - 0.014252	ki - 0.011105	un - 0.012249
9	Hex_Char_Count_7 - 0.032432	n2 - 0.014228	Digits_Ratio - 0.010799	Hex_Char_Count_3 - 0.012085
10	Max_Consecutive_Char_Count - 0.017846	Hex_Char_Count_4 - 0.012957	ma - 0.010620	il - 0.010725

For the illicit status model, features like Digits_Ratio, Max_Illicit_Substring_Length, and various Hex_Char_Count_X were highly influential. This suggests that the numerical composition, the presence of long illicit substrings, and the frequency of specific hexadecimal characters within an onion address are strong indicators for identifying criminality. For main class, Predicted_Illicit_Status emerged as the most important feature, highlighting the hierarchical model's effectiveness in leveraging prior predictions. Other important features included First_Char_Val, Last_Char_Val, and specific bigrams. Similarly for sub class, Predicted_Illicit_Status, First_Char_Val, and Last_Char_Val were prominent, along with Has_Illi_Keyword and Unique_Char_Count. Finally for language, First_Char_Val was the top feature, followed by Max_Illicit_Substring_Length, Last_Char_Val, and various bigrams (e.g., ug, ro, un, and il). These top features across all targets indicate that the models are leveraging a combination of structural, statistical, and n-gram features to make predictions. The consistent appearance of Predicted_Illicit_Status as a top feature in subsequent hierarchical models underscores the value of the cascading approach.

The second algorithm utilized was Support Vector Machine (SVM). The training classification report indicated reasonable performance for both classes. However, the testing report revealed a pronounced decline in performance for the "ILLICIT" class compared to RF, with very low precision at 0.21, recall at 0.17, and F1-score at 0.19. The SVM Illicit_Status testing confusion matrix further solidified this by showing that while 1472 "BENIGN" addresses were correctly identified, a significant number of "ILLICIT" addresses at 304 were misclassified as "BENIGN", and only 63 were correctly identified as "ILLICIT". This indicates a strong bias towards the majority "BENIGN" class, similar to RF, but with a more severe impact on identifying illicit activities. It is important to note that for the language prediction, the model could only be trained within a reasonable timeframe when relying solely on the inherent class_weight = 'balanced' parameter rather than the more computationally intensive SMOTETomek or other oversampling/undersampling techniques. This limitation highlights the sensitivity of SVMs to dataset size as the generation of synthetic samples significantly increases the computational burden for these algorithms. Table 3 presents the top 10 features of the SVM models.

Table 3. ML Model Feature Importance for Support Vector Machine.

Top Features	Illicit_Status	Main_Class	Sub_Class	Lang
1	gg - 0.002774	am - 0.006880	Letters_Ratio - 0.023687	Address_Entropy - 0.001505
2	Digits_Ratio - 0.002319	Has_Consecutive_Repeat - 0.004469	Digits_Ratio - 0.007622	Unique_Char_Count - 0.001505
3	Letters_Ratio - 0.002319	og - 0.003909	Hex_Char_Count_3 - 0.007392	vi - 0.001140
4	35 0.002251	bt - 0.003878	Has_Consecutive_Repeat - 0.005727	la - 0.001140
5	Max_Illicit_Substring_Length 0.002203	oi - 0.003544	Predicted_Illicit_Status - 0.005467	es - 0.001140
6	da - 0.002137	ik - 0.003540	222 - 0.004711	ab - 0.001140
7	in - 0.001951	ms - 0.003501	og - 0.004444	tl - 0.001060
8	am - 0.001782	Hex_Char_Count_3 - 0.003440	ra - 0.004424	st - 0.001047
9	se - 0.001736	or - 0.003353	22 - 0.003628	Hex_Char_Count_6 - 0.001040
10	sp - 0.001724	le - 0.003264	wo - 0.003386	Predicted_Illicit_Status - 0.000893

For the illicit status model, features like specific bigrams, Digits_Ratio, Letters_Ratio, and Max_Illicit_Substring_Length were the most influential. This suggests that while numerical composition and substring patterns are important, specific short character sequences also play a significant role for SVMs. Regarding main class, Has_Consecutive_Repeat and various bigrams were prominent once more. For sub class, Letters_Ratio, Digits_Ratio, Hex_Char_Count_3, Has_Consecutive_Repeat, and Predicted_Illicit_Status were among the top features. Finally for language, Address_Entropy, Unique_Char_Count, and various bigrams were highly ranked, along with Predicted_Illicit_Status appearing at a lower rank. These top features for SVM models indicate a reliance on a diverse set of structural, statistical, and n-gram features, similar to RF, but with some differences in the specific n-grams and the relative importance of features like Has_Consecutive_Repeat.

The third algorithm utilized was XGBoost. The training classification report demonstrated very high performance for both "BENIGN" and "ILLICIT" classes, with nearly perfect recall for "BENIGN" and high precision for "ILLICIT". The testing classification report showed a similar pattern to RF with strong performance for "BENIGN" but a significant drop in recall at 0.08 and F1-score at 0.14 for the "ILLICIT" class. The XGBoost illicit status model testing confusion matrix confirmed this with 1676 "BENIGN" addresses correctly classified, but 338 "ILLICIT" addresses misclassified as "BENIGN", and only 29 correctly identified as "ILLICIT". This implies that despite its overall high accuracy, XGBoost also exhibits a strong bias towards the majority "BENIGN" class in its predictions for illicit status. Table 4 presents the top 10 features of the XGBoost models.

Table 4. ML Model Feature Importance for XGBoost.

Top Features	Illicit_Status	Main_Class	Sub_Class	Lang
1	el - 0.010637	se - 0.025549	222 - 0.015025	First_Char_Val - 0.029001
2	222 - 0.009223	ix - 0.019241	Predicted_Illicit_Status - 0.009696	ug - 0.016292
3	og - 0.005795	Predicted_Illicit_Status - 0.008868	pa - 0.008112	pm - 0.010350
4	tc - 0.005631	un - 0.006664	zp - 0.006527	xo - 0.009284
5	Hex_Char_Count_2 - 0.005543	en - 0.006499	ms - 0.006039	a5 - 0.008354
6	le - 0.005470	ot - 0.006230	gn - 0.005837	un - 0.007299
7	Max_Illicit_Substring_Length - 0.005319	li - 0.006221	as - 0.005779	dn - 0.007290
8	tb - 0.005243	ib - 0.005493	2y - 0.005657	s4 - 0.007006
9	c3 - 0.004983	t6 - 0.005356	dl - 0.005322	tt - 0.006938
10	Hex_Char_Count_6 - 0.004856	bl - 0.005061	7j - 0.005216	Hex_Char_Count_1 - 0.006873

These top features show similarities with both RF's and SVM's top features. The important attributes for the illicit status model included various bigrams as well as Hex_Char_Count_2, Max_Illicit_Substring_Length, and Hex_Char_Count_6. This suggests a strong reliance on specific n-gram patterns and character counts for illicit status prediction. Likewise for the main class model, bigrams in addition to Predicted_Illicit_Status were highly ranked. Concerning sub class, bigrams along with Predicted_Illicit_Status were prominent once again. Finally for language, First_Char_Val was the top feature, followed by the bigrams ug, pm, xo, un, dn, s4, and tt. Hex_Char_Count_1 was also listed. The consistent appearance of bigram features as top indicators across various targets for XGBoost underscores its ability to leverage granular character sequence patterns.

4.2 Final Model Evaluations

The final evaluation phase involved a direct comparison of the three algorithms (RF, SVM, and XGBoost) across all four prediction targets using the consistent held-out test data of the DUTA-10K dataset with the engineered features. Figures 4 and 5 visually summarize this comparison, presenting the overall accuracy and F1-weighted scores which relates the precision and recall, respectively.

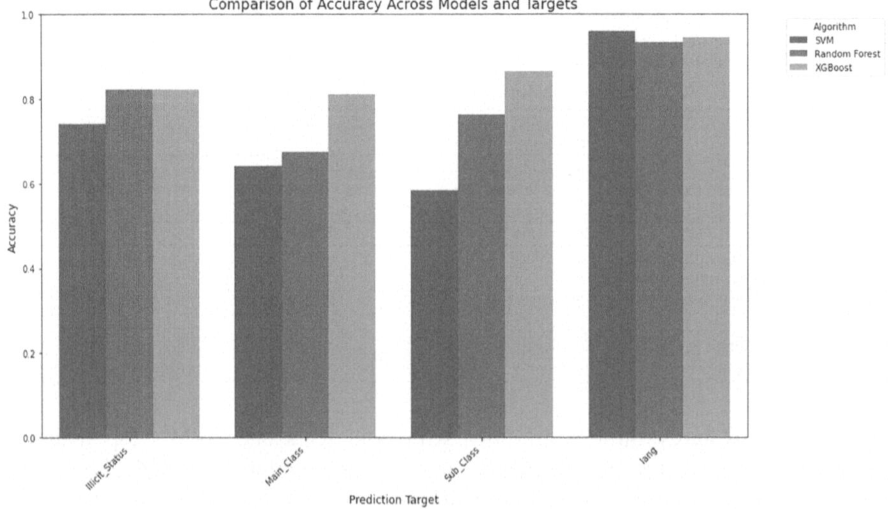

Fig. 4. Model Accuracy Comparison Graph

As observed in Table 1, XGBoost generally achieved the highest accuracy across all targets, with RF performing very similarly, often trailing by a small margin. SVM consistently showed lower accuracy values compared to the ensemble methods. For illicit status and language classification, all algorithms demonstrated relatively high accuracy with above 0.7 for SVM and above 0.8 for RF/XGBoost. This denotes a strong ability to predict these categories based on the address of a site. However, for main class and sub class, the accuracy values were significantly lower for all three algorithms. This highlights the increased complexity and potential challenges with these multi-class, highly imbalanced targets where a single entry can fall under a multitude of different labels that are distributed unevenly, making it more difficult for models to effectively learn.

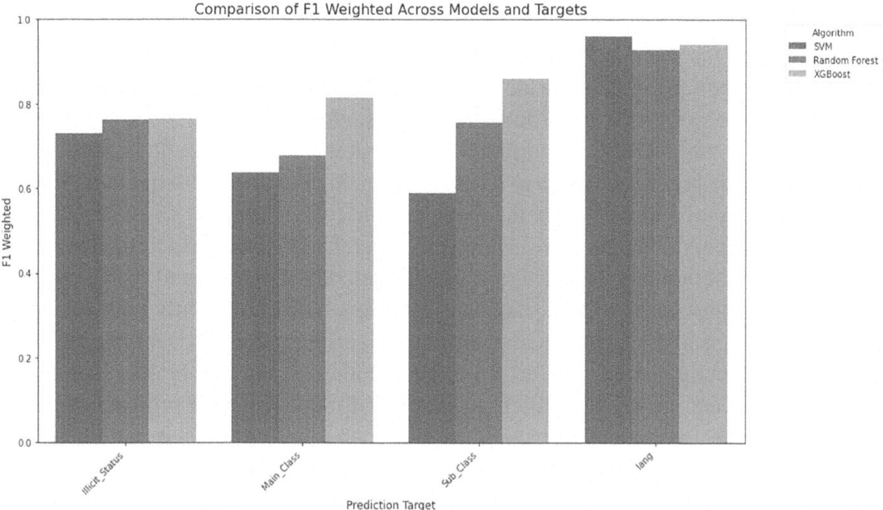

Fig. 5. Model F1-Score Comparison Graph

These final plots regarding precision, recall, and F1-score reinforce the previous trends seen in the accuracy comparison. XGBoost and RF generally outperformed SVM across all scores for weighted precision, weighted recall, and F1-weighted scores by a not insignificant margin, ranging from a few points away to over 0.2 for the sub class models. The exception to this is with the language models where SVM performed slightly better, but still not to the degree that it was outpaced by XGBoost for main class and sub class. The F1-weighted score, which balances precision and recall, provides a more robust measure for the imbalanced datasets. The lower precision, recall, and F1-scores observed for main class and sub class models, particularly for the minority classes, which were not explicitly shown here but can be inferred from overall low scores, are likely attributable to the high number of distinct classes and the severe class imbalance within these targets. With many categories where some have very few instances, it becomes inherently more challenging for models to learn robust decision boundaries for every class.

5 Conclusions and Future Work

This section discusses valuable insights gleaned from the results by looking at individual performances of the models as well as any overarching trends that emerged. It also dives into potential ways in which the work of this research could be continued in future work, ranging from straightforward to relatively complex.

5.1 Conclusions

This research aimed to develop machine learning models capable of classifying Dark Web onion addresses based on limited metadata, specifically predicting their illicit status, main class, sub class, and language. The methodology involved a systematic approach that began with the cleaning and formatting of the DUTA-10K dataset, followed by extensive feature engineering from the onion addresses themselves, and finally, the training and evaluation of hierarchical models using Random Forest (RF), Support Vector Machine (SVM), and XGBoost algorithms. The hierarchical structure allowed the output of the illicit status prediction to serve as an additional feature for subsequent models to improve overall performance when the models could reliably predict criminality.

The results generated indicated a varied but generally promising performance across the different prediction targets and algorithms. The models demonstrated strong capabilities in predicting primarily illicit status and language with consistently high accuracy, precision, recall, and F1-scores across all three algorithms during development. This suggests that the engineered features from onion addresses are highly effective in distinguishing between benign and illicit activities and identifying the primary language of a Dark Web site, as these sites have some underlying patterns in their naming shown through aspects such as the top features of the models often including n-grams of various characters.

In contrast to this, the prediction of main class and sub class proved to be significantly more challenging. While training accuracies for these targets were high, testing accuracies were notably lower, signifying a degree of overfitting on the training data. This performance disparity is likely a direct consequence of the inherent characteristics of these targets where there is a large number of distinct classes along with severe class imbalances where many categories have very few entries, as shown by the dataset statistics and the need to use one-hot encoding to congregate labels with very few entries to help narrow focus. The increased complexity of learning from a wide range of classes, some of which are sparsely represented, makes it difficult for models to generalize effectively to unseen data. The lower precision, recall, and F1-scores for these multi-class predictions are partially attributable to this challenge, as the instance of more than two labels to choose from inherently increases the complexity of the learning task while also helping to reflect multifaceted real-world scenarios where there are often more than just 2 potential options.

The analysis of top features across all of the constructed models revealed that there was no single feature or feature set that was consistently heavily weighted. Instead, the models leveraged a wide variety of features including First_Char_Val, Last_Char_Val, Max_Illicit_Substring_Length, various Hex_Char_Count_X features, and numerous two-character n-grams. This diverse set of important features suggests patterns in the beginning and end of addresses, the presence of specific illicit substrings, the frequency distribution of hexadecimal characters, and the occurrence of short character sequences are all highly discriminative for classifying Dark Web activities that would be considered unlawful. The consistent appearance of Predicted_Illicit_Status as a top feature in the main class, sub class, and language models further validates the effectiveness of the hierarchical modeling approach by demonstrating that the initial illicit status prediction provides valuable context for subsequent classification layers even as a single additional feature to learn from.

Regarding algorithm performance, XGBoost consistently produced the most effective models overall and typically achieved the highest scores across all metrics and targets. RF performed very similarly, often falling slightly behind XGBoost. SVM, despite showing reasonable performance for illicit status and language, consistently yielded lower scores compared to the ensemble methods. A notable observation was SVM's computational limitations. The SVM model for the language prediction, and likely for other multi-class targets, could only be trained within a reasonable timeframe when relying solely on its inherent class_weight = 'balanced' parameter instead of SMOTE-Tomek. In addition to this, it had much longer development time for the other prediction type models when compared to RF and XGBoost, even when they were all structured with the established 10 fits from 5 parameter combinations with 2-fold cross validation. This is primarily due to SVM's high computational complexity which can scale quadratically or cubically with the number of samples and is significantly exacerbated by the large number of synthetic samples generated by oversampling techniques like SMOTE and SMOTETomek. Outcomes such as this show how tree-based models like Random Forest and XGBoost are often preferred for large, high-dimensional, and imbalanced datasets due to their ability to scale much better and be less sensitive to the number of samples.

Furthermore, preliminary experimentation showed that increasing the number of randomly sampled hyperparameter combinations did not lead to substantially better performance but still dramatically increased the runtime, even more so for SVM and XGBoost. This made it so higher n_iter values were unreasonable for these algorithms unless there is a sizeable amount of both time and computational power available. The superior performance in illicit status and language predictions can likely be attributed to the binary nature of illicit status which can be simpler to learn, as well as the high prevalence of English ('EN') language entries which makes the language classification more similar to a binary problem (English vs. Non-English) for a large portion of the data along with the addresses containing characters from the English alphabet.

5.2 Future Work

To further enhance this research, several avenues for future work are proposed here.

The first potential method is the integration of live feeds and web crawling as previous studies have done. While this study focused on static dataset analysis, future work could involve integrating the models with live Dark Web feeds and web crawling capabilities. This would allow for real-time threat intelligence gathering and prediction in order to provide more timely and actionable insights.

Another approach that has been considered is collecting data from deeper sources. By expanding data collection efforts to more elusive areas such as dark web forums where threat actors often gather and communicate, richer and more nuanced features for model training could be produced. This would require addressing significant technical, ethical, and legal challenges as was discussed previously in the literature review.

A more straightforward way to expand on this study would be the exploration of alternative ML algorithms and architecture. Investigating other advanced ML algorithms or deep learning architectures could potentially yield higher performance. This is especially relevant for the multi-class and multi-label prediction tasks like the main class and sub class classifications where there is a wide array of potential labels an address can fall under.

Another avenue could be using advanced techniques to handle the data imbalances. Future work could delve into more sophisticated imbalance handling techniques tailored for multi-class, highly imbalanced datasets. Custom loss functions that penalize misclassifications of minority classes more heavily could also be explored.

Building upon what did well, more involved ensemble and stacking models could be utilized. Developing more complex ensemble or stacking models that combine the predictions of the current best-performing algorithms of RF and XGBoost could lead to further performance gains by leveraging the strengths of each individual model.

Lastly, external validation with new datasets could be conducted. Validating the developed models on entirely new and distinct Dark Web datasets would further assess their generalizability and robustness in real-world scenarios.

Overall, while the main goal of this research has been largely successful in being able to derive meaningful CTI from limited data through machine learning, there are still ways that this research can be taken further to help broaden its potential use. Whether through the straightforward expansion of the dataset or incorporating more complex techniques such as live feeds, the data science field is an always growing environment with a variety of tools available to make noteworthy contributions to CTI, cybersecurity, and beyond.

References

1. Topic: Cyber Threat Intelligence (CTI), Statista (2024). https://www.statista.com/topics/10666/cyber-threat-intelligence-cti/#topicOverview. Accessed 14 June 2025
2. Sangher, K.S., Singh, A., Pandey, H.M., Kumar, V.: Towards safe cyber practices: developing a proactive cyber-threat intelligence system for dark web forum content by identifying cyber-crimes. Information **14**(6), 349 (2023). https://doi.org/10.3390/info14060349. Accessed 3 June 2025
3. Sun, N., et al.: Cyber threat intelligence mining for proactive cybersecurity defense: a survey and new perspectives. IEEE Commun. Surv. Tutorials 25(3), 1 (2023). https://doi.org/10.1109/comst.2023.3273282. Accessed 30 May 2025
4. Schlette, D., Böhm, F., Caselli, M., Pernul, G.: Measuring and visualizing cyber threat intelligence quality. Int. J. Inf. Secur. **20**(1), 2025 (2020). https://doi.org/10.1007/s10207-020-00490-y.accessedJun.1
5. Ainslie, S., Thompson, D., Maynard, S., Ahmad, A.: Cyber-threat intelligence for security decision-making: A review and research agenda for practice. Comput. Secur. **132**(132), 103352 (2023). https://doi.org/10.1016/j.cose.2023.103352. Accessed 1 June 2025
6. Sarker, I.H.: Machine learning for intelligent data analysis and automation in cybersecurity: current and future prospects. Ann. Data Sci. **10**(1), 1473–1498 (2022). https://doi.org/10.1007/s40745-022-00444-2. Accessed 2 June 2025
7. Saeed, S., Suayyid, S.A., Al-Ghamdi, M.S., Al-Muhaisen, H., Almuhaideb, A.M.: A systematic literature review on cyber threat intelligence for organizational cybersecurity resilience. Sensors **23**(16), 7273 (2023). https://doi.org/10.3390/s23167273. Accessed 30 May 2025
8. Kolluri, V.: An extensive investigation into guardians of the digital realm: ai-driven antivirus and cyber threat intelligence. Int. J. Adv. Res. Interdisciplinary Sci. Endeavours **1**(2), 71–77 (2024). https://doi.org/10.61359/11.2206-2407. Accessed 1 June 2025

9. Basheer, R., Alkhatib, B.: Threats from the dark: a review over dark web investigation research for cyber threat intelligence. J. Comput. Networks Commun. **2021**, e1302999 (2021). https://doi.org/10.1155/2021/1302999. Accessed 3 June 2025
10. Zhou, Y., Tang, Y., M. Yi, C. Xi, and H. Lu, "CTI View: APT Threat Intelligence Analysis System," Security and Communication Networks, vol. 2022, pp. 1–15, Jan. 2022, https://doi.org/10.1155/2022/9875199. Accessed 30 May 2025
11. Koloveas, P., Chantzios, T., Alevizopoulou, S., Skiadopoulos, S., Tryfonopoulos, C.: inTIME: A Machine Learning-Based Framework for Gathering and Leveraging Web Data to Cyber-Threat Intelligence. Electronics **10**(7), 818 (2021). https://doi.org/10.3390/electronics10070818. Accessed 2 June 2025
12. Gao, P., et al.: Enabling efficient cyber threat hunting with cyber threat intelligence. IEEE Xplore, 01 April 2021. https://ieeexplore.ieee.org/document/9458828. Accessed 3 June 2025
13. Al-Nabki, M.W., Fidalgo, E., Alegre, E., Fernández-Robles, L.: ToRank: identifying the most influential suspicious domains in the Tor network. Expert Syst. Appl. **123**, 212–226 (2019). https://doi.org/10.1016/j.eswa.2019.01.029. Accessed 13 June 2025
14. Shin, G.-Y., et al.: Dark Side of the Web: Dark Web Classification Based on TextCNN and Topic Modeling Weight. IEEE Access **12**, 36361–36371 (2024). https://doi.org/10.1109/access.2023.3347737. Accessed 12 June 2025
15. Sennad, M., Ellaky, Z., Benabbou, F.: Enhancing DarkWebActivities Classification Using Embedding Methods, pp. 1–8, May 2025. https://doi.org/10.1109/iraset64571.2025.11008275. Accessed 12 June 2025
16. Nabki, A., Fidalgo, E., Alegre, E., Chaves, D.: Supervised ranking approach to identify infLuential websites in the darknet. Appl. Intell., July 2023. https://doi.org/10.1007/s10489-023-04671-9. Accessed 12 June 2025
17. IBM, "Random Forest," Ibm.com, Oct. 20, 2021. https://www.ibm.com/think/topics/random-forest. Accessed 24 June 2025
18. scikit-learn, "1.4. Support Vector Machines," Scikit-learn.org (2018). https://scikit-learn.org/stable/modules/svm.html. Accessed 24 June 2025
19. GeeksforGeeks: Difference Between Random Forest and XGBoost, GeeksforGeeks, Apr. 15, 2024. https://www.geeksforgeeks.org/machine-learning/difference-between-random-forest-vs-xgboost/. Accessed 24 June 2025
20. Al Nabki, M.W., Fidalgo, E., Alegre, E., de Paz, I.: Classifying Illegal Activities on Tor Network Based on Web Textual Contents. ACLWeb, Apr. 01, 2017. https://aclanthology.org/E17-1004/. Accessed 21 Jun 2025
21. SMOTETomek — Version 0.11.0: imbalanced-learn.org. https://imbalanced-learn.org/stable/references/generated/imblearn.combine.SMOTETomek.html. Accessed 25 Jun 2025

Detecting Deviation in User Activity Using Supervised Learning

Ciera Miller, Kemon Bynum, and Cheryl Hinds(✉)

Norfolk State University, Norfolk, VA 23504, USA
{c.miller126045,k.l.bynum}@spartans.nsu.edu, chinds@nsu.edu

Abstract. Cyberattacks continue to pose a serious threat to organizational and national security, with recent years showing a dramatic rise in the number and complexity of these attacks. In 2025 alone, ransomware attacks increased by 126%, averaging nearly 2,000 incidents per week worldwide [2]. Many of these attacks involve advanced techniques such as phishing, credential theft, and AI-driven exploitation, which make it harder for traditional security systems to detect threats. To address these evolving risks, it is critical to explore new methods that can detect malicious activity even when attackers appear as legitimate users. In this research, we developed a behavior-based anomaly detection system that uses supervised machine learning to identify abnormal login activity. The tool analyzes user behavior patterns such as login time, location, IP address, and device type, and flags suspicious deviations that may indicate account compromise or insider threats. We implemented and evaluated three machine learning models Random Forest, Logistic Regression, and Neural Networks to determine which approach is the best fitting in detecting anomalous user activity. This system will help organizations enhance their cybersecurity posture by proactively identifying unusual behavior patterns, offering earlier warnings of potential attacks, and ultimately reducing the impact of cyber intrusions.

Keywords: Supervised Learning · User Activity · Data Science

1 Introduction

Cyberattacks are a growing threat to network security, often launched through external malware such as trojans, worms, and viruses [1]. 2025 has witnessed an unexpected increase in cyberattacks, with ransomware attacks rising by 126% globally and an average of 1,925 per week for organizations [2]. These attacks have grown in complexity, frequently involving multi-stage infiltration techniques such as phishing campaigns, credential theft, and privilege escalation. Recent trends also indicate a surge in AI-powered attacks, where adversaries automate reconnaissance and exploit discovery using machine learning models [3]. Additionally, the growing number of connected devices and remote access points has widened the attack surface, making it increasingly difficult for traditional perimeter-based security to remain effective. As organizations shift more resources to cloud-based infrastructure and hybrid work environments, cybercriminals

I. Roy et al. (Eds.): CAINE 2025, CCIS 2740, pp. 336–348, 2026.
https://doi.org/10.1007/978-3-032-14893-3_23

are exploiting these changes with more targeted and persistent attacks [4]. This rising threat landscape highlights the urgent need for adaptive, behavior-aware systems capable of detecting subtle signs of intrusion before significant damage occurs.

In response to these evolving threats, security researchers and practitioners are increasingly turning to behavior-based detection methods, which leverage consistency in how legitimate users interact with systems. By modeling typical usage patterns, it becomes possible to identify deviations that may signal credential misuse, insider threats, or advanced persistent attacks. Modern digital systems rely heavily on users performing routine tasks within enterprise networks, cloud infrastructures, and online services. Users typically follow consistent activity patterns; unusual behavior can signal a potential attack [1]. Users typically exhibit consistent behavioral patterns such as logging in at predictable times and using the same devices, locations, and Ip addresses. These behaviors form a baseline that can be observed and statistically modeled. For example, an individual logs into their secure account with a specific user ID, location, IP address, and device (e.g., phone, tablet, laptop). Additionally, the individual has a typical login count and duration that contributes to a broader user profile. For instance, consider someone who regularly logs into a Netflix account from Norfolk, Virginia, using a specific device and IP address, typically three times a day for a total of five hours. By establishing this consistent pattern of behavior, we can detect deviations such as a login from an unfamiliar location or unusual usage patterns and raise a flag for potential security concerns.

As cyber threats continue to formulate and grow, attackers continue to exploit the trust embedded in these behavioral patterns. Many attacks involve techniques that involve mimicking users, such as stealing or forging their credentials. Attackers that mimic user credentials make it difficult for detection systems to differentiate between genuine and malicious users. This may result in attacks such as credential thefts, brute force login attempts, and insider threats. These attackers operate under the disguise of valid users, bypassing traditional security mechanisms and blending into the background of routine activity.

To mitigate these increasing attacks, many organizations have turned to anomaly detection systems, which are designed to flag deviations from established behavioral norms. These systems operate by learning what constitutes "normal" activity across various user metrics such as login time, location, frequency, and device type, then identifying outliers that differ significantly from these learned patterns. Unlike signature-based systems that rely on known attack patterns, anomaly detection methods are capable of uncovering novel or previously unseen threats. Common techniques include statistical modeling, rule-based thresholds, and unsupervised machine learning algorithms such as clustering or isolation forests. While these systems offer promise in identifying subtle irregularities, they often face challenges in distinguishing between benign anomalies and actual threats, especially in dynamic environments where user behavior naturally evolves over time.

Given the limitations of traditional anomaly detection systems, there is growing interest in adaptive learning systems that can evolve with user behavior. This research proposes a novel approach to detecting deviations in user activity using Supervised

Learning. Supervised learning is a class of machine learning algorithms that requires labeled examples (or data) of normal and malicious activity.

1.1 Project Motivation and Importance

The importance of this project includes identifying potential cyberattacks early by monitoring user activity instead of relying solely on known malware signatures.

Traditional signature-based detection systems are limited in their ability to recognize novel or sophisticated attacks that do not match predefined patterns [5]. By focusing on behavioral deviations, this approach allows for the detection of anomalous activity even when attackers use valid credentials or legitimate access points, which is particularly important in combating insider threats and account takeover incidents [6]. Additionally, the system can provide immediate notifications of suspicious behavior, enabling faster response and mitigation efforts before significant damage occurs. This proactive monitoring enhances organizational resilience by identifying threats in real time and minimizing the period between breach and detection. Ultimately, behavior-based detection strengthens cybersecurity defenses by complementing existing tools and helping secure modern environments where threats continuously evolve in complexity and stealth [7].

In this research, we will use data science techniques to develop a supervised learning system that models individual login behavior and detects deviations indicative of potentially abnormal or suspicious activity when the login is authenticated. Rather than relying on static rules or signature-based detection methods, this system will learn user specific behavioral baselines over time and continuously adapt as user patterns evolve. Additionally, this research will implement various supervised learning models, including Random Forest, Logistic Regression, and Neural Networks. These algorithms are suitable for both classification and regression tasks. The objective is to evaluate and compare their performance to determine the most effective model for addressing the problem. The system will provide risk-based responses to each login event, aiming to improve the speed and accuracy of anomaly detection while reducing false positives. This adaptive approach enhances user-specific security monitoring and contributes to the broader goal of proactively identifying credential misuse and insider threats in dynamic network environments.

The rest of the paper is organized as follows: Section 2 introduces related work. The methodology is given in Sect. 3. Finally, Sect. 4 will any references used in our research.

2 Related Work

This section provides an overview of prior work related to detecting deviations in user activity.

There has been some prior research conducted on detecting anomalies with user login behavior. The researcher's goal in [5] and [6] was to predict and identify any possible anomalies related to user login behavior. This was conducted using two unsupervised algorithms: isolation forest and local outlier factor (LOF). In conclusion, it was found that isolation forest algorithms would be more accurate compared to other algorithms since it predicted the highest number of anomalies with a low number of false positives in each dataset. A limitation of this research is that it does not list the features used. In other words, it does not list what defines user login behavior (e.g. timestamps, location, user id, etc.).

In paper [1], the author proposed a two-stage framework for insider threat detection that focuses on user behavior modeling and anomaly detection. The first usage uses normalized event sequences to capture user activity, while the second employs a Long Short-Term Memory (LSTM) network to learn temporal patterns in these sequences. Anomalous behavior is identified by calculating the reconstruction error based on the LSTM's hidden layer activations, with thresholds set using mean squared error. This approach enables more nuanced profiling of user activity compared to traditional methods that rely on aggregated or shallow features. The authors evaluate their model using the CERT v4.2 dataset, which has over 32 million events from 1,000 users across 502 days. Their best-performing LSTM configuration achieves an area under curve of 0.97, outperforming a baseline Support Vector Machine (SVM) model, indicating that the sequential modeling of user behavior significantly enhances detection capability.

The model's use of LSTM to capture sequential dependencies aligns well with the nature of insider threats, which often unfold over time rather than through isolated events. The shift from static feature vectors to dynamic sequence modeling represents a meaningful advance. However, the framework has limitations. It is trained and tested on a synthetic dataset, which may not fully represent the complexity and noise of real-world environments [1]. Furthermore, the model focuses on only a subset of event types (logon/logoff, device usage, and HTTP access), potentially overlooking other critical indicators of insider behavior, such as command-line activity, lateral movements, or privilege escalation attempts. The authors highlight that many existing models rely on aggregated features or shallow representations, failing to capture the sequential nature of user activity [1]. Traditional methods like SVMs and decision trees often lack the temporal modeling capabilities required to detect subtle behavioral shifts [1]. Additionally, much of the existing literature focuses on external threats (e.g., malware or website code injections), which do not adequately address the unique dynamics of insider attacks. While some user behavior analytics (UBA) systems are cited, they generally remain at a conceptual level and lack concrete implementation or evaluation strategies. Moreover, neither the paper nor the referenced works discuss adversarial evasion tactics, such as log manipulation or mimicry of normal behavior, which could be exploited by sophisticated insiders to bypass detection.

The authors in [8] designed and implemented a scalable system for detecting anomalies in database connection logs using unsupervised machine learning techniques. Recognizing the challenges of processing massive volumes of unstructured log data in real time, they developed a pipeline that combines ELK (Elasticsearch, Logstash, and Kibana) for real-time data ingestion and visualization with Apache Spark and Hadoop for large-scale batch processing. The architecture collects database connection logs across thousands of hosts and feeds them into a processing system that extracts features and applies clustering and classification algorithms, such as k-means and one-class SVM, to detect outliers without requiring labeled data.

The system is built to operate continuously, allowing organizations to monitor abnormal behaviors as they emerge across complex, distributed environments. By leveraging the ELK stack for near real-time insights and Spark for distributed computation, the authors demonstrate how log data can be analyzed efficiently and visualized meaningfully. Although the paper focuses more on the system's design than on detailed experimental results, it highlights the potential of combining traditional big data tools with unsupervised learning to detect anomalous login patterns and improve security monitoring at scale.

In Hansol Park's study [9], he presents a machine learning-based system designed to detect malicious user activity within Border Gateway Protocol (BGP) network traffic by identifying abnormal routing behavior. The approach leverages a range of techniques, including Natural Language Processing (NLP) to convert BGP path data into numerical features, Synthetic Minority Over-sampling Technique (SMOTE) to address class imbalance, and a variety of classification models such as Neural Networks, Long Short-Term Memory (LSTM) networks, and One-Class Support Vector Machines (OC-SVM). By combining these methods, the system aims to effectively identify anomalies that indicate malicious intent without relying solely on predefined rules or labeled attack data.

According to [9], the results demonstrate the potential of machine learning in analyzing complex routing behaviors. The study is constrained by its dependence on synthetic datasets and the absence of a real-time deployment framework. This limits the system's ability to generalize to real-world network environments, where traffic patterns are more varied and unpredictable. Nonetheless, the integration of sequence based deep learning models like LSTM with traditional anomaly detection methods marks a significant step toward intelligent, behavior-based threat detection in Internet infrastructure. Future work could benefit from testing live BGP data and exploring adaptive learning techniques for evolving network threats [9].

In paper [10], Benazzouza et al. proposed a hybrid model for identifying malicious users in cooperative cognitive radio networks by developing a robust stacking ensemble and deep learning approach. To detect spoofing and interference from malicious participants, they employed a multi-layer stacking model using chaotic compressive sensing for feature extraction, combined with ensemble classifiers to distinguish between genuine and malicious observations. This model achieved approximately 97% accuracy in identifying up to 50% of users being malicious, demonstrating strong resilience against varied adversarial behaviors. In parallel, they introduced a deep learning-based spectrum sensing component, where scalogram (time-frequency) representations of primary user signals were fed into convolutional neural networks. This component showed a high detection probability and low false alarm rates, while remaining computationally efficient with fewer training epochs. Despite promising results in simulation, their approach faces limitations in real-world deployment. The authors concluded that the reliance on synthetic or simulated data for training and validation raises concerns about the method's generalizability to live network environments where signal noise, propagation dynamics, and user behaviors may significantly differ [10]. Additionally, the study lacks an implemented real-time detection system, leaving open questions regarding latency, scalability, and adaptability under live traffic loads. Nevertheless, this work meaningfully advances the field by integrating compressive sensing, ensemble stacking, and deep-scalogram-based CNNs, suggesting promising avenues for adaptive, hybrid machine learning frameworks in future cognitive radio security deployments.

The authors in [3] propose a user behavior modeling approach to detect insider threats by applying unsupervised anomaly detection algorithms. They utilize the CERT R6.2 dataset to construct three types of behavioral representations: daily activity summaries (e.g., logins, website visits), topic distributions from email content using Latent Dirichlet Allocation (LDA), and communication network graphs extracted from weekly email interactions. To detect anomalies without labeled attack data, the authors apply one-class classification methods—including Gaussian density estimation, Parzen window estimation, PCA, and k-means clustering to learn patterns of normal behavior and flag deviations as potential threats. Their approach demonstrates strong performance across modalities, achieving over 90% detection rates for daily activity summaries and up to 98% for email topic analysis, with near-perfect detection of anomalous communication behavior in several cases [3].

According to the authors, the study has some limitations. First, each behavior modality is modeled separately, and the anomaly scores are not fused, which may reduce the system's effectiveness in capturing complex cross-modal threat patterns [3]. Second, their evaluation is conducted offline in batch format, preventing real-time detection that would be critical in active enterprise environments [3]. Finally, while the CERT dataset offers rich synthetic data, it does not fully represent the variability and noise found in real-world insider threat scenarios, which may affect the generalizability of the results [3]. Despite these constraints, the work makes a meaningful contribution by showing how unsupervised learning can be applied to diverse user behavior features for insider threat detection, and it offers a strong foundation for future research focused on real-time, multi-modal analysis.

3 Methodology

This study employed a mixed-methods approach combining synthetic data generation, data preprocessing, supervised machine learning techniques, and model evaluation to analyze abnormal user behavior based on login patterns. Our methodology is structured around the creation of a realistic but artificial user dataset, preparation of this data for modeling, selection and implementation of algorithms, and comparative evaluation of the results. Below is a detailed account of the technical and procedural elements of the methodology.

3.1 Framework and Library Utilization

The implementation heavily relied on Scikit-learn for its wide support of classification models, preprocessing tools, and performance metrics. Additional libraries included pandas and numpy for data manipulation, matplotlib and seaborn for visualization, and regex for string preprocessing. The Faker library was instrumental in generating synthetic yet realistic behavioral data, allowing us to explore the feasibility of behavior-based anomaly detection without compromising user privacy.

3.2 Synthetic Dataset Creation

To simulate user login activity, we developed a synthetic dataset using Python in a Google Colab environment. The dataset was generated using the **Faker** library, which allowed us to fabricate realistic login records while avoiding privacy issues associated with real user data. A constant user ID was used to simulate activity from a single user, generating a total of 10,000 login entries. These entries included structured attributes such as login time, IP address, city, session duration, and device type.

We established a "normal behavior profile" by assigning a fixed IP address and city to 8,000 out of 10,000 records (80%) to represent normal behavior. The remaining 2,000 records (20%) were used to simulate anomalies by using alternative IPs, cities, and irregular session durations. The 20% were varied using alternative IPs and cities to simulate anomalous behavior, such as remote logins or travel. Login durations were randomly assigned within a range of 1 min to 2 h and expressed as start and end times, later converted to total minutes for quantitative analysis.

To incorporate behavioral indicators, we transformed raw time and location data into features that could signal suspicious activity. For instance, login timestamps were converted into a 24-h numeric format, and logins occurring outside the 8:00 AM–5:00 PM window were flagged. Similarly, deviations from the known IP and city were marked as potentially suspicious. Sessions exceeding 90 min were also flagged. These conditions were aggregated into a final binary label— **"Abnormal vs. Normal"**—used for supervised model training.

3.3 Data Cleaning and Preprocessing

The dataset was cleaned and transformed to ensure compatibility with machine learning algorithms. Redundant fields such as raw timestamps and temporary flags were dropped. Categorical variables—namely city, IP address, and device type—were encoded using **one-hot encoding** to convert them into numerical format while mitigating multicollinearity via category drop-first techniques. This step ensured the dataset was in a structured and numerically stable format for algorithmic input.

We also performed **feature scaling** to normalize the input data, particularly important for models such as Logistic Regression and Neural Networks that are sensitive to variable magnitude. Quantitative attention was given to maintaining the balance between feature complexity and computational efficiency.

3.4 Algorithm Selection and Implementation

We adopted a **supervised learning paradigm** to classify login behaviors, as each record was labeled as either normal or abnormal. We selected three distinct classification models to evaluate performance across complexity levels:

1. **Logistic Regression** – A linear baseline classifier used to provide a reference point due to its simplicity and interpretability.
2. **Random Forest** – An ensemble method leveraging multiple decision trees to handle non-linearity and feature interactions while reducing overfitting.

3. **Neural Network (MLPClassifier)** – A feedforward multi-layer perceptron with ReLU activation, optimized using early stopping and regularization to detect complex, nonlinear patterns.

All models were implemented using the **Scikit-learn** framework, which provided tools for model training, cross-validation, and performance evaluation. To mitigate class imbalance, we employed stratified sampling during the train-test split and considered oversampling techniques where appropriate.

3.5 Model Training and Evaluation

Each model was trained on the same preprocessed training set and evaluated using a held-out test set. Evaluation metrics included **accuracy, precision, recall, F1-score**, and **confusion matrix analysis**—providing both quantitative (numeric metrics) and qualitative (error type) insights into performance.

- The **Random Forest** model achieved a strong balance of precision and recall with an accuracy of **90.45%**, indicating robust generalization and low misclassification rates.
- The **Neural Network** model demonstrated the highest accuracy at **99.05%**, though such high performance on synthetic data raised concerns of **overfitting** and potential poor generalizability to real-world datasets.
- The **Logistic Regression** model underperformed at **74.45%**, struggling especially with false positives and negatives in the minority class, highlighting its limitations in capturing complex behavioral patterns.

4 Results and Discussion

This section presents the outcomes of the model evaluation. It focuses on key performance metrics and the analysis of the confusion matrix. The evaluation metrics such as accuracy, precision, recall, and F1 score are used to assess the model's effectiveness across the dataset. In addition, the confusion matrix provides a detailed view of the model's classification performance by showing true positives, true negatives, false positives, and false negatives. Together, these results offer a comprehensive understanding of the model's strengths and limitations. Thus, forming the basis for later discussion and interpretation. The confusion matrix variables are as follows:

- True Positive (TP): correctly identifies deviations in user activity (the model flags a login as abnormal, and it truly is abnormal)
- True Negative (TN): Correctly identifies normal user behavior (the model classifies a login as normal, and it truly is normal)
- False Negative (FN): Fails to identify abnormal behavior (the model classifies an abnormal login as normal).
- False Positive (FP): Incorrectly flags normal behavior as abnormal (the model classifies a normal login as suspicious). This could lead to unnecessary alerts or disruptions for legitimate users.

Based on these variables, the evaluation metrics in the first category used in this research can be defined as:

- Accuracy: (TP + TN)/(TP + TP + FP + FN) [10]
- Precision: (TP)/(TP + FP) [10]
- Recall: (TP)/(TP + FN) [10]
- F1 Score: 2 * (Precision * Recall)/(Precision + Recall) [21]

The evaluation results for each model are presented and discussed below using Figs. 1, 2, 3, 4, 5 and 6, which include classification reports and confusion matrices. Each figure is referenced directly to explain model behavior and performance.

As shown in Figure 1, the Logistic Regression model performed poorly with an F-1 score of 0.68 and a major imbalance between all four classes in the report.

```
Accuracy : 0.7395

Classification report
              precision    recall  f1-score   support

           0       0.47      0.21      0.29       509
           1       0.77      0.92      0.84      1491

    accuracy                           0.74      2000
   macro avg       0.62      0.56      0.56      2000
weighted avg       0.70      0.74      0.70      2000
```

Fig. 1. Classification Report for Logistic Regression

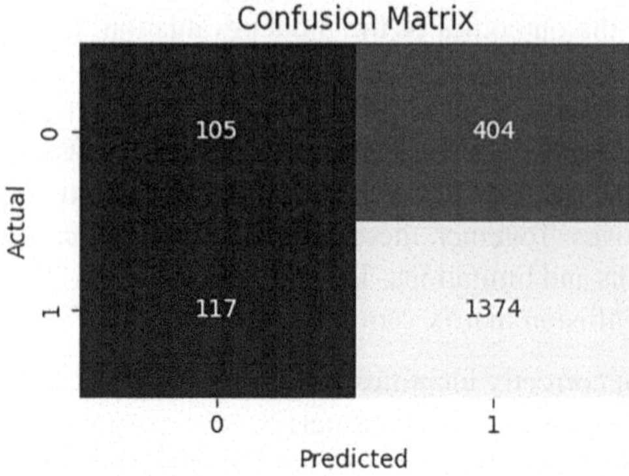

Fig. 2. Confusion Matrix for Logistic Regression

Figure 2 confirms that the Logistic Regression model performed the weakest, with an accuracy of 74.45%. It particularly struggled with the minority class (suspicious behavior), achieving low recall and precision for class 0. The confusion matrix confirmed this, showing a high rate of misclassification for suspicious activity. This underperformance is likely due to the model's linear nature, which is less effective at capturing complex relationships present in user behavior data.

```
Accuracy: 0.9045

Classification Report:
              precision    recall  f1-score   support

           0       0.91      0.90      0.90      1003
           1       0.90      0.91      0.90       997

    accuracy                           0.90      2000
   macro avg       0.90      0.90      0.90      2000
weighted avg       0.90      0.90      0.90      2000
```

Fig. 3. Classification Report for Random Forest

Figure 3 presents the classification report for the random forest model, achieving an F-1 score of 0.91.

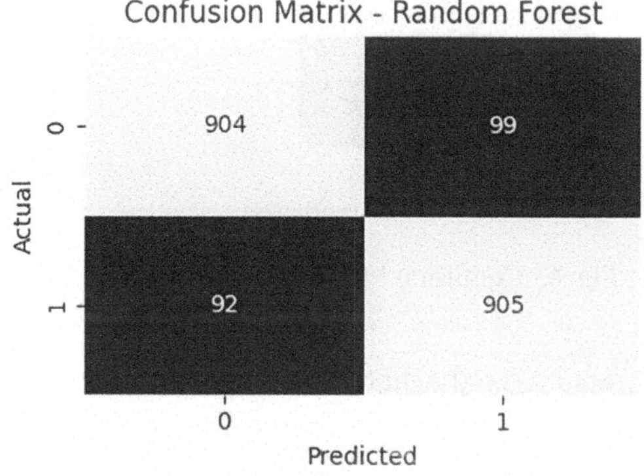

Fig. 4. Confusion Matrix for Random Forest

The Random Forest model emerged as the most balanced and reliable. It achieved an accuracy of 90.45%, with precision and recall values closely aligned across both classes. As shown in Fig. 4, the confusion matrix showed a well-distributed classification with low false positives and false negatives, indicating that the model generalized well without overfitting. These results suggest Random Forest is capable of consistently identifying both suspicious and normal behaviors in a balanced manner.

Figure 5 shows the Neural Network model's classification results, nearing perfect accuracy which suggests potential overfitting.

Figure 6's confusion matrix supports this concern, showing almost no classification errors. The Neural Network (MLPClassifier) achieved the highest accuracy at 99.05%, along with excellent precision and recall. However, the near-perfect scores raised concerns about overfitting, as the model may have memorized patterns in the synthetic

```
Neural Network (MLPClassifier) Evaluation
Accuracy: 0.9905

Classification Report:
              precision    recall  f1-score   support

           0       0.97      1.00      0.98       508
           1       1.00      0.99      0.99      1492

    accuracy                           0.99      2000
   macro avg       0.98      0.99      0.99      2000
weighted avg       0.99      0.99      0.99      2000
```

Fig. 5. Classification Report for Neural Networks

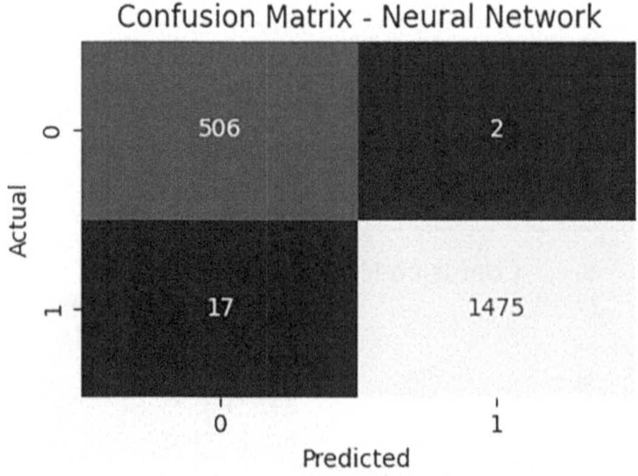

Fig. 6. Confusion Matrix for Neural Networks

dataset rather than learning generalizable behavior patterns. While powerful, this model may not perform as well on real-world data unless further regularization and validation steps are applied.

After training and testing each model, we compared their performance to determine which was most effective at identifying abnormal user behavior. All models were evaluated using the same synthetic test dataset and assessed using metrics such as accuracy, precision, recall, F1-score, and confusion matrices.

Figures 1, 2, 3, 4, 5 and 6 display the accuracy of the specific model and their evaluation metrics which include f-1 score, precision, and recall. Overall, as shown in Fig. 7, Random Forest was the best-performing model due to its balanced metrics, robustness to overfitting, and better handling of class imbalance, making it more suitable for deployment in practical scenarios involving user behavior analysis.

When compared to prior research, our Random Forest model achieved similar or improved results. For instance, Khan [8] used an unsupervised approach and reported lower precision due to high false positives. In contrast, our supervised approach using Random Forest achieved 90.45% accuracy and strong precision/recall values, due to the inclusion of time and location-based features.

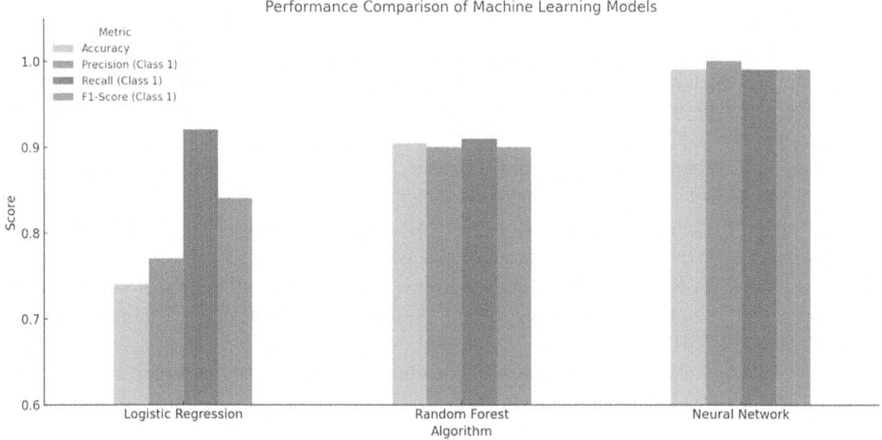

Fig. 7. Performance Comparison of Models

Additionally, while Kim et al. [1] reported an AUC of 0.97 using LSTM on insider threat detection, our neural network achieved 99.05% accuracy. However, due to the synthetic nature of our dataset, this high accuracy may not translate directly to real-world performance, as their dataset (CERT v4.2) is more extensive and realistic.

We selected Random Forest, Logistic Regression, and Neural Networks because they are commonly used for classification problems involving structured data and offer different strengths in terms of interpretability, performance, and flexibility. While unsupervised methods like Isolation Forest were used in prior work [5], our choice of supervised methods allowed for more precise evaluation against labeled behavior.

5 Future Work

While this study demonstrated promising results using synthetic data, future work will focus on expanding the model's applicability to real-world environments. One major enhancement would involve collecting and analyzing real user activity logs from organizational systems, ensuring privacy and ethical considerations are addressed. This would allow the models to learn from more diverse and nuanced patterns of behavior, ultimately improving their robustness and generalizability. Additionally, incorporating time-series analysis could help capture temporal anomalies and trends that static features may overlook. Future iterations may also explore the integration of unsupervised learning techniques for anomaly detection, which could flag previously unseen or rare behaviors without requiring labeled data. Lastly, deploying the best performing model in a live environment and evaluating its performance in real-time would provide valuable feedback on its practicality, efficiency, and adaptability under dynamic conditions.

Acknowledgements. This research was supported by NSF Grant 2205537.

Disclosure of Interest. There are no competing interests.

References

1. Alshehri, A., Khan, N., Alowayr, A., Alghamdi, M.: Cyberattack Detection Framework using machine learning and user behavior analytics. Tech Science Press. https://www.techscience.com/csse/v44n2/48277/html
2. Breached Company: Major Cyber Attacks 2025: A Comprehensive Analysis of the Year's Most Devastating Data Breaches and Ransomware Incidents, 28 May 2025. https://breached.company/major-cyber-attacks-2025-acomprehensive-analysis-of-the-years-most-devastating-data-breaches-andransomware-incidents/
3. Kim, J., Park, M., Kim, H., Cho, S., Kang, P.: Insider threat detection based on user behavior modeling and anomaly detection algorithms. Appl. Sci. **9**(19), 4018 (2019). https://doi.org/10.3390/app9194018
4. Rai, A., Morshed, M.: Adversarial machine learning in cybersecurity: a survey. IEEE Access **12**, 11245–11259 (2024)
5. Mukherjee, B., Heberlein, L.T., Levitt, K.N.: Network intrusion detection. IEEE Network **8**(3), 26–41 (1994)
6. Egele, M., Brumley, D., Fratantonio, Y., Kruegel, C.: An empirical study of credential reuse on the web. In: Proc. 23rd USENIX Security Symp., pp. 243–255 (2014)
7. Axelsson, S.: Intrusion detection systems: a survey and taxonomy. Technical Report, Chalmers University of Technology (2000)
8. Khan, K.: User Login Behavior Analysis and Anomaly Detection using Machine Learning. Medium, Oct 13 2024. https://medium.com/@khurram.khan_91792/user-login-behavior-analysis-andanomaly-detection-using-machine-learning-7e6f0f7e50ae
9. Park, H., Kim, K., Shin, D., Shin, D.: BGP dataset-based malicious user activity detection using machine learning. Information **14**(9), 501 (2023). https://doi.org/10.3390/info14090501
10. Benazzouza, Ridouani, M., Salahdine, F., Hayar, A.: A novel prediction model for malicious users detection and spectrum sensing based on stacking and deeplearning. Sensors **22**(17), 6477 (2022). https://doi.org/10.3390/s22176477
11. IBM, "What Is Machine Learning (ML)?," IBM Think, 22 Sept. 2021. https://www.ibm.com/think/topics/machine-learning
12. MathWorks, "Train Reinforcement Learning Agent for Simple Contextual Bandit Problem," Reinforcement Learning Toolbox Documentation, MathWorks. https://www.mathworks.com/help/reinforcementlearning/ug/contextualbandit.html
13. Cold start problem, ScienceDirect, Topics in Computer Science. https://www.sciencedirect.com/topics/computer-science/cold-startproblem
14. Zhang, L., Wang, T., Kim, H.: Security challenges in cloud-enabled IoT networks. IEEE Internet Things J. **9**(8), 6542–6555 (2022)
15. Mehta, S., Kothuri, P., Garcia, D.L.: Anomaly Detection for Network Connection Logs. arXiv preprint arXiv:1812.01941, 1 December 2018. https://arxiv.org/abs/1812.01941 [cs.LG]

An Obstacle-Adaptive A^* Heuristic for Path Planning: A Comparative Analysis with Repulsion-Augmented BFGS in Complex Environments

Vomsheendhur Raju[1], Deepak Dawar[2](✉), and Akhlaqur Rahman[3]

[1] North Dakota State University, Fargo, ND 58102, USA
vomsheendhur.raju@ndus.edu
[2] Miami University Hamilton, Hamilton, OH 45011, USA
daward@miamioh.edu
[3] Engineering Institute of Technology, Melbourne, VIC 3000, Australia
akhlaqur.rahman@eit.edu.au
https://vom1124.github.io/Website/,
https://miamioh.edu/regionals/profiles/deepak-dawar.html,
https://www.eit.edu.au/instructor_stories/akhlaqur-rahman/

Abstract. Pathfinding algorithms are fundamental to autonomous navigation, traditionally computing optimal routes based on defined start and goal locations using heuristic cost functions. While effective, these algorithms become computationally intensive in dynamic environments due to the need for exhaustive search. Gradient-based optimization methods, such as Broyden-Fletcher-Goldfarb-Shanno (BFGS), offer a reactive alternative by making local decisions based on the current position; however, they are susceptible to local minima and increased computational burden in complex cost landscapes, particularly when second-order derivative information is required. This paper proposes an enhanced A^* algorithm through the introduction of a custom heuristic function designed to reduce computational load and facilitate reactive navigation. The proposed heuristic, along with three well-established heuristics, is evaluated against an augmented BFGS approach that integrates a repulsive force field to improve obstacle avoidance. Performance metrics include path length and path behavior near obstacles. Experimental results demonstrate that the A^* algorithm with the proposed heuristic achieves obstacle avoidance performance comparable to BFGS, outperforming traditional heuristics. These findings suggest the potential for the enhanced A^* to support reactive robotic navigation, motivating further exploration towards hybrid strategies that leverage the strengths of both global search and local optimization methods.

Keywords: Hybrid Path Planning · A^* Heuristics · Obstacle Avoidance · Reactive Navigation

I. Roy et al. (Eds.): CAINE 2025, CCIS 2740, pp. 349–364, 2026.
https://doi.org/10.1007/978-3-032-14893-3_24

1 Introduction

Pathfinding algorithms are designed to determine the most efficient route between a starting point and a destination within a given spatial configuration. These algorithms play a critical role in a wide range of applications, from maze-solving games to autonomous navigation systems in robotics. Most pathfinding techniques are modeled as graph search problems, where the environment is discretized into nodes and edges. A search proceeds by expanding nodes based on specific cost functions or heuristics until the goal is reached. The efficiency and accuracy of these methods often depend on the underlying graph structure and the strategy used for node expansion. Traditional search methods such as Breadth-First Search (BFS) and Depth-First Search (DFS) offer foundational approaches but suffer from sub-optimality in practical scenarios [11,18]. BFS explores all nodes at the current depth before moving deeper, while DFS may follow an unfavorable branch far into the graph before backtracking. Neither approach guarantees the shortest path, and both are sensitive to the graph's topology and node exploration order. Furthermore, these algorithms often require additional logic to detect and handle revisited nodes, adding a layer of complexity for real-time implementation in robotic systems [14].

Alternative strategies have been developed to enhance the efficiency of pathfinding. One such approach involves Voronoi-based roadmaps, which discretize the environment into occupied and unoccupied regions, enabling binary navigation decisions that are particularly effective in static settings such as factory automation or surgical robotics [5,21]. However, the limitations of graph-based methods in dynamic or unknown environments have engendered interest in gradient-based approaches. These methods operate by iteratively updating the robot's state using local gradient information, eliminating the need to retain extensive state histories and thus offering advantages in memory-constrained systems. Despite this, their performance is often compromised by sensitivity to step size and susceptibility to local minima, which can hinder robustness in complex, non-convex environments [12].

Among gradient-based techniques, the Broyden-Fletcher-Goldfarb-Shanno (BFGS) algorithm stands out for its balance between computational efficiency and optimization performance. As a quasi-Newton method, BFGS approximates the Hessian matrix through gradient evaluations rather than computing it explicitly, significantly reducing computational cost and improving numerical stability, particularly in problems with large parameter spaces or noisy gradient information [3,7,15]. Unlike classical Newton's method, which requires functions to be continuous and possess well-defined second derivatives, BFGS can handle nonsmooth or irregular cost functions. Furthermore, it exhibits superlinear convergence under appropriate conditions, making it well-suited for a wide range of optimization problems. In mobile robotics, BFGS has shown significant promise for generating smooth and collision-free paths within SLAM (Simultaneous Localization and Mapping) frameworks [6,19]. Nonetheless, the method's performance can be affected by initialization sensitivity and numerical instabilities during iterative updates, which may degrade the quality of the Hessian approximation and impair convergence [4,16]. Moreover, as with other gradient-based approaches, BFGS remains susceptible to becoming trapped in local minima, limiting its ability to guarantee globally optimal solutions in complex or highly non-convex environments [1,2].

Recent advances in reinforcement learning and applied artificial intelligence for robotic navigation often employ simplified pathfinding algorithms to mitigate computational demands in known environments. By integrating graph theory with environmental knowledge, informed graph search methods can be tailored to specific navigation problems. Well-known examples of such algorithms include Dijkstra's, A^*, and D^*. Notably, the A^* algorithm can be considered an extension of Dijkstra's method, designed to guarantee optimal solutions provided an appropriate heuristic is employed. It operates by evaluating the true cost from the start node to the current node combined with an estimated cost from the current node to the goal, guiding the search towards minimizing this total cost. Due to its reliability, A^* has been widely adopted across various applications [8,10]. However, its performance is highly dependent on the choice of heuristic function; an unsuitable heuristic may result in incomplete solutions or non-termination, beyond merely increasing computational complexity. The algorithm's efficiency is closely tied to the number of nodes it explores. Importantly, A^* guarantees termination upon reaching the goal node, assuming the heuristic is consistent [20], ensuring that the search continues until a valid solution is found.

This paper proposes an improved heuristic function for A^* path planning and evaluates its performance against well-established heuristics commonly used in the literature. The proposed approach enhances pathfinding by modifying obstacle representation within the occupancy map and is benchmarked against an augmented Broyden-Fletcher-Goldfarb-Shanno (BFGS) algorithm that extends the method introduced by [19] through the integration of a repulsive force field(s). This force field controls the minimum proximity at which the robot initiates avoidance by dynamically activating the repulsive force as the robot enters the obstacle's Gaussian potential field. This approach addresses limitations of the original BFGS, where step size influences early detection—smaller steps enable earlier avoidance, whereas larger steps risk the robot climbing the obstacle's potential before reacting. By incorporating a user-defined proximity threshold into the weighted norm of the update direction, the augmented BFGS ensures preemptive and more efficient obstacle avoidance. The performance of both the augmented-BFGS and A^* (with all heuristics) are evaluated through simulation, using step count as the primary metric to assess path optimality and obstacle avoidance, allowing for controlled and reproducible analysis. This framework lays the groundwork for a potential hybrid strategy, where global A^* planning, or even locally constructed variants from real-time sensor data, could complement BFGS in adverse gradient conditions and inefficient escapes from local minima to enhance robustness and planning efficiency.

The paper is organized into four main sections, each containing focused subsections that detail key aspects of the proposed framework. Following the Introduction, Sect. 2 outlines the map construction process for both BFGS and

A^* algorithms, introduces the proposed augmentation to the traditional BFGS for improved obstacle clearance, and details the development of a novel heuristic function for A^*. The subsequent Sect. 3 evaluates the performance of these methods across increasingly complex environments, comparing path efficiency and obstacle avoidance behavior. Finally, Sect. 4 concludes the paper with a discussion on the implications of the results and potential directions for future work.

2 The Proposed Method of Implementing A^* Algorithm

This section details the procedures undertaken to evaluate and compare the performance of an augmented BFGS-based obstacle avoidance method and an enhanced A^* path planner incorporating improved heuristic design. The methodology is organized into three parts. First, the construction of the navigation environment is described, including the generation of two complementary map representations: a 2D occupancy grid for A^* path planning and a 3D topological cost surface with Gaussian potential fields for BFGS optimization. Next, the augmentation of the BFGS method is detailed, highlighting the obstacle modeling via Gaussian potentials and the integration of a repulsive force field to enable preemptive avoidance behavior. Finally, the improved A^* approach is introduced, focusing on the development and evaluation of an enhanced heuristic function benchmarked against conventional heuristics for effective navigation in cluttered environments.

2.1 Map Representations for BFGS and A^* Planning

Two distinct map representations are employed to support the different planning approaches. For the BFGS optimization, the workspace is initially modeled as a smooth parabolic (paraboloid) surface representing the cost landscape without obstacles. In this configuration, the start point is positioned at the highest level of the paraboloid, corresponding to the maximum cost, while the destination lies at the lowest level, representing the minimum cost. Obstacles are then introduced by superimposing Gaussian functions onto this surface, creating localized "hills" that raise the cost in their vicinity. These Gaussian potential fields effectively distort the parabolic cost landscape, shaping the gradient descent path to avoid the elevated obstacle regions. The cost landscape is composed of a parabolic base surface and Gaussian obstacle potentials based on the formulation presented in [19]. The resulting 2-D workspace $\mathcal{W} \in R^2$ is represented by coordinates $X = [x, y]^T$. The cost landscape guiding the robot is composed of a smooth quadratic function centered at the destination:

$$\Phi_{\text{base}}(X, X_d) = k(X - X_d)^T(X - X_d), \tag{1}$$

where $X_d = [x_d, y_d]^T$ is the goal and $k > 0$ determines the slope of descent. To incorporate obstacle information, each obstacle 'i' is modeled as Gaussian potential field:

$$\Phi_{\text{obs},i}(X) = \alpha_i \exp\left(-\frac{1}{\alpha_i}(X - X_{o,i})^T(X - X_{o,i})\right), \quad (2)$$

where $X_{o,i} \in \mathcal{W}$ is the obstacle center and α_i its influence weight. The complete cost landscape becomes:

$$\Phi_{\text{total}}(X) = \Phi_{\text{base}}(X, X_d) + \sum_{i=1}^{N} \Phi_{\text{obs},i}(X), \quad (3)$$

with N the number of observed obstacles or features. The cost landscape is visualized through surface and contour plots, employing color gradients to encode the cost magnitude along the vertical axis. This approach provides a clear depiction of the gradient descent trajectory and the influence of obstacle-induced potentials. Figure 1 presents both the initial parabolic surface and its alteration due to the added Gaussian obstacle fields.

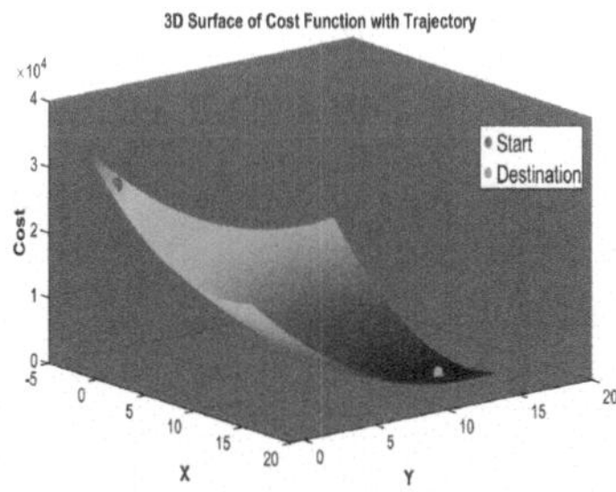

(a) Unobstructed parabolic descent path.

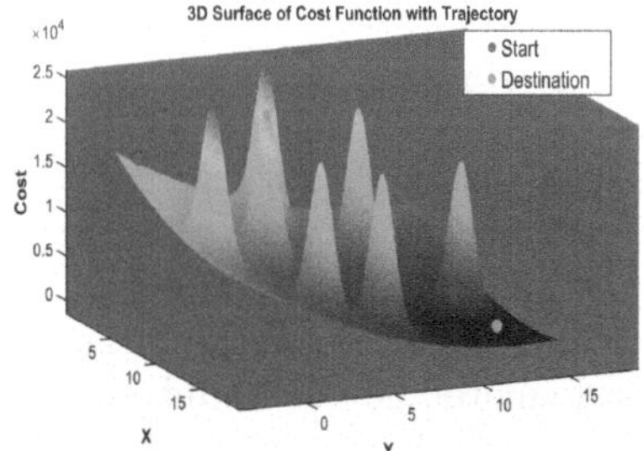

(b) Obstacle potentials deform descent path.

Fig. 1. BFGS cost landscape visualized with surface and contour plots.

In contrast, the A^* algorithm operates on a 2-D Cartesian occupancy grid map with a single-color grid representation. The workspace is discretized into cells with start and destination points defined as (x, y) coordinates, where obstacles are represented as approximate areas rather than precise shapes to reduce computational load from sensor data. Only obstacle width and height are used for classification, and all obstacles are treated as impassable. The grid cell size is set to approximately $1/100^{\text{th}}$ of the smallest obstacle dimension to ensure the robot's step size does not cause collisions, balancing considerations of robot size, speed, and computational capacity. Each obstacle is surrounded by a buffer zone, allowing the robot to approach safely and adjust its path before collision. Figure 2 illustrates the occupancy grid, showing start (red dot), destination (green dot), obstacles (cyan), and buffer zones (blue), with the start typically set close to the origin.

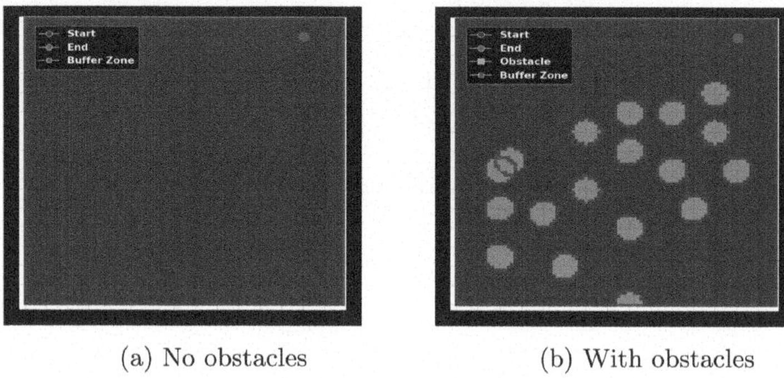

(a) No obstacles (b) With obstacles

Fig. 2. Occupancy grid with and without obstacles. (Color figure online)

2.2 Enhanced BFGS Path Planning via Potential Field Augmentation

Building on the cost landscape formulation detailed in Sect. 2.1, where the workspace is modeled as a parabolic descent surface combined with Gaussian obstacle potentials, this subsection presents the augmentation of the classical BFGS algorithm to improve convergence stability and obstacle avoidance. While the original method [19] effectively addressed local minima traps inherent in gradient-based navigation, it did not explicitly characterize the spatial vicinity within which traps are avoided or quantify the trade-offs involved in circumventing such traps. These limitations resulted in sensitivity to step size, obstacle size, and environmental complexity.

The proposed augmentation addresses these limitations through the integration of a user-tunable repulsive force field, conceptually rooted in classical artificial potential field methods, which have served as the bedrock for generating repulsive and attractive forces in robotic navigation since their introduction [9]. Such repulsive force formulations are widely recognized for enabling robots to maintain safe distances from obstacles by dynamically modulating cost or force influences based on proximity. By allowing dynamic adjustment of obstacle influence, this component enables the planner to balance efficient path convergence with conservative obstacle clearance tailored to the environmental context.

The enhanced BFGS approach involves three principal steps:

1. **Repulsive Force Augmentation:** A tunable repulsive force $\mathbf{F}_{\text{rep}}$ is integrated to actively modify the descent direction, enhancing obstacle clearance and mitigating local minima entrapment. This force accumulates contributions from all obstacles within a defined repulsion radius r_{rep}, computed as follows:

$$\mathbf{F}_{\text{rep}}(\mathbf{X}) = \sum_{i=1}^{N} \beta_i \, f_{\text{rep},i}(\mathbf{X}), \tag{4}$$

where each component is

$$f_{\text{rep},i}(\mathbf{X}) = \left\{ \left(\frac{1}{\|\mathbf{X} - \mathbf{X}_{o,i}\| + \varepsilon} - \frac{1}{r_{\text{rep}}} \right) \frac{\mathbf{X} - \mathbf{X}_{o,i}}{\|\mathbf{X} - \mathbf{X}_{o,i}\|}, \quad \text{if } \|\mathbf{X} - \mathbf{X}_{o,i}\| < r_{\text{rep}}, \text{otherwise}, \right. \tag{5}$$

with β_i controlling avoidance strength, $\mathbf{X}_{o,i}$ the obstacle position, and ε is a small regularization constant introduced to mitigate ill-conditioning in the inverse Hessian approximation (discussed later in the BFGS iterative procedure) and to ensure numerical stability during division operations. The total repulsive force is normalized to provide a direction for adjusting the descent path.

2. **Descent Direction Computation:** At each iteration 'k', the descent direction $p^{(k)}$ is formed as a normalized weighted sum of three components: the BFGS gradient descent $p_{\text{grad}}^{(k)}$, the repulsive force $\mathbf{F}_{\text{rep}}(X^{(k)})$, and the direct goal attraction $p_{\text{goal}}^{(k)}$. This combination balances convergence, obstacle avoidance, and goal orientation:

$$p^{(k)} = \frac{w_{\text{grad}}\, p_{\text{grad}}^{(k)} + w_{\text{rep}}\, \mathbf{F}_{\text{rep}}(X^{(k)}) + w_{\text{goal}}\, p_{\text{goal}}^{(k)}}{\left\| w_{\text{grad}}\, p_{\text{grad}}^{(k)} + w_{\text{rep}}\, \mathbf{F}_{\text{rep}}(X^{(k)}) + w_{\text{goal}}\, p_{\text{goal}}^{(k)} \right\| + \varepsilon}, \tag{6}$$

where

$$p_{\text{grad}}^{(t)} = -H^{(k)} \nabla \Phi_{\text{total}}(X^{(k)}), \quad p_{\text{goal}}^{(k)} = \frac{X_d - X^{(k)}}{\|X_d - X^{(k)}\| + \varepsilon}.$$

Here the $H^{(k)}$ is the inverse Hessian matrix, $\nabla \Phi_{\text{total}}(X^{(k)})$ is the gradient vector, and the weights $w_{\text{grad}}, w_{\text{rep}}, w_{\text{goal}}$ regulate the influence of each component. Once the descent direction is computed, the robot's actual step length $\Delta X^{(k)}$ of the iteration is defined as:

$$\Delta X^{(k)} = \mu^{(k)} \cdot p^{(k)}, \tag{7}$$

where $\mu^{(k)}$ is a user-defined step size parameter for varying the magnitude of the descent direction.

3. **Iterative BFGS Solution:** The path optimization is solved iteratively using the BFGS quasi-Newton method, which avoids direct computation of the full Hessian matrix, instead approximating its inverse using gradient evaluations. The next iteration's inverse Hessian $H^{(k)}$ which updates the path estimate computed using the standard BFGS formulation [7,13] based on all previous path updates and gradient differences. The optimization then updates both the $H^{(k)}$ and the path estimate $X^{(k)}$ using the robot's step length $\Delta X^{(k)}$ computed earlier as:

$$\mathbf{X}^{(t+1)} = \mathbf{X}^{(t)} + \Delta X^{(k)}. \tag{8}$$

2.3 Proposed A^* Implementation

The efficiency and optimality of the A^* algorithm heavily depend on the choice of heuristic function guiding the search process. Several well-established heuristics are commonly used in path planning, each with unique characteristics suited to different environments and movement constraints. The following formulations summarize these classical heuristics for points $\mathbf{x} = (x_1, x_2, \dots, x_n)$ and $\mathbf{y} = (y_1, y_2, \dots, y_n)$:

- **Manhattan Distance:** Computes the sum of absolute coordinate differences, often used in grid-based environments with orthogonal movements only,

$$\text{Manhattan}(\mathbf{x}, \mathbf{y}) = \sum_{i=1}^{n} |x_i - y_i|. \tag{9}$$

- **Jaccard Distance:** A similarity measure based on set intersection over union, useful for comparing occupancy or feature vectors in map representations,

$$\text{Jaccard}(\mathbf{x}, \mathbf{y}) = \frac{\sum_{i=1}^{n} x_i \cdot y_i}{\sum_{i=1}^{n} x_i + \sum_{i=1}^{n} y_i - \sum_{i=1}^{n} x_i \cdot y_i}. \tag{10}$$

- **Canberra Distance:** A weighted sum of coordinate differences normalized by their magnitudes, which emphasizes relative differences and can better capture dissimilarities in sparse data,

$$\text{Canberra}(\mathbf{x}, \mathbf{y}) = \sum_{i=1}^{n} \frac{|x_i - y_i|}{|x_i| + |y_i|}. \tag{11}$$

These heuristics functions provide reasonable approximations of path cost but inherently possess limitations. The Euclidean distance, though not explicitly presented above, is frequently favored for its admissibility, as it guarantees that the estimated cost never exceeds the actual cost. However, this optimistic property often leads to significant underestimation in practical robotic navigation, where factors such as vehicle orientation and maneuverability impose additional constraints. In many scenarios, the shortest geometric path does not correspond to the most viable or efficient trajectory, especially in environments that require frequent or acute directional changes. In order to address these limitations and improve search performance, this paper introduces a refined heuristic formulation that incorporates a tie-breaking strategy inspired by examples in [17] and define an improved directionally informed heuristic. This formulation introduces cross-product-based directional vectors to emphasize alignment between the current-to-neighbor vector and the goal direction:

$$\boldsymbol{U}_{cross} = \hat{CN} \times \hat{NN} \tag{12a}$$

where 'CN' is the current node and 'NN' is the neighboring node.

$$d(x, y)_a = (U_{cross_x} - goal_x,\ U_{cross_y} - goal_y) \tag{12b}$$

$$d(x, y)_b = (start_x - goal_x,\ start_y - end_y) \tag{12c}$$

where 'd' is the distance and (x, y) is the coordinate locations of the corresponding path. Then the final heuristic cost to be implemented in A^* becomes

$$\text{Cost} = |\boldsymbol{d}_a \cdot \boldsymbol{d}_b| \text{ or } |\boldsymbol{d}_a \times \boldsymbol{d}_b| \text{ or } \min(|\boldsymbol{d}_a|, |\boldsymbol{d}_b|). \tag{12d}$$

The choice of the heuristic cost in Eq. (12d) is application-dependent, balancing directional accuracy with computational efficiency. This approach prioritizes maintaining a straight trajectory from the current node to the goal in obstacle-free regions while flexibly rerouting around obstacles when necessary. The vector alignment principle aligns with quaternion-based orientation control, where the robot rotates to face the destination before proceeding, reflecting the intended motion planning behavior this heuristic seeks to emulate.

The A^* algorithm implementation follows the classical best-first search paradigm, integrating multiple heuristic functions—including established metrics and the proposed heuristic—to guide efficient path planning. Figure 3 presents the flowchart detailing the key procedural steps, from node initialization and priority queue operations to heuristic-based node evaluation and path reconstruction. The diagram highlights the modular integration of different heuristics for comparative performance analysis.

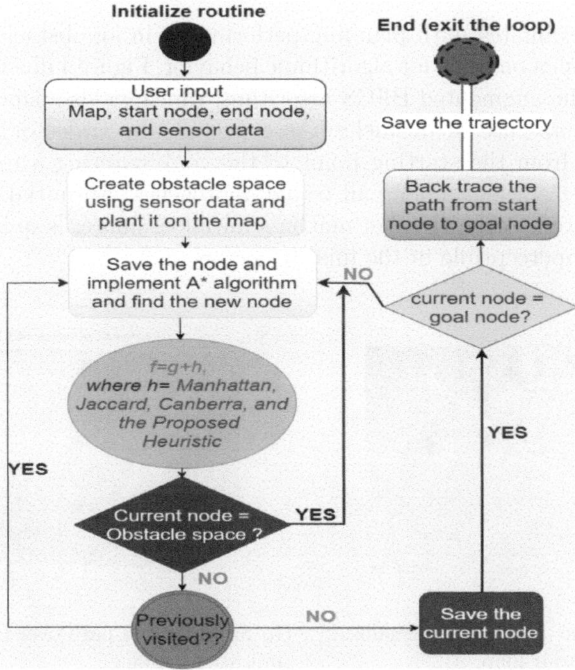

Fig. 3. Flowchart illustrating the modular implementation of the A^* algorithm.

3 Results and Discussion

This section begins by independently evaluating the performance of the augmented BFGS algorithm, followed by a comprehensive analysis of the A^* search method incorporating all heuristics introduced earlier (Sect. 2.3). Subsequently, the proposed improved heuristic within the A^* framework is compared against the augmented BFGS results to demonstrate its effectiveness. Finally, the proposed heuristic is benchmarked against established alternatives to underscore its relative advantages.

Three representative scenarios of increasing complexity are considered. The first scenario examines path planning in an obstacle-free environment to establish baseline behavior. The second scenario introduces obstacles incrementally as the robot moves toward the goal, allowing assessment of heuristic sensitivity and adaptive path planning. The final, most challenging scenario involves a cluttered environment with obstacles positioned similarly to those encountered previously, but of larger size, resulting in narrower passages and heightened navigational difficulty. These scenarios collectively demonstrate the robustness and comparative advantages of the proposed approach.

3.1 Scenario-1: Obstacle-Free Navigation Space

This scenario evaluates path planning performance in an obstacle-free environment to establish a baseline for algorithmic behavior. Figure 4 illustrates the path generated by the augmented BFGS algorithm, which yields a smooth parabolic descent in the absence of obstacles. As expected, the trajectory converges to a straight line from the starting point to the goal, covering a total of 72 unit steps—the most efficient among all tested approaches. An initial minor perturbation, attributed to the Hessian matrix's initial estimate, is quickly corrected and remains imperceptible in the final trajectory plot.

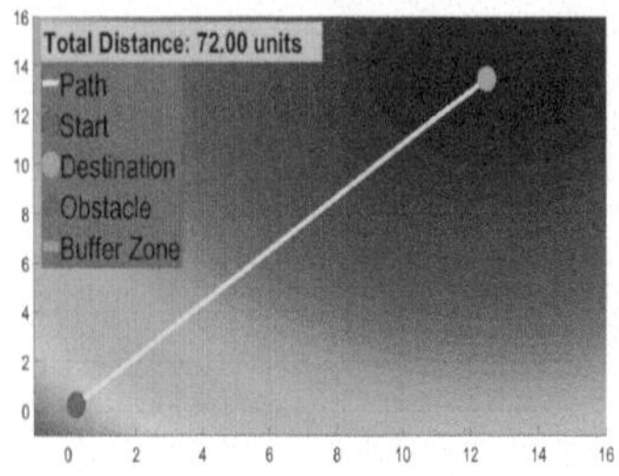

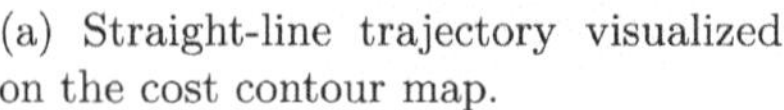
(a) Straight-line trajectory visualized on the cost contour map.

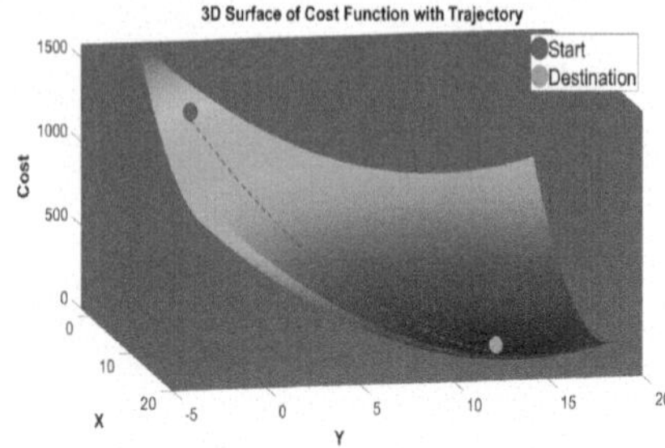

(b) Minimal-cost path over the smooth optimization surface.

Fig. 4. Augmented BFGS performance in an obstacle-free environment.

The A^* algorithm demonstrates comparable performance to the augmented-BFGS in the absence of obstacles, as illustrated in Fig. 5. Across all tested heuristics, a brief initial deviation from the optimal path was observed, which was corrected within a few steps. Both the Manhattan and the proposed heuristic produced a total path length of 74 unit steps; however, their trajectories differed qualitatively. As shown in Fig. 5(a), the Manhattan heuristic generated a zigzag path aligned with horizontal and vertical grid lines—a direct result of its underlying metric. In contrast, Fig. 5(d) highlights how the proposed heuristic quickly corrected its early deviation and continued with a nearly straight diagonal path to the goal. This difference, despite identical step counts, can be attributed to the distance evaluation scheme in grid-based algorithms, where diagonal moves are often not penalized with higher costs. The behavior of the Jaccard heuristic is depicted in Fig. 5(b) It followed a similar general trajectory to the Manhattan heuristic but introduced a more pronounced initial deviation away from the goal, resulting in a slightly longer path of 76 unit steps. Among all tested heuristics, the Canberra heuristic performed the worst. As illustrated in Fig. 5(c), it produced a disjointed, circuitous path that deviated significantly from the shortest route, ultimately covering 111 unit steps. Such a trajectory would be unsuitable for real-world navigation due to its inefficiency and potential safety concerns in constrained environments. Overall, the proposed heuristic outperformed all others by yielding a near-optimal, smoother trajectory with minimal deviation, making it the most effective choice for grid-based navigation.

3.2 Scenario-2: Incremental Obstacle Introduction

Figure 6 illustrates the behavior of the augmented BFGS algorithm in the presence of obstacles. The descent begins smoothly, but the trajectory gradually deforms as repulsive gradients from nearby obstacles reshape the cost landscape. The path adjusts continuously, curving around each obstacle without abrupt turns. Despite a slight increase in overall path length (81 unit steps), the algorithm maintains trajectory smoothness and demonstrates robust adaptability in continuous space, consistently preserving a safe clearance around obstacle boundaries.

The A^* algorithm's response to the same scenario is shown in Fig. 7. The Manhattan heuristic generated a stepwise path that closely hugged obstacle boundaries, occasionally raising safety concerns, yet it resulted in the most efficient path length of just 79 unit steps, as shown in Fig. 7(a). On the other hand, the Jaccard heuristic exhibited delayed obstacle avoidance, resulting in more noticeable lateral deviations as depicted in Fig. 7(b). This behavior is attributed to its strong initial deviation away from the destination during early convergence. The Canberra heuristic once again produced erratic and non-intuitive routes, compromising efficiency as illustrated in Fig. 7(c). However, it was more conservative in obstacle avoidance, often adopting paths that encountered multiple obstacles in successive steps.

Conversely, the proposed heuristic demonstrated superior performance. As seen in Fig. 7(d), it adapted early to emerging obstacles, generating a direct,

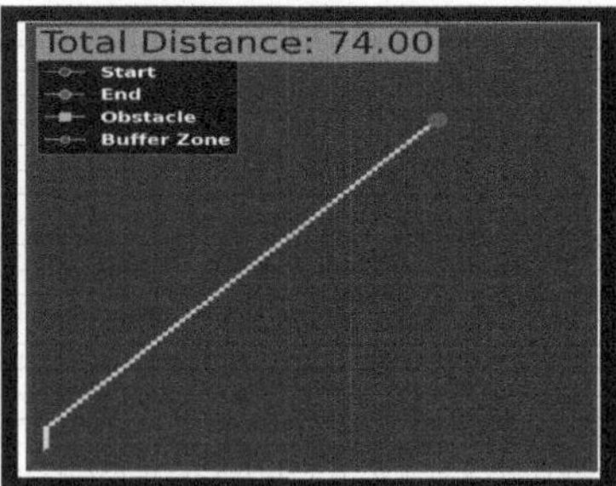

(a) Manhattan heuristic: grid-aligned and non-diagonal steps.

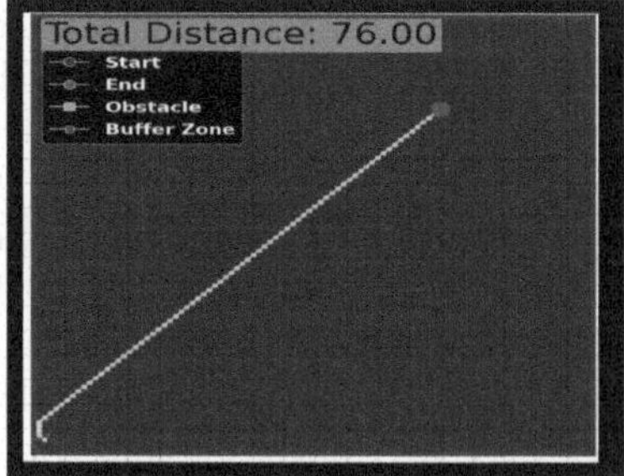

(b) Jaccard heuristic: early deviation before convergence.

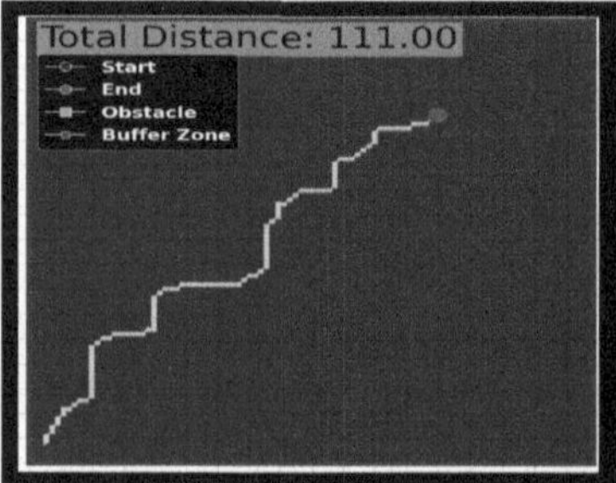

(c) Canberra heuristic: erratic and inefficient trajectory.

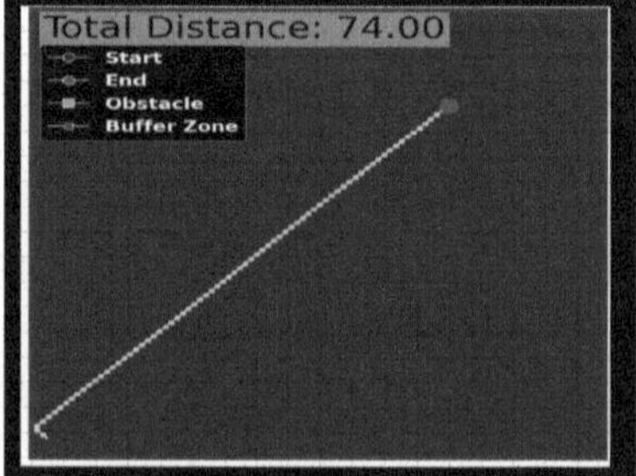

(d) Proposed heuristic: early correction and near-straight path.

Fig. 5. A^* path planning performance under different heuristics in an obstacle-free environment.

coherent path with fewer abrupt turns and minimal deviation from the optimal route. Most deviations occurred near the starting location and became insignificant once convergence was achieved. This indicates stronger spatial foresight and better alignment with the cost-to-go principle. It produced a path length of 79 unit steps, matching the Manhattan heuristic. In this scenario, both the Manhattan and proposed heuristics produced efficient paths compared to the augmented BFGS. However, BFGS excelled in terms of safety clearance due to its repulsive force field, maintaining a more cautious distance from obstacles. Essentially, while the BFGS maintained smoothness and the A* heuristics varied in their effectiveness, the proposed heuristic consistently delivered a direct, obstacle-aware path that closely approximated the continuous behavior of BFGS, outperforming all standard alternatives in both path quality and adaptability.

3.3 Scenario-3: Cluttered Environment with Enlarged Obstacles

In this scenario, algorithm performance is evaluated under a more demanding setting, characterized by enlarged obstacles that intensify navigational complexity. Figure 8 presents the trajectory generated by the augmented BFGS algorithm.

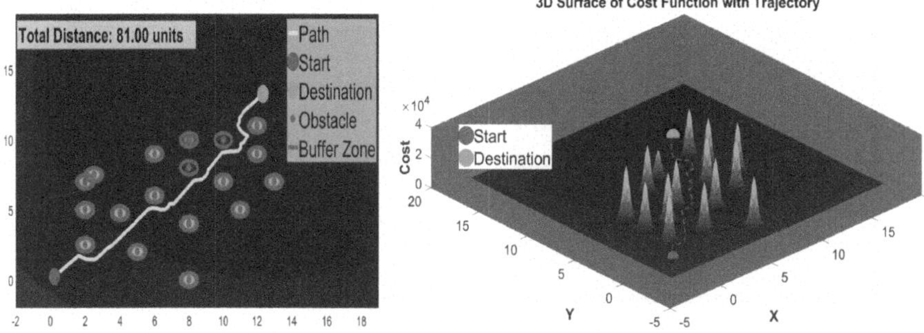

(a) Top-view contour plot of the cost landscape. (b) 3D topographical surface plot showing descent path.

Fig. 6. BFGS navigation in a workspace with incrementally introduced obstacles.

The robot begins with a smooth descent but progressively adapts its trajectory in response to stronger repulsive gradients from the larger obstacles. Despite the constrained environment, the path remains smooth and continuous, extending to 94 unit steps while maintaining safe clearance around all obstacles. The path adaptation closely mirrors human-like re-routing, highlighting the augmented-BFGS algorithm's responsiveness and fluidity in continuous space.

The A^* algorithm's response is shown in Fig. 9. Figure 9 shows the corresponding paths generated by the A^* algorithm. The Manhattan heuristic, shown in Fig. 9(a), followed a tightly bounded route that intersected several enlarged obstacles but achieved an efficient path length of 99 unit steps. The proposed heuristic, presented in Fig. 9(d), maintained a similarly efficient trajectory of 98 unit steps while proactively selecting routes with fewer critical encounters, demonstrating stronger spatial foresight. Figure 9(b) highlights the behavior of the Jaccard heuristic, which exhibited delayed obstacle avoidance and substantial lateral deviations. It initially converged into a region densely populated with obstacles, resulting in a total path length of 101 unit steps. Meanwhile, the Canberra heuristic, shown in Fig. 9(c), favored broad clearance from obstacles and adopted a more cautious, natural-looking trajectory. However, this overconservatism led to a significantly longer path of 126 unit steps, underscoring its inefficiency in densely cluttered spaces.

Overall, the proposed heuristic continues to offer the best balance between safety and efficiency among the A^* variants. It proactively avoids critical obstacle encounters and maintains near-optimal path lengths even in cluttered conditions. While the augmented BFGS demonstrates superior adaptability, smoother trajectories, and enhanced safety margins through continuous-space reactivity, the grid-based A^* approach with the proposed heuristic remains highly effective, delivering competitive performance under increased spatial complexity. The augmented BFGS, enhanced with repulsive barrier fields, ensured smooth and safe trajectories, but like other gradient descent-based methods, it can be vulnerable

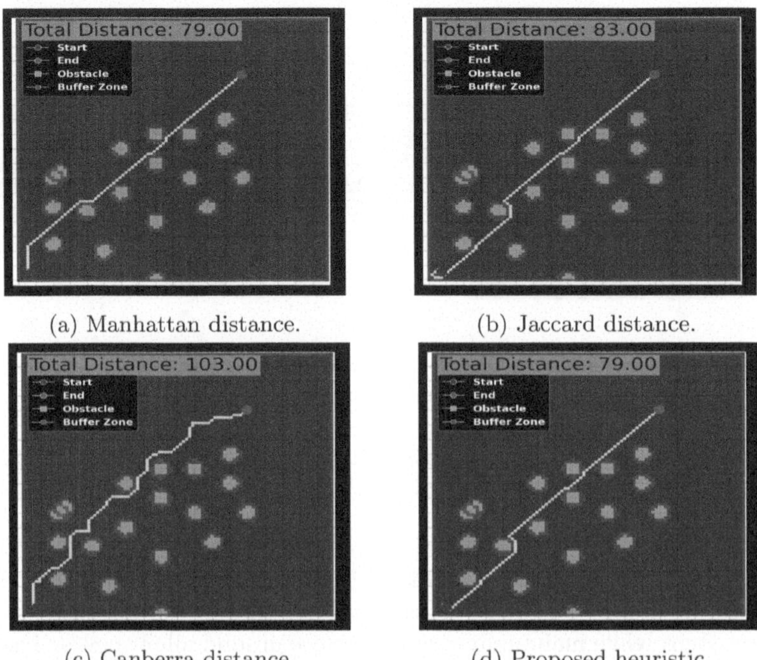

(a) Manhattan distance. (b) Jaccard distance.

(c) Canberra distance. (d) Proposed heuristic.

Fig. 7. Paths generated by A^* with different heuristics in an environment with incrementally added obstacles.

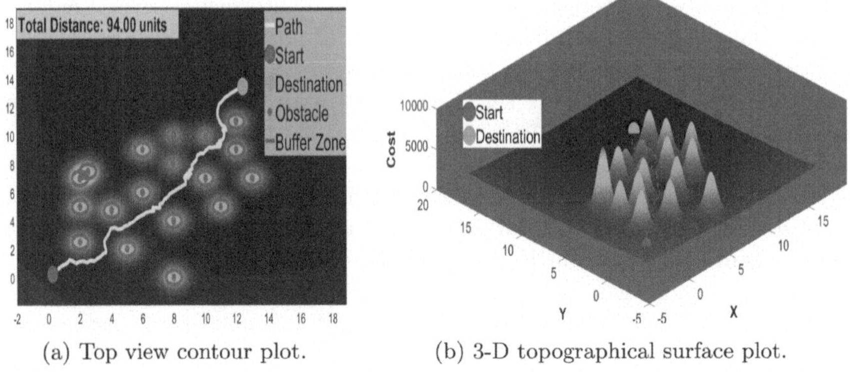

(a) Top view contour plot. (b) 3-D topographical surface plot.

Fig. 8. BFGS cost landscape and path trajectory in a cluttered environment with enlarged obstacles.

to local minima and challenges arising from ill-conditioned Hessian computations in dense or irregular environments.

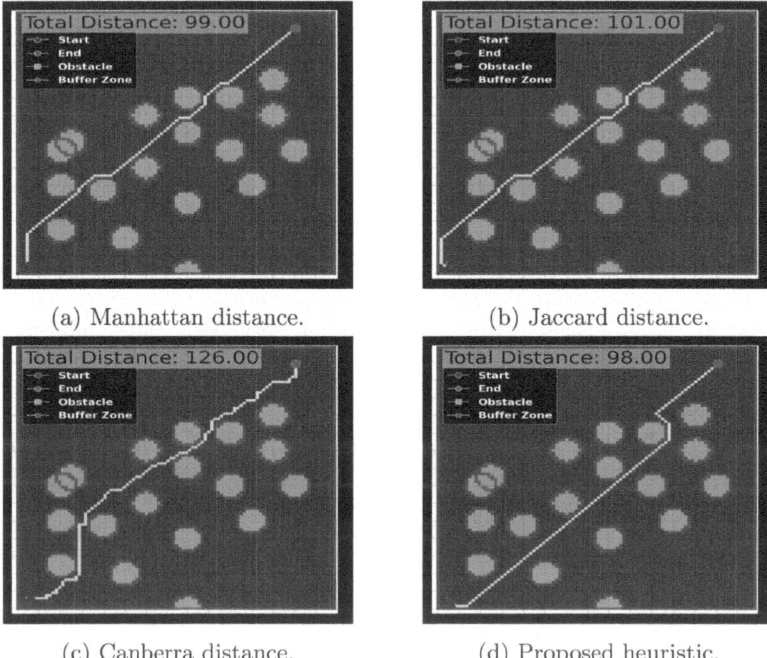

(a) Manhattan distance.

(b) Jaccard distance.

(c) Canberra distance.

(d) Proposed heuristic.

Fig. 9. Path planning using A^* heuristics in a cluttered environment with enlarged obstacles.

4 Conclusions

This paper aimed to develop and compare enhanced path-planning algorithms capable of handling both discrete and continuous navigation challenges in complex environments. A comprehensive evaluation of an improved A^* heuristic and an augmented-BFGS approach for path planning in complex environments was done. The proposed A^* algorithm demonstrates robust performance, generating near-optimal paths with efficient obstacle avoidance capabilities. The augmented BFGS, enhanced with repulsive barrier fields, ensured smooth and safe trajectories, but like other gradient descent-based methods, it can be vulnerable to local minima and challenges arising from ill-conditioned Hessian computations in dense or irregular environments. These results lay the foundation for a potential hybrid framework that combines the discrete efficiency of A^* with the continuous adaptability of BFGS, offering a resilient solution for real-time autonomous navigation in dynamic and complex settings.

Disclosure of Interests. The authors have no competing interests to declare that are relevant to the content of this article.

References

1. Bertsekas, D.P.: Nonlinear Programming, 2nd edn. Athena Scientific, Belmont (1999)
2. Boyd, S., Vandenberghe, L.: Convex Optimization. Cambridge University Press (2004)
3. Broyden, C.G.: The convergence of a class of double-rank minimization algorithms 1. General considerations. IMA J. Appl. Math. **6**(1), 76–90 (1970). https://doi.org/10.1093/imamat/6.1.76
4. Byrd, R.H., Lu, J.S., Nocedal, J., Zhu, C.: A limited memory algorithm for bound constrained optimization. SIAM J. Sci. Comput. **16**(5), 1190–1208 (1994)
5. Choset, H., et al.: Principles of Robot Motion: Theory, Algorithms, and Implementations. MIT Press, Cambridge (2005)
6. Dellaert, F., Kaess, M.: Factor graphs for robot perception. Found. Trends® Robot. **6**(1-2), 1–139 (2017). https://doi.org/10.1561/2300000043
7. Fletcher, R.: Practical Methods of Optimization. Wiley, Hoboken (2013). https://books.google.com/books/about/Practical-Methods-of-Optimization.html?id=_WuAvIx0EE4C
8. Foead, D., Ghifari, A., Kusuma, M.B., Hanafiah, N., Gunawan, E.: A systematic literature review of A∗ pathfinding. In: Procedia Comput. Sci., pp. 507–514. (2021). https://doi.org/10.1016/j.procs.2021.01.034
9. Khatib, O.: Real-time obstacle avoidance for manipulators and mobile robots. In: Proceedings of 1985 IEEE International Conference on Robotics and Automation, vol. 2, pp. 500–505 (1985). https://doi.org/10.1109/ROBOT.1985.1087247
10. Koenig, S., Likhachev, M., Furcy, D.: Lifelong planning A^*. Artif. Intell. **155**(1), 93–146 (2004). https://doi.org/10.1016/j.artint.2003.12.001
11. Konolige, K., Chou, K.: Markov localization using correlation (1999). https://api.semanticscholar.org/CorpusID:13817822
12. Kuffner, J.J., LaValle, S.M.: RRT-connect: an efficient approach to single-query path planning, vol. 2, pp. 995–1001 (2000). https://api.semanticscholar.org/CorpusID:17124403
13. Lam, A.: BFGS in a Nutshell: An Introduction to Quasi-Newton Methods (2021). https://towardsdatascience.com/bfgs-in-a-nutshell-an-introduction-to-quasi-newton-methods-21b0e13ee504. Accessed 23 May 2025
14. LaValle, S.M.: Planning Algorithms. Cambridge University Press (2006). http://planning.cs.uiuc.edu/
15. Nocedal, J.: Updating quasi newton matrices with limited storage. Math. Comput. **35**(151), 951–958 (1980). https://doi.org/10.1090/S0025-5718-1980-0572855-7
16. Nocedal, J., Wright, S.J.: Numerical Optimization. Springer (2006)
17. Patel, A.: Heuristics (2023). https://theory.stanford.edu/~amitp/GameProgramming/Heuristics.html. Accessed 23 May 2025
18. Platform, N.G.D.: Graph Search Algorithms - Developer Guides (2023). https://neo4j.com/developer/graph-data-science/graph-search-algorithms. Accessed 23 May 2025
19. Raju, V., Selekwa, M.F.: On the mapping problem in SLAM approaches for autonomous robot navigation. ASME Digit. Collect. (2022). https://doi.org/10.1115/IMECE2021-70452
20. Russell, S.J., Norvig, P.: Artificial Intelligence: A Modern Approach. Pearson Education, Inc. (2010)
21. Solanki, P.B., Reddy, G.H.V., Mukerjee, A.: Voronoi diagram based roadmap motion planning. Technical report, IIT Kanpur Institute (2013)

Author Index

A
Abdelahad, Corina 126
Alshoshan, Abdullah I. 32, 42
Anreddy, Sujan 54

B
Barati, Mehdi 244
Becerra-Vivanco, Paulina 265
Bhattacharya, Ansuman 171
Bynum, Kemon 336

C
Castro-Pastrana, Lucila I. 265
Chen, Bernard 189
Chowdhury, Md Minhaz 145
Cueva, Rony 277

D
Dahal, Samana 100
Dawar, Deepak 349
Debnath, Narayan 126, 212

F
Farías, Andrés Alejandro 212
Farías, Andrés Francisco 212

G
Garis, Ana Gabriela 212
Ghosh, Saurav 112
Gomez, Reinaldo 126
Graham, Jonathan 321
Griffin, Terry 223
Gupta, Bidyut 100
Gurazada, Pavankumar 3

H
Haque, Mehmuna 100
Hinds, Cheryl 321, 336

J
Jaekel, Arunita 160
Jung, Byungkwan 189

K
Kallam, Yogeshvar Reddy 299
Kambhampaty, Krishna 299
Khanduja, Sarthak 160
Kumar, Suman 189

L
Li, Zhouzhou 24

M
Maity, Moutusy 3
Mathers, Garrett 223
Miller, Ciera 336
Mitra, Reshmi 100, 112
Mittal, Akshay 69
Montejano, Germán Antonio 212
Montejano, German 126

N
Nguyen, Linh 189
Nunna, Satya Krishna 299

P
Padmanabhuni, Sriya 299
Pagidoju, Ravi Teja 87
Panchumarthi, Lovely Yeswanth 299
Pant, Nipesh 112
Pappy, Melissa 189
Parchuri, Lavanya 299
Parvathaneni, Dharani 171
Pastor, Piero 277

R
Rahimi, Nick 244
Rahman, Akhlaqur 349
Raju, Vomsheendhur 349

I. Roy et al. (Eds.): CAINE 2025, CCIS 2740, pp. 365–366, 2026.
https://doi.org/10.1007/978-3-032-14893-3

Riesco, Daniel 126, 212
Rosas-Romero, Roberto 265
Roy, Indranil 100, 171
Roy, Prasanta Kumar 171

S

Saei, Saviz 54
Sayani, Praveen Kumar 309
Sha, Zhanxin 24
Stringfellow, Catherine 223
Sugathan, Krishnendu Kottakkal 171

T

Taylor, Derek 244
Tupia, Manuel 277

U

Upadhyay, Hemant 112

W

Ward, Trenton 321
Wohlbach, Cameron J. 145

Y

Yao, Yicai 24
Yuan, Juefei 24

The manufacturer's authorised representative in the EU is Springer Nature Customer Service Centre GmbH, Europaplatz 3, 69115 Heidelberg, Germany. If you have any concerns regarding our products, please contact ProductSafety@springernature.com

Printed and bound by CPI Group (UK) Ltd, Croydon, CR0 4YY
07/07/2026
02160906-0010